Integrated Analytical Systems

Series Editor:
Radislav A. Potyrailo
GE Global Research Center
Niskayuna, NY

For other titles published in the series, go to
www.springer.com/series/7427

Margaret A. Ryan · Abhijit V. Shevade
Charles J. Taylor · Margie L. Homer
Mario Blanco · Joseph R. Stetter
Editors

Computational Methods for Sensor Material Selection

Ⓐ Springer

Editors

Margaret A. Ryan
California Institute of
Technology
Jet Propulsion Laboratory
4800 Oak Grove Drive
Pasadena CA 91109-8099
USA
mryan@jpl.nasa.gov

Abhijit V. Shevade
California Institute of
Technology
Jet Propulsion Laboratory
4800 Oak Grove Drive
Pasadena CA 91109-8099
USA
abhijit.shevade@jpl.nasa.
gov

Charles J. Taylor
Seaver Chemistry
Laboratory
Pomona College
645 N. College Avenue
Claremont CA 91711
USA
chuck.taylor@pomona.edu

Margie L. Homer
California Institute of
Technology
Jet Propulsion Laboratory
4800 Oak Grove Drive
Pasadena CA 91109-8099
USA
mhomer@jpl.nasa.gov

Mario Blanco
California Institute of
Technology
Molecular Simulations
Center
Pasadena CA 91125
USA
mario@wag.caltech.edu

Joseph R. Stetter
KWJ Engineering
8440 Central Avenue
Newark CA 94560
USA
rstetter@kwjengineering.
com

ISBN 978-0-387-73714-0   e-ISBN 978-0-387-73715-7
DOI: 10.1007/978-0-387-73715-7
Springer Dordrecht Heidelberg London New York

Library of Congress Control Number: 2009929915

Printed on acid-free paper

Springer is part of Springer Science+Business Media (www.springer.com)

Series Preface

*In my career I've found that "thinking outside the box"
works better if I know what's "inside the box."*
Dave Grusin, composer and jazz musician

*Different people think in different time frames: scientists
think in decades, engineers think in years, and investors
think in quarters.*
Stan Williams, Director of Quantum Science Research,
Hewlett Packard Laboratories

*Everything can be made smaller, never mind physics;
Everything can be made more efficient, never mind
thermodynamics;
Everything will be more expensive, never mind common sense.*
Tomas Hirschfeld, pioneer of industrial spectroscopy

Integrated Analytical Systems

Series Editor: Dr. Radislav A. Potyrailo, GE Global Research, Niskayuna, NY

The book series *Integrated Analytical Systems* offers the most recent advances in all key aspects of development and applications of modern instrumentation for chemical and biological analysis. The key development aspects include (i) innovations in sample introduction through micro- and nanofluidic designs, (ii) new types and methods of fabrication of physical transducers and ion detectors, (iii) materials for sensors that became available due to the breakthroughs in biology, combinatorial materials science, and nanotechnology, and (iv) innovative data processing and mining methodologies that provide dramatically reduced rates of false alarms.

A multidisciplinary effort is required to design and build instruments with previously unavailable capabilities for demanding new applications. Instruments with more sensitivity are required today to analyze ultra-trace levels of environmental pollutants, pathogens in water, and low vapor pressure energetic materials in air. Sensor systems with faster response times are desired to monitor transient in vivo events and bedside patients. More selective instruments are sought to analyze specific proteins in vitro and analyze ambient urban or battlefield air. For these and many other applications, new analytical instrumentation is urgently needed. This book series is intended to be a primary source on both fundamental and practical information of where analytical instrumentation technologies are now and where they are headed in the future.

Looking back over peer-reviewed technical articles from several decades ago, one notices that the overwhelming majority of publications on chemical analysis has been related to chemical and biological sensors and has originated from Departments of Chemistry in universities and Divisions of Life Sciences of governmental laboratories. Since then, the number of disciplines has dramatically increased because of the ever-expanding needs for miniaturization (e.g., for in vivo cell analysis, embedding into soldier uniforms), lower power consumption (e.g., harvested power), and the ability to operate in complex environments (e.g., whole blood, industrial water, or battlefield air) for more selective, sensitive, and rapid determination of chemical and biological species. Compact analytical systems that have a sensor as one of the system components are becoming more important than individual sensors. Thus, in addition to traditional sensor approaches, a variety of new themes have been introduced to achieve an attractive goal of analyzing chemical and biological species on the micro- and nanoscale.

Preface

Sensor arrays for chemical vapor sensing, frequently known as electronic noses, have grown in popularity over the last two decades. The relative simplicity of design and small size, leading to ease of use, make electronic noses very appealing for applications such as process control monitoring, environmental monitoring and medical diagnosis. Since the introduction of the concept of an electronic nose in the 1980s, starting with work on arrays of metal-oxide vapor sensors, there has been a significant increase in research into sensing materials and the use of arrays. Today, there are several journal articles a month discussing evaluation and selection of sensing materials as well as associated work such as transduction methods, measurement circuitry, data analysis, and sampling methods.

As approaches to designing and using sensing arrays have become more mature, and as applications of the technology have grown, it has become increasingly important to tailor the sensor materials in an array to the selected application. From the early days of research and demonstration, work has moved to focused applications, which require attention to selection of types of sensing materials as well as to selection of specific sensors within a type. Empirically derived models and first-principles computer simulations are playing an increasingly important role in our understanding of the interactions between sensing material and analytes, where the sensing materials may be polymers, metal oxides, self-assembled monolayers (SAMs), or biologically based materials.

In general, selection of sensors for a sensing array is a three-step process. First, a transduction method and a class of sensing material appropriate to that method are selected. Second, specific sensing materials within the selected class are evaluated as candidates for inclusion in the array. Third the sensing materials which will make up the array are selected. We have focused this volume on the second and third steps, selecting and evaluating specific sensing materials in order to select the elements in an array. This volume covers methods which have been used successfully in the construction of full sensing devices as well as emerging methods which show promise, with a particular emphasis on computational and statistical approaches to materials and array evaluation and selection.

We begin this volume with an introductory chapter focused on experimental methods for evaluation and development of chemical sensors and sensor arrays. Chapter 1 begins with a discussion of the mammalian model of olfaction and how it has inspired the array-based approach to chemical sensing. It goes on to establish the issues that must be considered in developing sensing materials and sensing arrays, such as sensor feature space, sensor orthogonality, geometries, and transducers. Chapter 1 also discusses the issues underlying the design of experiments and sensor evaluation, and finally, the use of experimental data in arriving at an endpoint in the evaluation process. This introductory chapter lays the groundwork for all the approaches discussed in this volume; that is, it establishes an approach to planning what to do once we have determined which sensors to test.

These computational approaches to sensor and array evaluation and selections are divided into three parts (1) First-Principles Methods of Materials Evaluation and Selection, (2) Multivariate and Statistical Methods of Materials Evaluation and Selection, and (3) Methods for Array Selection and Optimization.

Part One, Chaps. 2–6, discusses First-Principles Methods of Materials Evaluation and Selection. The general goal of developing a model of sensor performance based on first principles is not to replace existing experimental methods or knowledge-based methods of sensing material selection, but to complement these by providing quantitative approaches which can be used to prioritize the selection of new materials. First-principles design methods are being developed which can be used to plan rational modifications in the structure and function of a sensing material. This design methodology allows us to develop a theoretical understanding of the sensing material and analyte system and to predict their interactions. The predictions can then be put to an experimental test.

First-principles calculations include quantum mechanical, molecular dynamic, and structural approaches. These methods have focused primarily on developing fundamental electronic and atomic level descriptions of materials to provide insight into chemical interactions of materials with target analyte(s). Quantum mechanical techniques are discussed in Chaps. 2, 3, and 6. Molecular dynamics or atomistic techniques and statistical mechanical and multiscale approaches are discussed in Chaps. 3, 4, and 6. Chapters 3 and 6 describe a method which relies on both quantum mechanical and molecular dynamics approaches for screening sensors for their response to specific analytes. De novo structure-based design of receptors for selective chemical sensors as described in Chap. 5 applies fundamental information about structure and bonding as a basis to search for host architectures that are highly organized to form a complex with a guest molecule.

Chapter 2 uses application examples to illustrate the use of Density Functional Theory and electronic transport modeling based on nonequilibrium Green's Function in modeling carbon nanotube-based nano-electromechanical sensors and the gas-sensing properties of carbon nanotubes and metal-oxide nanowires.

Chapters 3 and 6 show that a combination of quantum mechanics with first-principles molecular dynamics can afford a great deal of information that is useful in designing and selecting materials for specific analytes.

Chapter 4 investigates the predictions of sensor responses using Grand Canonical Monte Carlo simulations. This method is used to predict the degree of sorption of analyte into polymers by calculating partition coefficients of alcohols, aromatics, ketones, esters, alkanes, and perfumes for typical gas chromatography films and compares predicted values with experimental values.

Chapter 5 presents an overview of a computer program, HostDesigner, that has been created to allow the de novo structure-based design of receptors that are structurally organized for complexation of small ionic and molecular guests. The methodology applies fundamental information about structure and bonding as a basis to search for host architectures that are highly organized for guest complexation.

Part Two, Chaps. 7–9, discusses Statistical and Multivariate Methods for Materials Evaluation. In this section, the work of various laboratories that have taken a combined theoretical and experimental approach to problems in vapor sensing and identification is discussed. Statistical and multivariate methods include semiempirical approaches, such as combinatorial approaches, Quantitative Structural Activity Relationships and Quantitative Structure Property Relationships, and calculation of solvation energy relationships. Many of these approaches have been developed to elucidate mechanistic aspects of sensing material activity. These approaches can, however, also be used to guide selection of materials. As array-based chemical sensing is still a relatively young field, many of the computational methods for sensor selection are still in a developmental phase.

Chapter 7 covers the experimental technique of high throughput (HT) screening, which applies combinatorial strategies to screen large sets (tens and hundreds) of sensing materials. This topic is discussed in greater detail in a companion volume in this series.

Chapter 8 discusses a statistical and multivariate method for correlating sensor response with molecular descriptors using a combination of Quantitative Structural Activity Relationships and Quantitative Structure Property Relationships. This approach develops statistically validated models of sensor response based on experimentally developed data.

Chapter 9 shows how an understanding of solubility interactions informs the selection of polymers to obtain chemical diversity in sensor arrays and obtain the maximum amount of chemical information, using principle components analysis to analyze array data. This chapter also discusses new chemometric methods which have been developed to extract chemical information from array responses in terms of solvation parameters serving as descriptors of the detected vapor.

Part Three, Chaps. 10–12, Designing Sensing Arrays, considers the computational and experimental methods that have been used together to select the components of an array designed to detect a particular analyte set.

Statistical methods based on experimental data have been used successfully to optimize an array; statistical methods may also be used with data simulated in the computational approaches discussed in Part I or with sensing data analyzed by methods discussed in Part II. The process of selecting the components of an array considers both type and identity of sensing materials, the optimum number of sensors to be used in an array for a particular set of analytes, and how the responses of sensors will be treated in data analysis.

Chapter 10 presents a generic approach for designing sensor arrays for a given chemical sensing task. This chapter describes a correlation-based metric used to assess the analytical information obtained from chemiresistors as a function of operating temperatures and material composition combined with a statistical dimensionality-reduction algorithm to visualize the multivariate sensor response obtained from sensor arrays.

Chapter 11 discusses an iterative approach to statistical evaluation of experimental responses of candidate materials for a sensing array by developing parameters which are used to evaluate sensor performance. These three parameters are used to compute a measure of sensor suitability for inclusion in an array designed to detect a given set of analytes.

Chapter 12 discusses a hybrid sensor array, a multimodal system that incorporates several sensing elements and thus produces data that are multivariate in nature and may be significantly increased in complexity compared with data provided by single-sensor-type systems. In this chapter, various techniques for data preprocessing, feature extraction, feature selection, and modeling of sensor data are introduced and illustrated with data fusion approaches that have been implemented in applications involving data from hybrid arrays.

Finally, we close with some thoughts on future directions for work in developing computational approaches to sensor evaluation. There are several computational approaches, which have been used to design and evaluate select materials for chemical sensors. Computational methods also include use of statistical and computational approaches to characterize measured and experimentally observed analyte-sensing material interactions and sensing material responses to the presence of analyte. With the increasing use of sensing arrays, computational approaches offer complementary information to that developed through experimental approaches.

Contents

Contributors

Joseph R. Stetter
Ecosensors TTD, KWJ Engineering Inc., 8440 Central Ave, Newark, CA, Suite 2D
94560, USA

Amitesh Maiti
Lawrence Livermore National Laboratory, Livermore, CA 94551, USA

Mario Blanco
Division of Chemistry and Chemical Engineering, California Institute of Technology, BI 139-74, Pasadena, CA 91125, USA

Abhijit V. Shevade
Jet Propulsion Laboratory, California Institute of Technology, Pasadena, CA 91109, USA

Margaret A. Ryan
Jet Propulsion Laboratory, California Institute of Technology, Pasadena, CA 91109, USA

Takamichi Nakamoto
Graduate school of Science and Engineering, Tokyo Institute of Technology, Japan

Benjamin P. Hay
Chemical Sciences Division, Oak Ridge National Laboratory, Oak Ridge, TN 37831-6119, USA

Vyacheslav S. Bryantsev
Division of Chemistry and Chemical Engineering, California Institute of Technology, Pasadena, CA 91125, USA

Michael C. McAlpine
Department of Mechanical and Aerospace Engineering, Princeton University, Princeton, NJ 08544, USA

James R. Heath
Division of Chemistry and Chemical Engineering, California Institute of Technology, BI 139-74, Pasadena, CA 91125, USA

Radislav A. Potyrailo
Chemical and Biological Sensing Laboratory, Materials Analysis and Chemical Sciences, General Electric Company, Global Research Center, New York, 12309 Niskayuna, USA

Vladimir M. Mirsky
Department of Nanobiotechnology, Lausitz University of Applied Sciences, Senftenberg, Germany

Margie L. Homer
Jet Propulsion Laboratory, California Institute of Technology, Pasadena, CA 91109, USA

Hanying Zhou
Jet Propulsion Laboratory, California Institute of Technology, Pasadena, CA 91109, USA

Allison M. Manfreda
Scientific Investigation Division, Los Angeles Police Department, Los Angeles, CA 90032, USA

Liana M. Lara
Jet Propulsion Laboratory, California Institute of Technology, Pasadena, CA 91109, USA

Shiao-Pin S. Yen
Jet Propulsion Laboratory, California Institute of Technology, Pasadena, CA 91109, USA

April D. Jewell
Tufts University, Chemistry Department, 62 Talbot Avenue, Medford MA 02155, USA

Kenneth S. Manatt
Jet Propulsion Laboratory, California Institute of Technology, Pasadena, CA 91109, USA

Adam K. Kisor
Jet Propulsion Laboratory, California Institute of Technology, Pasadena, CA 91109, USA

Jay W. Grate
Pacific Northwest National Laboratory, Chemistry and Materials Sciences Division, P.O.Box 999, WA 99353, Richland, USA

Michael H. Abraham
Department of Chemistry, University College London, 20 Gordon Street, 1H 0AJ, London, WCIH OAJ, UK

Barry M. Wise
Eigenvector Research, Inc., 3905 West Eaglerock Drive, Wenatchee, WA 98801, USA

Baranidharan Raman
Chemical Science and Technology Laboratory, National Institute of Standards and Technology, MD 20899, Gaithersburg, USA

Douglas C. Meier
Chemical Science and Technology Laboratory, National Institute of Standards and Technology, MD 20899, Gaithersburg, USA

Steve Semancik
Chemical Science and Technology Laboratory, National Institute of Standards and Technology, MD 20899, Gaithersburg, USA

Kirsten E. Kramer
Cognis Corporation, Cincinnati Innovation Concept Center, 4900 Este Avenue, Building 53 Cincinnati, OH 45232-1419 (513) 482-2242

Susan L. Rose-Pehrsson
Naval Research Laboratory, Chemistry Division, Code 6181, Washington, DC 20375-5342, USA

Kevin J. Johnson
Naval Research Laboratory, Chemistry Division, Code 6181, Washington, DC 20375-5342, USA

Christian P. Minor
Naval Research Laboratory, Chemistry Division, Code 6181, Washington, DC 20375-5342, USA

Part I
First Principles Methods For Materials Evaluation

Chapter 1
Introduction: Experimental Methods in Chemical Sensor and Sensor Array Evaluation and Development

Joseph R. Stetter

Abstract Sensors are devices, sensor arrays are collections of sensors, and it is through experimentation and computation that we obtain the knowledge we need to make useful analytical measurements. Gas and liquid chemical sensor arrays provide a new multidimensional analytical technique not unlike Gas Chromatography, Liquid chromatography, or GC/MS [gas chromatography mass spectrometry]. Exciting possibilities for advanced analytical measurements are emerging with the development and use of chemical sensor arrays. The multidisciplinary nature of sensor development and the diversity of the types of sensors, analytes, and applications provide a rich venue for research and development as well as the complex issues that lead to lively debates. Progress in developing arrays for analytical purposes is coming from applying new knowledge about biosystems that use sensor arrays, advanced predictive chemical computational capabilities, and significant increases in experimental materials and methods. The protocols for the experimental understanding of sensor arrays provides the foundation for present strategies and future models that will enable realization of the contributions of sensor arrays to analytical measurement science and technology.

Acronyms and Definitions

Analyte	Substance or chemical constituent whose identity or quantity is determined by conducting the analytical procedure
ANN	Artificial neural network
Ar	Argon
atm	Atmosphere (pressure)

J.R. Stetter
Ecosensors TTD, KWJ Engineering Inc., 8440 Central Ave, Newark, CA, Suite 2D 94560, USA
e-mail: jrstetter@kwjengineering.com

M.A. Ryan et al. (eds.), *Computational Methods for Sensor Material Selection*,
Integrated Analytical Systems,
DOI 10.1007/978-0-387-73715-7_1, © Springer Science+Business Media, LLC 2009

A_s	Analytical sensitivity
BAW	Bulk acoustic wave
C	Capacitance
CGS	Combustible gas sensor
Chembio	Chemical–biological
CI	Chemical interface
Cl_2	Molecular chlorine
cm^3	Cubic centimeter
CO	Carbon monoxide
CO_2	Carbon dioxide
CPS-100	Chemical Parameter Spectrometer – 100
E	Electromotive Force or Voltage
GC	Gas chromatography
H_2	Hydrogen
HCN	Hydrogen cyanide
H_2S	Hydrogen sulfide
I	Current – charge per unit time
IMCS2	International Meeting on Chemical Sensors 2
IR	Infrared
K or k	Sensitivity – signal per unit concentration
KNN or k-NN	k-nearest neighbor
L	Liter
LOD	Limit of detection
M	Mass
mL	Milliliter
MOSES II	Laboratory electronic nose by Lennertz
MS	Mass spectrometry
mV	Millivolt
nA	Nanoampere
N_2	Nitrogen
Ne	Neon
NH_3	Ammonia
NO_2	Nitrogen dioxide
O_2	Oxygen
OR	Olfactory Receptor – a G-receptor protein used in olfaction
pA	Picoampere
ppb	Parts per billion – by volume
ppq	Parts per quadrillion
ppt	Parts per trillion
R	Resistance – ohms
S	Sensor signal
SAW	Surface acoustic wave
SPME	Solid-phase microextraction
SSTUF	Shared sensor testing user facility

TAS	Total analytical system
TCD	Thermal conductivity sensor
TIC	Toxic industrial chemical
TIM	Toxic industrial material
VOC	Volatile organic compound
Z	Impedance

1.1   Introduction

At the turn of the nineteenth century, Lord Kelvin (Sir William Thompson) said, "To measure is to know," and "If you can not measure it, you can not improve it." These thoughts are timeless. For those of us developing sensors and sensor arrays and applying them to new and increasingly difficult and complex analytical problems, these quotations exemplify the two reasons we make experimental measurements with sensors and arrays. First, measurements allow us to understand the sensor principles of operation and the mechanisms of their response to analytes so that we can develop new sensors and improve the old ones. Second, experimental measurements allow us to calibrate sensors, evaluate their performance, use them effectively, and rely on them for quantitative results even in life-threatening and critical health-monitoring situations. Sensors, arrays, and the measurements they enable, help us work and improve our quality of life.

Every sensor array consists of individual sensor elements, either discreet or integrated, and all the analytical information we obtain comes from the sensors' signals. Therefore, all the information that characterizes either the concentration and/or the molecular identity of the analyte, a situation, or a simple or complex chemical environment is created by the experiment that generates the sensor data. Sensory experiments thus demand stringent control and the specific experimental setup and procedure are intricately tied to the data quality and hence the precision, accuracy, and validity of the outcome or analysis.

The first law of analytical measurement is: "a measurement is useless without a report of the error." That is, every experimental result should be reported with a quantitative statement of the uncertainty. To be fair, uncertainty boundaries are frequently implied in experimental reports, but for the most precise work or for accurate comparisons of sensors, specific error analysis is critical.

Constructing a workable sensor array and applying it with confidence are extremely difficult if the sensors that make up the array have not been fully characterized and understood. Not characterizing the sensors and yet building an array is akin to an architect's building a bridge without knowing the strength of his materials. It is thus crucial that a sensor's performance under probable use conditions be known or anticipated. Understanding the signal(s) from the sensing system, especially their error sources, drift characteristics, and failure modes is essential for accurately interpreting the signals over short- and long-time intervals.

In developing a focused analytical method, therefore, three issues need to be addressed (1) the invention and development of new sensors, (2) the invention and development of new sensor arrays, and (3) the application of the arrays. These three issues are best addressed from three perspectives, each of which can constitute a separate project (or a separate phase of a larger project). Integrating sensor technology, a sensor array, and an analytical method, constitutes both a challenging and a complex systems problem.

What data are needed about sensors and arrays for a given problem? How should the data be gathered and used experimentally in each case? Where does the most critical information come from, and how is it created in the first place? These questions are the topic of this chapter. In this regard, the "Edisonian" or experimental approach to R&D will complement theory vis-à-vis chemical sensors. Experiments are required to obtain the chemical sensor array data to improve on theory, create models, and to guide the design of the experiments themselves.

1.2 Chemical Sensing: Inspiration from Biomimetics and Biology

Natural sensing processes provide biology-inspired and biomimetic examples of the use of sensor arrays and, of course, computation in chemical sensor array research. In mammals, gas sensor arrays in the nose, liquid sensor arrays in the tongue, light sensors in the eye, and mechanical sensors in the ear involve chemical transduction processes that are part of a larger, intricate sensory system. Artificial sensors and arrays, today often called electronic noses (e-noses), e-tongues, etc., mimic biological organs, often relying on the actual biological molecules and the biological processes of sensing. Understanding the mammalian sense of smell (Fig. 1.1) provides a useful backdrop for chemical sensor array research – research that is increasingly amalgamating knowledge and techniques derived from physics, math, chemistry, biology, and engineering.

Our understanding of mammalian olfaction is expanding rapidly. Recent publications [1–3] illustrate this merging of the realms of biological and physical sciences and indicate that creating the bionic or at least the cybernetic nose is not very far away off. Additional research using authentic olfactory receptors or other bio-derived molecules as elements in sensors for target analytes is also being conducted [4, 5]. Such work parallels the extensive R&D in immunosensors, where biological molecules or fragments thereof are being used as molecular recognition elements [6, 7].

1.2.1 Understanding the Mammalian Model

Inspiration from biology has led to experimental models for sensing structures, processes, and components, including those for gas sensor arrays. Studies of the

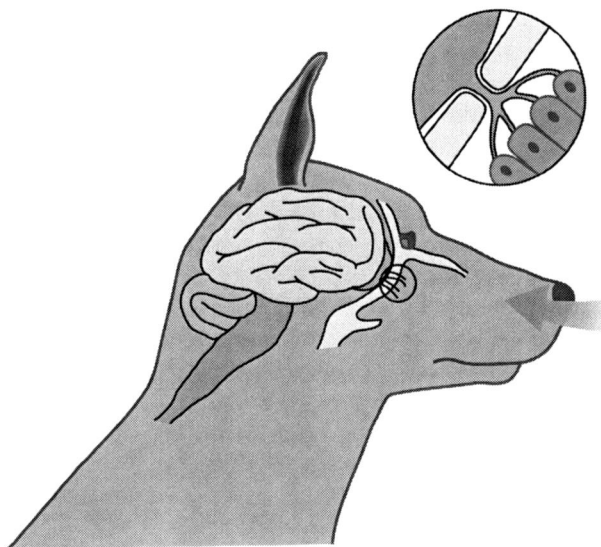

Fig. 1.1 Dog olfactory system

characteristics of mammalian olfaction have provided background information for experimental aspects of sensing with arrays. For example, the human nose contains as many as 350 types of olfactory receptors (ORs), and dog, rat, and mouse noses contain more than a 1,000 ORs. In this regard, throughout this chapter, we often use the dog olfaction model in exemplifying various sensor aspects. Thousands of each type of receptor are spatially dispersed over the olfactory epithelium in an inter-mixed, albeit not random, fashion. However, signals from each type of OR all meet at the same location in the olfactory bulb and from there, a "movie" of impulses (more study is needed to resolve the temporal/spatial aspects of this signal) is sent to the brain for interpretation. The nasal sampling system preconditions the tempera-ture, humidity, and particulate matter in the gas sample.

A dog's initial sniff brings the gas sample into the nose in about 0.2 seconds through a channel to the back of the throat where it meets a larger chamber and wafts slowly over the olfactory epithelium containing the ORs. The flow of the sniffed sample over the mucus depletes mucus-soluble odorant molecules from the sample and concentrates odorants using odorant-binding proteins. The binding event of an odorant to an OR triggers a signal in the ON (olfactory neuron); these neurons provide a nonlinear array of signals to the glomeruli (a cluster of nerve fibers in the olfactory bulb) that has the spatial and temporal qualities to encode the signals' variations, representative of the original sample's interactions. The working of the dog's mammalian olfactory system [8, 9] helps illustrate each of three complex components in a sensor array: *sampling (preparation/separation)*, *sensing (transport/binding/detection)*, and *subsequent pattern analysis*.

In a bionic-, cybernetic-, or electronic-nose, the sampling component allows for the separation, conditioning, and presentation of the sample to the detection layer either in a constant or a variable manner. The detection process and equipment can include filtration materials, such as is provided by the mucus in the mammalian nose, and can be configured so as to geometrically and chemically bias the detection system for individual molecular or supramolecular target analytes. We do not yet understand the entire role of the mucus layer in mammalian olfaction, but we do know it is a complex mixture containing molecules like the odorant-binding proteins that help in trapping, concentrating, or otherwise facilitating transport or binding of the target odor molecules. We also know that signal conditioning and chemical amplification occur in the detector layer of a mammal's nose and that the nose also contains multiple detectors or a sensor array. The olfactory analysis in mammals includes the collection of the sensory outputs into an odor record and the use of that record to identify and quantify the currently sniffed molecular presentation or odor. The analysis of sensor array patterns is performed by computers in an artificial nose but is, as yet, a poor mimic of the mammalian brain.

Systems engineering aspects to the analysis of an odor with a "nose" is revealed by consideration of the synergistic operation of the parts. Mammalian sensors apparently have "off," "on," and "partially on" qualities in their levels of response for a given molecular stimulus. "Off" is a particularly important experimental state, and one that is often ignored in the construction of artificial sensor arrays [10]. Artificial sensors can also have several partially "on" states including "on positive" and "on negative" states [11]. Sensory transmission to the brain is complex and considered largely digital with temporal and spatial qualities [8]. Mechanically, the nose contains a complex gas sampling system [11], and the sample changes composition and temperature as it enters the nose and passes over the olfactory epithelium (sensor array). In interpreting olfactory signals as odors, memory and connections to other brain areas such as the limbic system are used. For example, one way of training a dog to respond to a target odor entails using a reward to link the target directly to a response from the limbic system. Moreover, results from zoology experiments show that training a dog in odor recognition can result in a sensitivity increase that is an order of magnitude greater [12] than that of an untrained dog.

Where does this increase in sensitivity come from, given the sensor array is thought to be similar and of the same sensitivity in both trained and untrained animals? Possible answers include improved sensory feedback from the brain or a more trained and efficient cerebral algorithm resulting from something akin to weighting vectors. Alternatively, the number or activity of a certain specific OR may change, or signals from certain types of samples may be suppressed. Thus, whether the sensitivity increase from training is due to improvements in computational capability or to physical changes, or to both is still debated.

We can cite examples of physical and neuronal changes that affect odor perception in mammals. Bodily influence on mammalian olfactory performance is exemplified in the well-known heightened sense of smell for some odorants in pregnancy and the depressed sense of smell in people undergoing chemotherapy. Accident victims with brain trauma can temporarily or permanently lose all or some of their sense of smell. In any case, the sensory capability in the physical hardware of the sensory system can be affected in many ways, and the brain's plastic computational capability can be changed to enhance or destroy the effective sensory experience. Hence, using the mammalian model affords us with parallel choices in constructing artificial sensory arrays, and we should make full use of our knowledge of these systems to create increased experimental possibilities.

1.2.2 Extrapolating from the Mammalian Model

We can make experimental progress in sensory problems by examining biology in two ways. First, we can create a hypothesis about the manner in which the biological system works and then emulate it in engineered hardware and software, with the resulting system evaluated and studied experimentally. If we achieve the expected result, we may be on the way both to understanding the biological system as well as to improving our own sensor array performance. Second, we can continue to unravel the mechanism of cellular signaling and interconnections at the molecular level. This second task is daunting, but progress has been and is being made at a rapid pace in using receptors and cloning receptors for use in artificial sensing.

We live in an odor-rich world in which an initial analog sensory interaction is converted to a digital code both in mammals and artificial noses. Each sample is characterized by a high degree of chemical or molecular variation, interactions of different molecules with the sensors are often unique, and the signal information created by the sensor(s) has the ability for extremely high and diverse information content. We need to better understand each of these issues from the experimental perspective. This issue of "richness" in sample and signal can be called the *experimental diversity*, and we address it in Sect. 1.3.

1.3 Experimental Diversity in Chemical Sensors and Sensor Arrays

In this section we examine the diversity or dimensionality of the problem and experimental hardware and methodology. The sample has diversity and the array of different signals create the dimensionality. Experimental sensor arrays are not yet able to gather and use all available data, data features, and available signal dimensions. However it is good to consider chemical sensor array experiments as "imaging" in the various dimensions that *are* available: space (x, y, z), time, chemical and biochemical composition, and concentration dimensions. Feature space for imaging purposes is an abstract space wherein each dimension can be considered a coordinate and n-dimensions result in an n-dimensional feature space. Thus, any property of an odor/molecule can be used as a dimension or feature in n-dimensional feature space. A molecule can be sorted on the basis of molecular weight, polarity or dipole moment, electrochemical activity, or other chemical property in chemical feature space. Therefore, *feature space*; that is, an abstract space where each sample or molecule (or "pattern/element" or "feature") is considered a point in n-dimensional space, with its dimension determined by the number and value of "the patterns" or "the features" used to describe the sample or molecule." The importance of the chemical and biochemical dimensions of sensors becomes more apparent when we consider that each molecule in a sample that is different from another molecule contributes to the sample's molecular diversity and that each molecular property being measured by the sensor–molecule interaction is considered a dimension – whether it is electrochemical activity, partitioning into a polymer matrix, or the molecular mass of the analyte.

1.3.1 Sample Diversity

Let us consider sample diversity first (see Table 1.1). Although perfect analysis of a gas sample is not yet possible, consideration of the molecular diversity of

Table 1.1 Molecular diversity in a single breath

Volume of a normal breath (L)	0.5
Molar volume at standard temperature and pressure (L)	22.4
Moles in a breath	0.022321429
Molecules per mole	6.02E + 23 (Avogadro's number)
Molecules in a breath	1.34357E + 22

Breath constituents	**Molecules**	**Total molecules**
78% N_2	1.05E + 22	1.04813E + 22
20% O_2	2.69E + 21	1.31688E + 22
1.9% H_2O at about 80% relative humidity	2.55E + 20	1.34241E + 22
400 ppm CO_2 (0.04%)	5.37500E + 18	1.34294E + 22
5 ppm CO (0.0005%)	6.71875E + 16	1.34295E + 22
500 ppb for 150 VOCs each	6.71875E + 15	1.34295E + 22
600 ppt for 100 unknowns	6.71875E + 12	1.34295E + 22
500 at femtomolar ($10e^{-15}$)	6.71875E + 09	1.34295E + 22
X at attomolar ($10e^{-18}$)	10,000 ?	
X at zeptomolar ($10e^{-21}$)	10 ?	

Avogadro's number is the number of molecules in a gram-mole of any chemical substance. In the table entries E = 10, and " + number" value following E = the exponent of 10. That is, for Molecules in a breath, 1.34375E + 22 = 1.34375×10^{22}

a single breath sample is instructive in order to consider analytical complexities. In one normal human breath about 500 mL of air is exhaled from the lung, which has about a 4.8-L capacity. If we knew the exact chemical composition of this 500 mL sample at 1 atm and room temperature, we would need to specify the number, identity, and position of about 10^{22} molecules in the 500 cm^3 space. Even though most of the molecules are oxygen and nitrogen, there are a dozen other gases present at the ppm level, including CO_2, Ar, and CO; and at least 150–1,200 common organic materials, which have been measured by gas chromatography and mass spectrometry (GC/MS) on breath samples at the ppm and ppb level [13, 14]; plus possibly many more such as ammonia (NH_3), hydrogen cyanide (HCN), or hydrogen (H_2) that are not easily chromatographed along with breath volatile organic compounds (VOCs). A few techniques are able to peer into the window below ppb levels, but literally thousands and thousands of different compounds could be present at the ppt, femtomolar, and attomolar levels, including those produced by human metabolic processes or from the environment (e.g., some explosives have a vapor pressure of 10^{-14} atm and such compounds could be present if the solid particles and their vapors are present and inhaled and exhaled by humans). In Table 1.1, as we sum the molecules, when we get to the attomolar level, there are only 10,000 molecules in our sample of each type, and so even 10,000 types will only add 10^8 molecules to our total of more than 10^{22} and so we would need 10^{11} types to bring our cumulative total to account for each molecule in the breath. It is clear, current analysis of human breath can still have many unreported compounds at the ppm, ppb, and lower levels.

In fact, the true chemical diversity in one breath could easily be 10^{17} different types of molecules illustrating the immense composition and concentration diversity in a single breath. And 10^5 molecules or more of each of billions of different chemicals are likely to be present. Even 10,000 molecules are clearly sufficient concentration to measure with today's best analytical techniques and with some sensors. When analysis of a complex mixture, e.g., human breath, reports only a few hundred analytes present, we are most assuredly missing much information! Sample diversity is a daunting problem and immense opportunity for today's sensor experimenters. How will we achieve our "gedanken" experiment of perfect analysis, i.e., to specify the location and identity of each molecule even in a single cc of sample?

1.3.2 Experimental Diversity in Sensor Arrays

From a statistical perspective, we need at least as many dimensions in our experiment as we have in our real sample, otherwise the method may not have the statistical capacity to perform an analysis. We now know how many dimensions we might expect from our sample. The next question is how many dimensions can we create with a sensor array experiment?

The dimensionality or features of a sensor array experiment can be estimated by considering its constituent parts of the sensor array and their function, according to (1.1) [15]:

$$\begin{array}{ccccccc} \text{Materials} & \times & \text{transducers} & \times & \text{structures} & \times & \text{methods} & = & \text{features}/\text{dimensions} \\ 10^8 & \times & 10^2 & \times & 10^3 & \times & 10^8 & = & 10^{21} \end{array} \quad (1.1)$$

1.3.2.1 Types of Sensor Feature Space

Equation (1.1) is a bit like the Drake equation from astrophysics (an equation that allows quantification of the factors that determine the number of extraterrestrial civilizations in our galaxy with which we might come into contact) in which we can only estimate each term to provide a gross indication of the dimensionality of the imaging space that can be created by sensor arrays [15]. In (1.1), we estimate that there are about 10^{21} features in sensor array space due to the four different properties of materials, transducers, structures, and methods that produce differentiating molecular interactions/signals in sensors. *Material space* includes the elements from the periodic table that can be applied singularly or combined in many organic, inorganic, and/or biochemical compounds and composites to make sensors. *Transducer space* is divided into the different forms of energy used to transduce each sensor signal, with each providing a different class of sensor (see Fig. 1.2) for which there are many types. The classes of sensors are based on transducers for light (radiant), heat (thermal), charge (electronic), chemical (electrochemical), mechanical (mass and force), and magnetic energy. Combinations of these classes provide for, conservatively, 100 types of sensor responses. The structure and geometry of the sensor materials also profoundly influence the sensory response of those materials (e.g., the material can consist of a thin or thick film on a mechanical or optical transducer). Finally, the richest source of diversity, limited only by the creative mind of the developer, is that created by operational method. For example, a sensor can be heated to constant or variable temperatures and with a high or low sample flow rate yielding differentiating chemical responses. In addition, there are many electronic influences on sensor signals. It is the interplay of all these diverse possibilities that results in the immense number of features that exist in chemical or biochemical space and any given sensor array can contain a large number of these features. The examples shown in Fig. 1.2 summarize these aspects of sensor diversity organized by the structure and the process used to create the dimension.

1.3.2.2 Orthogonal and Nonorthogonal Features

Not all features are "orthogonal" (i.e., not correlated with one another, strictly in mathematics two vectors must be perpendicular to be orthogonal, but here we will use

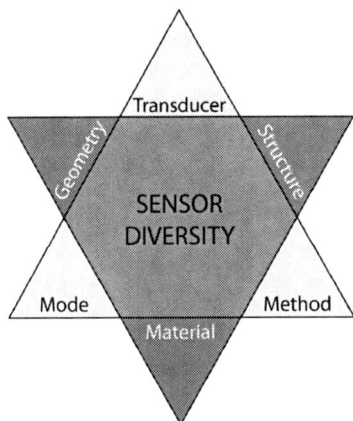

- **Transducers**
  - Mechanical — mass/force, piezoelectric
    - BAW, SAW, cantilever
  - Optical — absorption, emission
    - Fiber optic, fluorescence
  - Electrochemical/electronic — ions, e^-
    - R, C, I, E, transconductance
  - Thermal — catalytic, calorimetric
    - Pellister, thermopile, pyroelectric
  - Magnetic — Hall effect, paramagnetic
- **Method/Mode**
  - Chemistresistor, variable temperature
  - Chromatography separation/ temperature cycle
  - Amperometric, cyclic voltage
  - Sensor, spatial array
- **Material, Structure, Geometry**
  - SnO_2, NiO, WO_3, polymer, composite
  - Nanoparticles, crystal, amorphous
  - Dense, thin film, embedded electrodes

Fig. 1.2 Sensor diversity comes from design, hardware, and processes

the term non-rigorously to infer a degree of non-relatedness). For example, electronic conductivity and mass loading could have the same concentration dependence on a given sensor and hence be correlated and thus not be orthogonal. If responses are not correlated (i.e., if conductivity is exponential in concentration and mass loading is linear in concentration), they would have at least some degree of orthogonality.

In an array experiment, the orthogonality of the sensory responses is important because it implies a higher information content for the array of signals. Creating orthogonal parameter space with sensors can be challenging, but is necessary for effective analytical results. Experimentally, more often than not, a one-to-one relationship does not exist between the number of sensors in an array and dimensionality in the sense of independent noncorrelated or partly correlated responses of the data. A single sensor can be operated by many methods, including cyclic heating, cyclic voltage application, and pulsed current, and these combinations can produce multidimensional data rapidly. Sensors can also combine transducer platforms as is done in spectro-electrochemistry to produce multidimensional data from a single sensor/method.

An early heterogeneous gas sensor array [16, 17] used four sensors and a heated catalytic combustible gas sensor (CGS) in a synergistic operation to detect virtually all gaseous compounds at ppm to percent levels. The chemical parameter spectrometer (CPS-100) instrument that implemented this early pre-electronic nose sensor array era gathered 16 channels of data from four sensors operated in four modes. These 16 channels were not totally independent (orthogonal) for all analytes as shown by statistical analysis [17]. Later, a modulated concentration sensor was used to provide added features in the data that reflected the chemical composition [18]. This example of forming multidimensional data by using different sensors and processes is presented in more detail later to illustrate the impacts of experimental work on sensor array research.

A classic GC/MS experiment illustrates another form of the sensor array experiment. The GC separates hundreds of compounds in time, and the MS detects them in sequence. In this case, one MS detector provides a multidimensional mass spectrum of data for each compound, often doing so multiple times each second. Collection of the entire data set can take more than an hour for each GC sample, producing extremely large multidimensional data bases. Such data has been used to derive structure-activity relationships for, among many other things, drugs and to differentiate killer bees from European honey bees. Modern arrays, sensors, and sensing systems use many dimensions as well as scientific and engineering methods/techniques [19].

1.3.2.3   The Experimental Matrix Surrounding a Single Sensor

To understand "sensor-created" feature space, we need to look at the experimental matrix that can be obtained from a single sensor. A single sensor exposed to a step change or pulse of gas produces signal-vs.-time data. These data, in turn, are characterized by several features, including signal height, the area under the signal–time curve, the response and decay times, and the ratios of the several features. The many dimensions possible by combining just the material and transducer platforms and modes of operation are estimated at more than 10^8 combinations, which, in fact, appear to be a conservative estimate.

1.3.2.4   Transducers

Transducers can manifest any of the six types of energy exchange – thermal, mechanical, chemical, electronic, magnetic, and optical. Each transducer category can be considered a sensor class; each class can consist of many sensor types, with each type based on the parameter measured. Electronic transducers can measure resistance (R), capacitance (C), or impedance (Z). Mechanical transducers may respond to and their signal may reflect change in mass (M) or elastic properties or a combination thereof. Optical transducers measure a change in optical properties (e.g., absorption or emission of a specific wavelength of specific energy or the change in the refractive index). A conservative estimate of transducer parameter space is 10^2 dimensions because transducers can operate alone or in combinations.

1.3.2.5 Sensor Geometries and Structures

To produce differing responses, sensors can also be made with many geometries and structures. On a mechanical platform, thick sensors may produce slower but more sensitive responses per unit of analyte than do thin sensors. In addition, an observed sensor response may be due to more that one effect (e.g., simultaneous mass change and elasticity change in a polymer with increasing analyte concentration). Each of these effects can be the dominant response under a different set of conditions (e.g., sensor thickness, geometry, or formulation, or at a different analyte concentration). Such effects produce multidimensional data from sensor responses and allow "imaging" in chemical space when combined with the other dimensions of the structure.

1.3.2.6 Methods of Sensor Operation

By far the largest contribution to diversity in sensor response is the method of operation. Methods include internal and external modulation of sensor responses, combination of sensors into homogenous or heterogeneous arrays, and the production of hyphenated, multidimensional sensor responses from sensor systems of all kinds, various qualities, and any number of sensors [19]. Examples include heterogeneous arrays [17], modulation of input concentrations [18]; modulation of sensor operational variables like electrochemical bias [16], operation in nonsteady-state modes [20]; and discontinuous, cyclic, or pulse modes of operation. The highest information content is acquired when the sensor or sensor array can obtain parameters that contain noncorrelated (orthogonal) responses to the same analyte, the concentrations, or the target matrix. For example in spectroelectrochemistry, the data from electrochemistry can relate to the reaction of one functional group in a molecule [e.g., nitrogen dioxide (NO_2) reduction] and the spectroscopic information can relate to another part of that molecule (e.g., light absorption of the aromatic side chain). Both signals provide information about the molecular identity, and signal intensity relates to concentration. Differing concentration dependence of the two responses can add significant data dimensionality for example, if one parameter measurement functionality is linear and the other logarithmic and the method involves changes in concentration.

Experimental parameter space for sensors is extremely large and difficult to navigate, but can yield an enormous number of features in an array's sensory responses. The prospect of working in such a space experimentally with so many possibilities is daunting but also leads to the richness of this research and the possibility of significant analytical power from sensors. The important role of experimentation is thus to quickly screen choices and select promising practical approaches from the many possibilities. Combining experimentation with theory helps narrow choices for achieving different sensor responses by forming hypotheses about those responses. These hypotheses can guide and simplify experimental work. In sensor array development, a powerful paradigm for analytical advances is created when theory and experiment work together.

1.4 Experimentation to Create New Sensors

1.4.1 New Tasks for Sensors

Today, chemical and biochemical sensors are being tasked with providing information about more and more analytically-complex endpoints. We not only ask of the sensor/instrument how much substance in general is present (quantitative analysis), but also ask how much of a particular substance A is present (qualitative analysis). Instruments that contain sensor arrays are also tasked with answering other complex questions about product quality and environmental situations: Has this coffee been roasted? Has this cheese been adequately aged? Does this plastic have an off odor? Where does this toxic spill come from? Is this situation hazardous or toxic? What "type" of fire is beginning and in what "stage" is it? The endpoints desired are often chemically or biochemically complex, and demand a great deal, analytically from the sensory system.

We assume that the answers to these questions are contained in a sample's molecular diversity. The complex endpoint analyses needed are often analogous to finding a needle in a haystack (e.g., a ppb-level of benzene in air or does this package contain TNT?). Can a chemical sensor array conduct these analyses? We know that the analytical task is possible because a dog can do it, often astounding us by performing such experimental feats in only a few seconds with a sniffer, a biosensor array, and just a few ounces of gray-matter as its "computer." We know that the airborne information is transmitted in the molecular diversity of the sample and all we need is a suitable sampler, sensor array, and computational capability to answer such complex analytical questions. However, the performance of all sensor array systems is not equal and artificial noses have not been able to duplicate the feats of mother nature yet, although the gap is closing today between artificial and natural approaches.

Clearly, we have much more to learn from natural systems, particularly vertebrate and invertebrate, and even from insect olfactory organs. We can find the answers only through experiment. This last statement may be controversial. For example, what can we learn from theory and what can we learn from experiment? In many cases, doing the experiment and creating a device is the only way to understand something. In the early days of electricity, Michael Faraday sent a copy of his electric motor to his colleagues, not a paper about it, because the action of the device transmitted more information than could be written at the time. There is knowledge and information in the experiment itself. We shall not further argue here the merits of experiment vs. theory. As we know, they are complementary and valuable. However, experiment is the final arbiter and herein, we are emphasizing the role of experimentation.

1.4.2   Defining the Approach for Creating Sensors and Arrays

Chemical sensor array instruments (like the e-nose or e-tongue) have been created primarily to perform experiments. To understand these devices, we must break them down into their components, just as the experimental process itself can be divided into instructional steps. We need to follow specific steps to progress toward achieving a goal and, whether they are formal or informal, building an effective sensor array instrument that needs to include several distinct steps.

1.4.2.1   Step 1: Seek to Understand, from a Fundamental Viewpoint, the Problem/Goal

For any sensor array experiment, it is important to articulate as clearly as possible its specific goal or objective; otherwise, we cannot tell when we have achieved that which is wanted. Conducting a comprehensive literature search about the topic is typically required, and clearly stated goals with a specific and focused application in mind will narrow the often vast possibilities.

1.4.2.2   Step 2: Isolate and Identify Major Issues in Reaching the Goal

All available information on the topic needs to be distilled to isolate and identify relevant gaps in knowledge or technology. The major issues/problems that prevent the immediate design and implementation of a complete solution to the problem at hand should be listed. Step 2 can be conducted concurrently with Step 1, with the ongoing literature search and theoretical analysis used to identify and quantify the major issues as they are defined. Those issues will most likely consist of which of the materials, structures, components, processes, and algorithms need to be chosen, as well as a list of systems and interface problems.

1.4.2.3   Step 3: Develop a Strategy

A strategy consists of creative thought and conclusions about the possible approaches that will be required to reach the goal and overcome the major problems. A strategy should seek to bring typically disparate or often conflicting requirements into harmony. Sometimes, for instance, a solution may require both a fast sensor and an extremely sensitive sensor; if so, compromises will be required in which more sensitivity will require more time for sample collection. In this case, the strategy would need to incorporate anticipatory response time algorithms to meet the speed goals for sensitive, but slow sensors. This is but one simple example of strategy development.

1.4.2.4 Step 4: Prepare a Workable Plan

A workable, efficient, and effective experimental plan that outlines the best experimental path for implementing the strategy and getting to the goal is needed next. The experimental plan should be viewed as a living document that will change as execution of the experiment proceeds. What is learned along the way will be used iteratively to refine Steps 1–3 and update the plan. Careful design of experiments can save immense time and effort.

In an ideal world, budget and time would be unlimited; in the real world, the challenge is to frame the solution to the problem into a feasible plan. Doing so is often accomplished in a reality feedback session that makes up part of the strategy discussions. It always helps to get feedback from skilled colleagues.

1.4.3 Carrying out the Approach

During implementation, when an impasse is reached, and when we truly do not know what is the best experiment to do, the best advice is often to do something. Virtually any relevant and well-crafted experiment is better than no experiment and will provide guidance and direction so that progress can resume.

Scientists have developed chemical sensor array systems that produce complex data sets to detect or monitor specific target endpoints. The experimental work draws on two major sources:

1. The physical world of hardware consisting of components and devices.
2. The myriad methods the hardware uses to generate data.

Figure 1.3 illustrates the gaseous chemical sensor array experiment, showing the order of the hardware components and implementation of a simple method. Shown are:

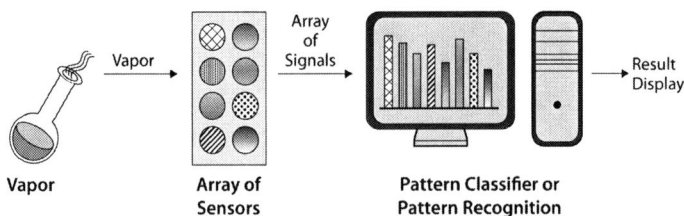

Fig. 1.3 Sampling system, sensor array, signal processing algorithm/display

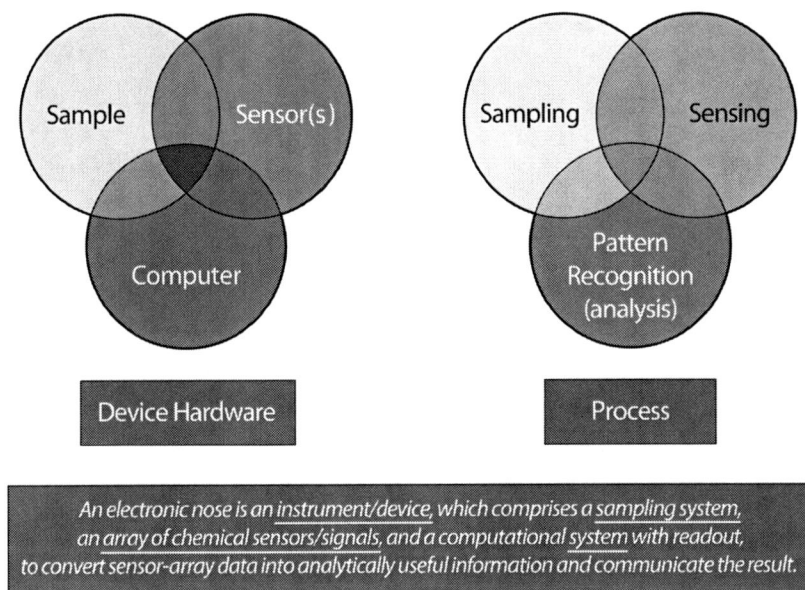

> **Sample** **Sensor(s)**
>
> **Computer**
>
> **Device Hardware**
>
> **Sampling** **Sensing**
>
> **Pattern Recognition (analysis)**
>
> **Process**
>
> *An electronic nose is an instrument/device, which comprises a sampling system, an array of chemical sensors/signals, and a computational system with readout, to convert sensor-array data into analytically useful information and communicate the result.*

Fig. 1.4 The e-nose is a device that implements an analytical process

1. The sample and sampling procedure.
2. The sensors and the sensing process that generate the raw data about the endpoint.
3. The analysis of the sensory data using a computational and/or pattern-recognition method with display of the result.

Of course, many creative systems of components can be used experimentally to implement the e-nose method. From the perspective of sensor R&D, it is helpful to clearly divide the parameter space being investigated into hardware and process components (see Fig. 1.4). This is especially true when trying to design experiments and then explain and interpret the resulting sensory data. We return to this theme of the division of issues later in the chapter. It is important to note that where *division* of issues is required for understanding data and experimental design, it is also imperative to *integrate* components as quickly as possible in the instrument development cycle in order to uncover interface issues and deal with them expeditiously.

1.4.3.1 Sampling and Samples

Samples can be gases or the gases above solids or liquids. Often, for the gas sensor or sensor array, all three phases exist together because the air is laden with particles of micrometer and sub micrometer size, as well as aerosols, molecules, and supramolecular clusters. The target for the sensor(s) analysis could be a specific

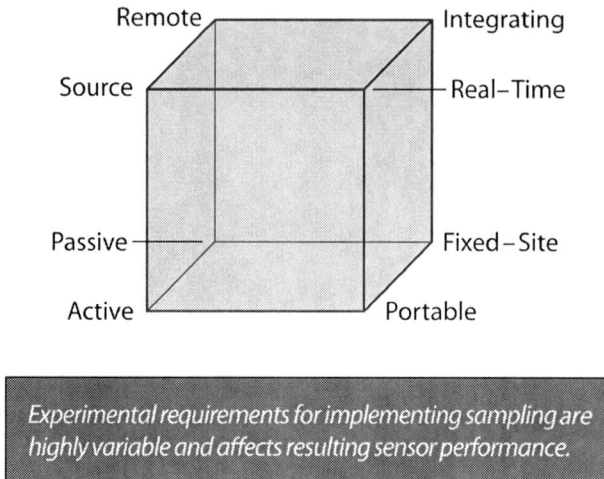

Experimental requirements for implementing sampling are highly variable and affects resulting sensor performance.

Fig. 1.5 Sampling methods are diverse because of the demands of applications

chemical analyte (e.g., gaseous chlorine [$Cl_2(g)$] or hydrogen sulfide [H_2S]) or a more complex endpoint or situation such as gasoline vapors, a fire, an explosive hazard, or diagnostic marker(s) for a disease we wish to detect.

The sampling cube shown in Fig. 1.5 illustrates the major dimensions of sampling. The sampling method is the experimentalist's first chance to purposively direct the sensing system toward the desired goal by eliminating interferences and enhancing sensitivity or other performance variables. Typical gas sensor arrays are designed to continuously sample an environment at a point, with the sample either actively drawn into the sensors or allowed to passively diffuse to the sensors. The sensor array can be portable, wearable, or installed at a fixed site. Fixed-site analyzers typically have more power available to them, and hence more sampling options for the instrumentation are possible using pumps. The sampling process can take place either in real time and/or be designed to collect sample over time (integrating). A real-time sensory readout can be integrated into the instrument, but a sampler that collects a specimen cannot provide time resolved data shorter than the collection interval.

Several types of samplers are commonly used, and together with sensors, can be utilized to implement a wide variety of methods targeted to include everything from finding clandestine burial sites [21, 22] or determination of environmental contamination [23, 24] to early diagnosis of lung cancer using human breath samples [25]. Samplers include the following types:

• Auto samplers
• Solid phase microextraction (SPME)
• Sorbent tubes
• Passive badges
• Canisters

- Bags
- Active pumps

The MOSES II is a sensor array instrument designed as a modular electronic nose [26] and was often used with an auto sampler capable of controlling the exposure of the array to precise aliquots of headspace vapors. The auto sampler was designed to hold a number of sample vials into which a sample of the materials is placed. The auto sampler robotically extracts an exact volume (e.g., 3 mL of air or headspace air at 35°C and 1 atm) for introduction into the sensor array. The auto sampler's robotic implementation allows significant sampling precision over manual methods. The SPME method involves collection of the sample by absorption into a polymer layer like a GC stationary phase with a retractable syringe needle coated with the absorbent. SPME is most often used with GC/MS analysis, but can also provide a repeatable method for collecting adsorbable organic materials in any analytical scheme. Sorbent tubes of variable size and shape and passive badges can be filled with a suitable sorbent that either adsorbs or absorbs chemicals of interest. Canisters and bags are evacuated vessels that can collect an entire sample, not just the part that is "sorbable"; however, the sample can decay or react before analysis can take place in the sensor array, and this is a limitation of any method involving significant time for sample transport between collection and analysis. Pumps or gaseous diffusion barriers provide real-time sampling systems and, if interfaced properly, let the sensor "see" the entire sample before it is degraded by the passage of time. The precision of the sampling method is important since it can limit the precision of the overall analytical method. Similarly, the accuracy of the method will be affected by the fidelity of the sampling scheme.

Appropriate sample collection and preconcentration can increase sensitivity by a factor of 100–1,000, can improve selectivity by biased collection of the relevant fraction of the sampled composition, and can improve the precision of the method through auto samplers and auto-injectors. Samplers separate a sample into sub-groups of molecular interest and reject those that are interferences or of no interest, which can vastly increase the selectivity of the overall analytical system.

For each application, therefore, the developer must consider the advantages and disadvantages of sampling systems that can be interfaced with the sensor array. Applications are quite varied: consider the Scent Transfer Unit™ (STU 100), which was designed to obtain a scent pattern and impart it to the "live" dog's olfactory sensor array [27]. Such a sampler could be used with any sensor or sensor array and impart similar operational advantages. The examples provided here reflect only a small part of the art of chemical sampling, and although working on the sampling often lacks the glamour of sensors, experimentally sampling provides the first chance and a most significant chance to endow the resulting sensor array instrument with advanced analytical prowess.

As with sensors, it is instructive to learn about sampling and about the interface of sampling with sensor arrays from nature, and in particular the dog [8, 9]. The dog's nose contains a "bio" sampler system and provides an example of an effective sampler interface with a sensor array system. A bloodhound's nose takes air into the center of the nose, and expels it through a slit at the side of the nose. The air sample is drawn in straight and fast through a channel that conditions both the temperature and relative humidity of the sample, and then follows a tortuous path in which it passes over the sensors in the olfactory epithelium. These sensory receptors lie beneath a soup of nasal mucus. As the sample is sniffed rapidly [0.2 s per sniff] and conditioned by the nasal cavity, a relatively slow and repeated exposure to the olfactory sensor array occurs. The sample, as it first enters the olfactory epithelium, is intact and homogenous, but as it progresses along the epithelium, materials soluble in mucus are depleted. This process results in a sample that is richer in insoluble material and light gases for the latter part of the epithelium. One would expect changes in the sensor response to vary as well, in such a system.

The entire canine sampling process produces a time- and location-dependent sensory response to an increasingly fractionated sample. The olfactory epithelium contains millions of receptors of thousands of different types interspersed in the epithelium in a spatially non-random manner. Each receptor type sends its signal to the same neuron, thereby gathering the signals from many ORs over time into a fewer number of nerve impulses that in turn are transmitted to the brain for processing. Although much of the olfactory process remains to be understood, observation indicates that the dog's brain, which weighs only a few ounces and uses much less power than the 250-W processor in our PCs, can perform enviable feats of detection and identification. The dog model teaches that it is possible to sniff an air sample and isolate and identify markers for specific and even complex endpoints, including a specific person's identity in the presence of many (but not all) interferences.

Experimentally a sampling system must be selected and designed using the parameter space that is appropriate for the analytes, endpoint, and sensor array hardware and software. The systems engineering and component interface issues surrounding sampling are important as well, as discussed in the next section.

1.4.3.2 Sensing, Sensors, and Sensor Arrays

Each of the three generic types of sensors – chemical, physical, and analytical systems - can be placed in an array to produce multidimensional data. Chemical sensors (see part 1 of Fig. 1.6) use a chemically reactive layer to interact with the analyte and produce an analytically useful response. Physical sensors (see part 2 of Fig. 1.6) can also be used for chemical analysis by measurement of a physical interaction of energy and the molecule. And sensor systems can be built (see part 3 of Fig. 1.6) that implement an analytical sensing process as may be represented by the lab-on-a-chip GC example illustrated or by a complex instrument like a mass spectrometer or ion mobility spectrometer. Each type of sensor requires the developer to approach the experimental research differently.

Chemical-biological (chembio) sensors can use optical, magnetic, electrochemical, electronic, mechanical, thermal, or magnetic transducers to produce the

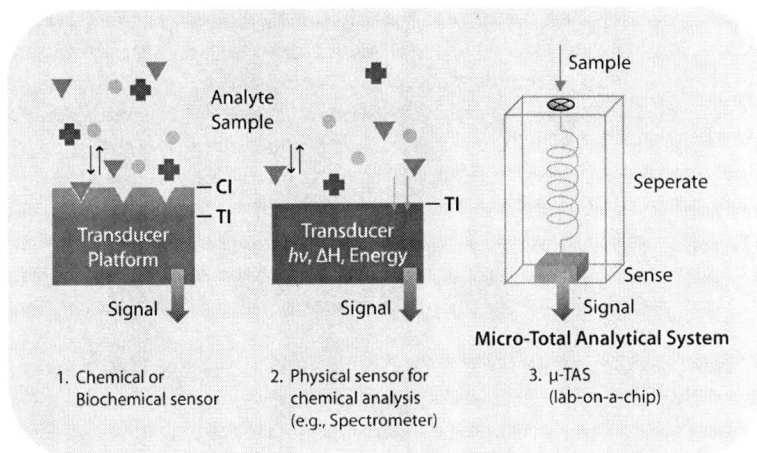

Fig. 1.6 The three designs for sensing that can be used to produce multidimensional data

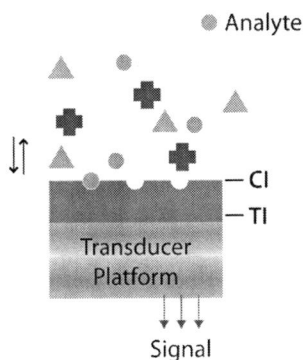

It is a small self-contained Integrated System of parts that because of

- A chemical interaction or chemical process between the analyte and the sensor's interfacial layer

- Transforms chemical or biochemical information into an analytically useful signal [32]

CI: *chemical interface*

TI: *transducer interface*

Fig. 1.7 What is a chemical sensor?

electronic signal that is proportional to molecular identity and concentration. The chemical interface consists of a coating or detection layer that chemically reacts with the analyte in some manner (Fig. 1.7).

The physical sensor [e.g., an infrared (IR) spectrometer or a thermal conductivity sensor (TCD), see part 2 of Fig. 1.6], does not chemically react as a part of the sensor with the analyte, but instead senses a change in energy transduction as the analyte absorbs, emits, or scatters energy. The IR spectrometer and TCD sensors are physical sensors that can respond to chemical changes in the environment. Chembio-sensors are fundamentally different from physical sensors because the sensor coating changes chemically, i.e., the coating actually is a different chemical or composite during sensing and after sensing because of its reaction with the analyte. A chembio-sensor's coating is a different material form during each phase of the sensing process, but must then return to prior conditions chemically for reversibility to be observed. In fact, virtually all of the analytically relevant performance characteristics of the chembio-sensor (e.g., sensitivity, selectivity, speed of response, stability) are determined by the thermodynamic and kinetic reaction characteristics of the chemical reaction between the coating and the analyte.

In a physical sensor (part 2 of Fig. 1.6), physical processes and fundamental transducer performances limit sensor performances in contrast to the chembio-sensor (part 1 of Fig. 1.6) wherein reaction with the coating often limits sensor response. While this process/mechanism of detection is not always relevant to the end user, it must clearly be understood by the sensor developer in order to tailor the sensor response to the sensing objective.

To address complex analytical problems in small sensor systems, total analytical systems (part 3 of Fig. 1.6) have been developed that combine sampling, separation, and sensing operations. Miniaturized instruments, including MS, IR, and other types of spectrometry and spectroscopy, are also widely developing as sensors. Such approaches to sensing can be considered sensing systems like the lab on a chip because they include a sample introduction and processing as well as the detection of some property of the processed molecular material. In MS, for example, the sample is introduced, ionized to create charged molecules or fragments, accelerated through an electric and/or magnetic field and then the charge is detected. In addition, because systems like GC/MS have become the "gold" standard for many analytical measurements, they are being miniaturized in order to be useful in tiny portable sensing systems.

All three types of sensors and sensor systems shown in Fig. 1.6 can be used to generate multidimensional data. Sensor arrays, single sensors operated under different conditions, or combinations of similar or different sensors and/or operating conditions produce the multidimensional data needed for solving chemical identification and quantification problems.

1.5   Sensor Evaluation and Information Content Generation

All of the information we eventually learn about a sample comes from the signals that the sensor(s) generate. Experimentally, our sensor or array needs to have three properties: precision, accuracy [28, 29], and validity [30]. Precision is needed to get the same response from the same stimulus each time, accuracy is needed so that the true values of the response are produced within the error bounds of the precision, and validity is needed to insure that the method measures the endpoint of interest in the situation required. Of these three analytical qualities, the first two are addressed as a matter of course in analytical work, but validity is often overlooked. Nonetheless, validity is critical because it assures us that the analytical method is measuring what is intended to be measured in the application without ambiguity or interference [30].

The parameters important in the evaluation of sensors can be thought of as the "five S's" of sensor characterization (see Fig. 1.8), and include Sensitivity, Selectivity, Speed of response, Stability, and Size/shape/cost. Sensor performance parameters, those that provide quantitative boundaries for the analytical capability, all derive their measured value from the sensor signal. These parameters are required and used to develop and compare sensors, and to interpret sensor results. Comparison of these parameters provides the tool for understanding the failure modes, knowing when sensor performance needs to be improved, and indicating the method by which it can be improved. We are essentially blind during sensor development without accurate determination of these sensor characteristics.

Sensitivity, k, is often confused with the sensor signal, S, or the limit of detection. Sensitivity is defined as the slope of the calibration curve, i.e., the slope of the signal vs. the concentration curve and has the units of signal per unit concentration. Figure 1.9, to be discussed later, provides an example of a linear calibration curve for S vs. $[X]$. If sensitivity is reported for a nonlinear sensor, the concentration at which the sensitivity was determined must also be reported. Analytical sensitivity, A_s, an analytically useful but less often used measure of sensitivity, is the sensitivity at a given concentration, divided by noise, s (also in signal units). Noise is measured in the same units as the signal and, if the sensitivity

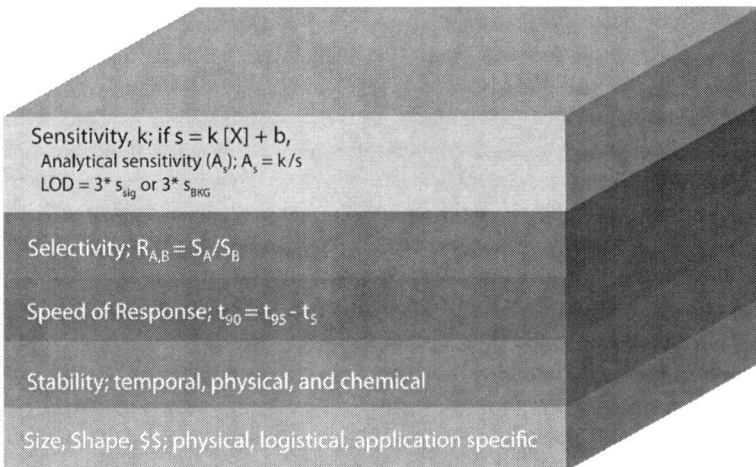

Sensitivity, k; if s = k [X] + b,
Analytical sensitivity (A_s); $A_s = k/s$
LOD = 3* s_{sig} or 3* s_{BKG}

Selectivity; $R_{A,B} = S_A/S_B$

Speed of Response; $t_{90} = t_{95} - t_s$

Stability; temporal, physical, and chemical

Size, Shape, $$; physical, logistical, application specific

All information comes from measurement of "S," the sensor signal recorded under various conditions and at various times.

Fig. 1.8 Characterization of Chemical Sensors

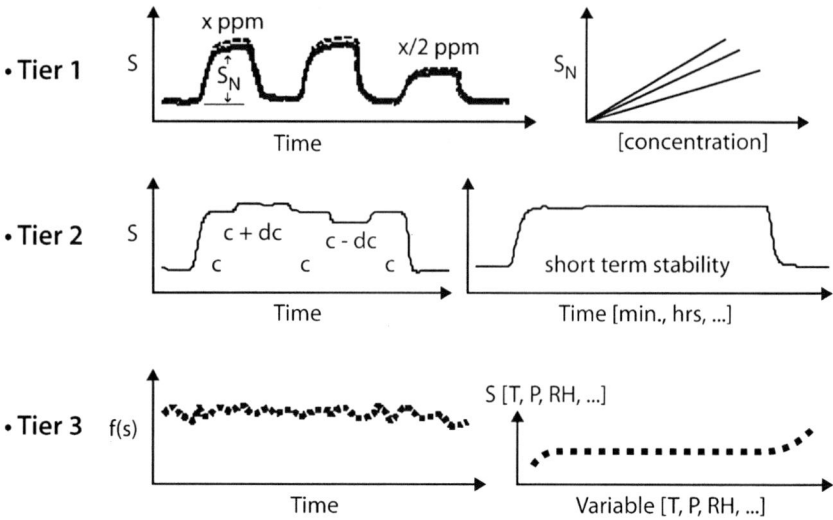

Fig. 1.9 Quantitative sensor evaluation

is in mV ppm^{-1}, and noise is in mV, the analytical sensitivity is in ppm^{-1} units. Analytical sensitivity measures the amount a signal must change before that change can be "sensed" by the sensor system at the analytical level the sensitivity is measured. Analytical sensitivity is useful for comparing the performance of sensors with different mechanisms such as comparing electrochemical sensors with heated metal oxide (HMOx) CO sensors [31]. In this example, the electrochemical sensor has a linear response and sensitivity is constant all the way to the detection limit while the HMOx response is a power law and increases at lower concentration. Comparison of the analytical sensitivity over a range of concentrations illustrates that the electrochemical sensor is more sensitive at higher concentrations, but the HMOx is more sensitive at lower concentrations [31]. The limit of detection (LOD) is typically taken as three times the noise expressed in concentration units. If the response is nonlinear, analytical sensitivity at a certain concentration can be used to determine which sensor is more sensitive to changes at the LOD concentration [31]. If a calibration curve is generated for more than one target analyte, selectivity can be calculated as a ratio of sensitivities. Again, it is important to report the concentration at which the selectivity is determined.

Sensor response and recovery times are often difficult to estimate, given that the exposure experiment must be able to produce virtually instantaneous changes in analyte concentration without causing other system upsets (e.g., changes in pressure, relative humidity, temperature) that may influence the sensor signal. It is often convenient to choose a time of response from 10–90% or 5–95% of the signal when comparing several sensors or analytes since these times can often be easily determined during an experiment.

Stability has many dimensions, and each must be evaluated independently. Under constant conditions of exposure, the temperature or the pressure can be varied to obtain the temperature coefficient of response or pressure coefficient of response, respectively. This concept can be expressed as a partial differential equation for the change in signal due to the analyte, S_A, and each influencing parameter where observed signal change, ΔS, is expressed in (1.2):

$$\Delta S = k[A] + [\Delta S/\Delta T]\Delta T + [\Delta S/\Delta P]\Delta P + [\Delta S/\Delta RH]\Delta RH + \text{etc.} + \ldots, \quad (1.2)$$

where ΔT is the change in temperature, ΔP is the change in pressure, ΔRH is the change in relative humidity, and so on. The total effect on the analyte signal is the sum of, to a first approximation, independent effects that must be compensated for in order to accurately determine the change in signal due only from the analyte concentration change. The effects of temperature and other variables must either be compensated for electronically, or this variable must be kept constant during measurement (e.g., $\Delta T = 0$). Sometimes it is possible to equip the sensor with a thermostat or to reduce temperature dependence by limiting the sensor signal by diffusion of the analyte. For diffusion-controlled signals, dependence on temperature follows a square root, $T^{-1/2}$, dependence rather than the exponential T dependence of reaction kinetics. The above equation is, of course, simplistic because it does not include the possibility that the temperature dependence of the signal might also be concentration dependent. Furthermore, the sensitivity, k, includes effects due to many experimental aspects of the sensor system including sensor design and structure as well as flow rate, electronics, and other operational parameters.

1.5.1 Protocols for Sensor Evaluation: Tiered Tests

The analytical characterization of a sensor or array can require many measurements. To expedite the evaluation process, a practical and tiered test system can be used (see Fig. 1.9) with testing implemented in the detail required by the goal of the experiment or particular sensor application. Although many test methods are possible and vary in their degree of efficiency, we have selected the following example to illustrate a typical protocol for evaluation of sensor performance. A simple reversible sensor for a simple molecular analyte is used as the example.

1.5.1.1 Tier 1 Testing

In first-tier testing (see Fig. 1.9) the sensor is exposed over time to a series of different controlled concentrations, selected to cover the range of interest, with periods of zero concentration in between each controlled concentration. This simple experiment generates what analytical chemists call the working curve for the sensor, more often called the calibration curve, and is obtained from the recorded data. From this simple calibration experiment, response and decay times, sensitivity, noise, and hence LOD, as well as short-term zero stability and span stability, can also be derived. We have assumed that the method of exposure of the sensor allows accurate determination of these parameters, i.e., the flow rate effects, temperature, rapid sensor chamber purge, and other experimental issues that have been dealt with and considered in setting up the apparatus for the calibration of the sensor.

The sensor calibration test is extremely valuable and useful. It can be used to screen sensor coatings and compare their performance, as well as to compare different formulations of the same sensor to select the "best" formulation as defined by any analytical objective such as the "most" sensitive, the "fastest" responding, etc. Moreover, this test allows comparison of different sensor classes vis-à-vis the same analyte [32] or mixture of analytes.

When this test is used for screening, the initial results quantitatively indicate if a sensor is "good" enough as is, for the intended application and whether a more thorough evaluation is warranted. Of course, to compute more accurate statistics from the calibration curve, seven or more points over the concentration range of interest would be preferred, but the complexity and time needed for doing so have to be balanced against the intended use of the resulting data (e.g., quick comparison screening of sensors, detailed quantitative characterization and calculation of engineering specifications, and/or a comparison of analytical techniques). The developer also needs to take into account that a parameter like "noise" in a measurement is a "system" property and not exclusively a property of the sensor. "Noise" types include sensor noise, electronic or pneumatic instrument noise, or even noise in the gas sampling system. Noting which type of noise is occurring is often important, especially when defining a sensor's LOD. For example, typical amperometric gas sensors [33–35] detect ppb levels with a signal of about 10 nA ppb^{-1}. If we look at the theoretical limit of current measurement, pA or lower, ppt or ppq concentrations would easily be measured with modern electronics. However, electrochemical systems with low enough background currents to allow ppq detection have yet to be developed.

Each of the variables in the experiment (even when temperature, pressure, and relative humidity of the sample are held constant), and each component in the experimental apparatus must also be considered in the overall error analysis. The formal treatment of systems with variable temperature, relative humidity, pressure, and composition during the recording of the sensor signal (e.g. the real world) is more complex and beyond the scope of this chapter, but is addressed in a similar manner to every other evaluation of sensor performance. The major point we emphasize here is that the tier 1 experiment must be well controlled to provide adequate accuracy of information on which to build our sensor characterization. Tier 1 is designed to be rapid and low cost but provide the information essential as a first level characterization for making the decision as to when and how to move forward with development of a sensor.

1.5.1.2 Tier 2 Testing

Tier 2 testing focuses on additional operational variables, including the sensor's analytical sensitivity and short-term stability. At some level, sensor performance becomes demanding for most applications. For example, an alarm may be wanted when 50 ppm of CO or 2 ppm of H_2S are reached - levels where toxic effects become significant. In these cases, analytical sensitivity and the short-term stability measurement can be designed to provide that information at the concentration of most interest. Tier 2 tests create a sensor response profile for exposure to the concentration of interest and then vary that profile by a chosen amount up, down, and back to the concentration of interest, and/or expose the sensor to a constant concentration for an extended period. These tier 2 tests allow calculation of analytical sensitivity, noise, and short-term stability. The tier 2 tests provide a second level of characterization for sensors that operate over the entire range of interest (tier 1) and allow screening of the sensors that will now perform at the critical levels of one or more critical functions. Also, short term stability is required before long term stability is considered. Furthermore, many variations of such tests are possible in order to suit testing to a particular sensing objective we might have for a sensor.

1.5.1.3 Tier 3 Testing

Finally, if a sensor can satisfy tier 1 and tier 2 basic needs, Tier 3 tests can be constructed to gather the tier 1 and tier 2 measurements for all of the variables relevant to the sensor's application, including temperature, pressure, time, relative humidity, matrix changes, and/or flow rate. When measurements are made over a period of time (e.g., once a day for 2 weeks, once a week for 12 months or more), we can compute the long-term drift/stability for an individual sensor or compare two sensors with one another [31]. A test over a temperature range allows computation of the temperature coefficient for span (the sensor's analytical signal to the analyte), background (the sensor's output or signal with no target analyte present), and response time variables, and, similarly, indicates the coefficient for any of these possible influences on sensor signal. The more measurements that are made, the more is known about the sensor's analytical characteristics and the easier it is to assess performance in a given application.

Using a tiered approach to sensor characterization, we can make important quantitative comparisons of sensor designs, structures, materials, and/or operating protocols with a cost-effective and time-efficient approach. The tiered approach also offers the advantage of being easily customized for the development project. Automation of the various steps also allows many sensors to be screened in a combinatorial approach. And if many sensors of the same type are evaluated, this approach can be used to assess yield in sensor manufacture and therefore aid in process or product development. In all cases, the measurement precision and accuracy should be reported along with the results.

Before a sensor can be introduced into any critical application such as a safety application on which life depends, acquiring tier 3 type data is a necessity. For example, a set of tier 3 tests that are made using sensors made from (1) different batches of the construction materials in order to span material variations, (2) different process batches or runs of the process to span process variations, and (3) multiple days of processing to span other variables in time such as different operators, tools or environmental variations can be used to assess a sensor production method and determine QA/QC requirements. In this case, the average performance of sensor to be provided by this process is characterized and understood completely. The characterized set of sensors also can undergo tier 1 and tier 2 testing each day for a week, then once a week for 3 months, and finally once a month for up to several years to determine specification over the sensor lifetime. In addition, sensor specifications need to be written to serve as boundaries for each material, part, assembly process, and, finally, the resulting anticipated responses under expected user conditions.

Severe environmental testing or accelerated lifetime testing may also be warranted, and a subgroup of sensors could be used to make Tier 1 measurements at different temperatures, pressures, and relative humidity. In the end, a complete data set for the intended purpose should exist in order to specify performance for the sensor in quantitative terms for variables such as sensitivity over short and long terms (i.e., short-term and long-term drift), and to specify a temperature compensation curve that can be implemented in hardware or software.

In summary, the key measurement in all sensor evaluations is the calibration curve, and its measurement over all of the variables, one at a time, provides the data for calculation of the analytical performance specifications. This approach may sound tedious, but even such a thorough evaluation of the sensor may be insufficient if the sensor has multiple degradation mechanisms or response properties that are not functionally independent. At any rate, the evaluation and characterization of a sensor should take the form required by the objective and be as comprehensive as required. The above is a discussion of a viable means for approaching the requirements for sensor characterization, comparison, and development.

1.5.2   Measuring Equipment

Specialized experimental equipment is needed to measure sensor signals under the many differing conditions. Of the many experimental setups that can be used to evaluate sensors, one example is the Shared Sensor Technology User Facility (SSTUF), designed and built by the Sensor Research Group of the International Center for Sensor Science and Engineering [36] and illustrated in Fig. 1.10. We here describe the SSTUF apparatus which is one of many sensor characterization systems that have been built. The SSTUF was designed in order that many types of user groups could test different chemical sensors and so that virtually all chem-

Fig. 1.10 SSTUF apparatus illustrating chamber for evaluation of chemical sensors and arrays

sensor characterization requirements could be met. Of course, it is not practical to design a system for characterization of "all" sensors under "all" possible practical conditions. However, the SSTUF is a good example of a versatile sensor evaluation and characterization tool.

The SSTUF has a sampling system whose gas-mixing system provides different concentrations of target analytes(s) in a matrix of background gases to the sensor. SSTUF operates over wide temperature ranges for studies ranging from sub-ambient to about 450°C or more. The sensor test chamber is sealed and can operate from sub-atmospheric to several atmospheres of pressure and has relative humidity control. The measurement of response time requires the design of a special interface that is designed specifically to hold each sensor within the SSTUF chamber. This holder for the sensors is sealed inside the chamber so that the inflowing gas rapidly purges the area around the sensor to allow fast sensor response times to be measured. Sub-millisecond response times can be measured in a separate chamber inside the SSTUF chamber. The computer-controlled SSTUF automatically gathers all of the data from the sensor with multiple feed-throughs that are pneumatic for gas changes and electronic for sensor reading, power, and control. The matrix gas mixture, pressure, temperature, relative humidity, and data logging for the sensors are computer-controlled [36].

If all that is needed is a signal-vs.-concentration curve, simpler systems can obviously be used, but the SSTUF is a good example of a sensor evaluation system that can complete virtually any performance test on any chemical sensor or series of sensors in a short time and implement a tiered sensor characterization scheme.

The SSTUF has been used to provide quick experimental answers to important sensor performance questions for many sensors in many collaborations to date.

Once sensor performance data has been obtained, we can combine the sensors into arrays and obtain sensor array data.

1.6   Experimentation, Artificial Intelligence, and Sensor Arrays

Gathering and interpreting sensor array data can be done in many ways. In principle, a sensor array experiment gathers data simultaneously from many sensors. And it should be possible to use the same principles that were used to characterize individual sensors to characterize sensor arrays and to determine an equivalent sensitivity, selectivity, stability, and response time performance characteristics for arrays. However, this task is not so straightforward. Consider the property of "signal drift" in the more complex dimensions of chemical imaging space associated with arrays. It is often difficult to understand intuitively and often difficult to handle multidimensional drift and/or noise, experimentally. Calculating the limit of detection for a sensor array when sensors have differing sensitivity to the target analyte or to a mixture is a case in point and has been discussed in the literature [37, 38]. In such cases, it is the pattern of responses that is sinking into the noise and elements of the pattern disappear with different detection limits. It is, therefore, important to specify the data set in which the pattern exists in order to specify the detection limit.

In retrieving both qualitative and quantitative information from patterns produced by sensor arrays, we need to make four assumptions related to the analytical result. We make these assumptions explicitly or implicitly when we use pattern recognition methods with sensor array data to achieve an analytical result. We assume that:

1. The measured response is related analytically to the endpoint we seek.
2. The endpoint can be uniquely represented by a set of responses.
3. The quantitative relationship of the response and endpoint can be discovered or quantified by the method we use.
4. The relationship can be extrapolated to unknowns and additional situations with a statement of the analytical uncertainty.

Assumption 1 seems obvious, but must be tested experimentally. Assumption 2 assumes that the sensors have a unique response compared to everything else that is possible for this application. Assumption 3 deals with the differing effectiveness of the various pattern recognition or AI methods and we know they differ because relative effectiveness has been demonstrated in numerous cases and improvement continues to be made in pattern recognition algorithms. Assumption 4 is the most frequently violated assumption in practice. There must be proof that the unknown, within some statistical uncertainty, belongs to the data set. Assumption 4 best works for comparative studies, i.e., we have a coffee that has been roasted to perfection, and then the unknown coffee is analyzed in exactly the same manner, and compared to the known pattern to determine similarity within some confidence interval. But it is not necessarily sufficient to calibrate the sensor array with standards in a lab and then use these responses to determine patterns for the same molecules as they are sampled in environmental or process stream samples. The environmental or process samples could contain compounds that interfere with the observation of the pattern either directly or indirectly.

1.6.1 Arriving at Complex Endpoints: Experimental Tests

As discussed above, people have long marveled at the feats the dog nose and brain sensory system can accomplish. Recent reports indicate that some trained dogs can detect bladder cancer by sniffing urine samples [39]. What the dog responds to is not known, however, it may simply be to trace nicotine in the subject's urine. If that is the case, the dog is not really analyzing the urine for bladder cancer per se, but indicating whether the subject is a smoker or not (the incidence of bladder cancer is, in fact, higher in smokers than nonsmokers). Such an indication would be an *association with*, not *direct detection of* a marker of the disease. Determining associations is not useless, but this example is given to highlight that we must test our assumptions experimentally and quantitatively to give them the appropriate scientific credibility. Anthropomorphic sensor arrays produce patterns, and we need to understand if the pattern matching with an analytical endpoint is a statistical association or a deductive fact.

A simple model is helpful here in understanding how to evaluate such complex systems experimentally and to obtain important performance insights from only a few targeted experiments. The model can be stated as shown in (1.3):

$$A \rightarrow B \rightarrow C \text{ or } A \rightarrow B \text{ and } A \rightarrow C \text{ or } A \rightarrow C \text{ and } B \text{ is present}, \qquad (1.3)$$

where A = agent/event; B = cancer/endpoint; and C = a marker compound.

If exposure to agent/event A causes cancer B, and the cancer causes marker compound C, then the detection of that marker is an indication of cancer. On the other hand, if A causes B but also causes C to appear, C is not directly related causally to the cancer we want to detect, B, but rather to the presence of A. And in the third case, A can cause C without any relation to endpoint B, which can still be present and there is no relationship between C and B.

In designing our experiment, therefore, we have to understand the situation at hand and clearly determine the exact associations of the variables and the desired endpoint. For a sensor array, this means determining the relationship of the array of sensory signals to the complex endpoint under study. In this regard, too much of the published literature in the electronic-nose field fails to clearly demonstrate associative or causative relationships in the data; as a result, much of e-nose capability is misguided and misapplied for its intended applications [10, 40].

The literature on pattern recognition is extensive, as are algorithms for different kinds - from vector machines [41] to simple KNN (i.e., *k*-nearest neighbor) statistical algorithms [42], to the construction of models as done using the net analyte signal approach [43]. Other chapters in this book discuss these algorithms for data treatment in more detail as does the open literature, and these topics are not covered here. However, we need to bear in mind that experimentation plays a vital role in the elucidation of the pattern recognition method to use, determination of how well it works (i.e., quantitation of error), and the discovery of potential pitfalls in data interpretation. These issues are addressed in Sect. 1.6.2.

1.6.2   Experimental Aspects of Pattern Recognition

The history of sensor arrays and pattern recognition consists of significant experimental work, perhaps much more experimental than theoretical work. Computational guidance was very much needed in the early days of this field. Early work conducted in 1980 by the author, addressed a complex analytical problem posed by the US Coast Guard. The USCG was, at that time, in charge of dealing with oil spills in the US coastal waters and hazardous chemical spills anywhere. The Coast Guard needed a portable instrument to use in detecting and identifying each of the more common 1,000 toxic, flammable, and otherwise hazardous chemicals shipped in US trade and commerce. Any of these could be involved in a spill or accident. These are the same toxic industrial materials (TIMs) and toxic industrial chemicals (TICs) that yet concern us today. Portable instruments for gas detection in health and safety [44] were available for compounds like combustibles, CO, and H_2S. However, the goals of detecting and identifying virtually every kind of chemical over a range that spanned ppb to percent levels were seemingly impossible tasks for a single low cost, portable, easy to use instrument package [42].

We assembled a team and began experimenting with sensor arrays that combined different kinds of sensors (e.g., a CGS in front of a CO and H_2S sensor), and made a surprising discovery: the combined sensors not only produced a distinct signal, but also a unique pattern of responses were obtained for the different molecular analytes. Furthermore, these different sensors constructively and destructively interfered with each other's signals. For example, by placing a CGS in the front of four electrochemical sensors, we created a sensor array that could detect ppb to percent levels of all types of gaseous analytes. The CGS, itself, could detect percent levels of virtually all vapor phase hydrocarbons, but the very low ppm level signal was too weak to be observed. However, even at ppm levels, the CGS reacted with the hydrocarbons to produce CO and other products that the electrochemical sensors readily detected at very low ppm and sub-ppm levels. This sensor array proved sensitive to virtually all gases/vapors of chemical compounds except inert gases (e.g., Ar, Ne) [11, 16, 42, 45–48]. Moreover, because the sensory responses differed for virtually all of the compounds, a compound could be identified by its "fingerprint" [16, 49–56]. The fingerprint or pattern was akin to a mass spectrum or IR spectrum for the molecule, and we developed a library of responses and we called this technique "chemical parameter spectrometry" or CPS because the fingerprints were the array of chemical sensor responses for the analyte on each of the different sensors in the sensor array.

This early 1980s experimentation created a heterogeneous sensor array that provided orthogonal and complementary sensor responses, and demonstrated the synergy within arrays of sensors that result in an entirely new information content in its combined signals – the signal from the electrochemical sensor due to a reaction taking place at a previous sensor. We later optimized this sensor array into one or more reactive filaments followed by portable low-cost sensors [10, 11, 16, 17, 42]. A homogenous array, e.g., use of just the electrochemical sensors, would allow sensing a variety of target analytes, including any oxidants and/or reductants, but not have the added possibilities of a synergistically operating sensor array with high information content.

This early 1980s experimental sensor array began the exploration of the multi-dimensional feature space that Göpel [15] described almost 20 years later. A similar technique was used years later for detection of chlorinated hydrocarbons [20, 57]. This early experimental work led to a patented instrument, demonstrated to the Coast Guard in 1984, called the chemical parameter spectrometer (CPS-100) that detected and identified more than 100 compounds [55, 58]. This work led to the fabrication of the world's first instrument with a sampling system, heterogeneous sensor array, and pattern recognition algorithm for the purpose of identifying and quantifying unknown compounds in the environment with a portable instrument [45, 47–49, 51–56]. The ensuing demonstrations and experiments led to a patent and spin-off instrument that employed this unique approach for field analytical detection and identification of chemicals. The experiments were later extended to liquids and solid materials [58, 59] to allow identification of fertilizers and plastics, as well as energetic materials like explosives [57, 60–63]. In this early work, examples of creating the chemical imaging space were evident long before formal discussions of chemical imaging and the work predated even the term "electronic nose" clearly illustrating the powerful synergy among sensors and the high dimensionality of the feature space that can be created by sensors [64, 65]. All of this work was performed experimentally before the data existed upon which the theory began to be developed.

Formal pattern recognition methods were in their early stages at this time, but application of what will now seem simple statistical methods to the sensor array data led to additional surprising results. Chemometrics was just being developed and immediately began to provide powerful methods for visualizing the multidimensional data created by sensors. Chemometrics and chemical sensor arrays seemed a natural combination. The chemometrics of the day made it possible to visualize some of the chemical feature space produced by sensor arrays, and we could begin to appreciate the "imaging" power of a sensor array that is only being quantified now.

In this early experimental work, it was also important to begin to understand the surprising power of comparison of the complex chemical space that these sensor arrays created. Experiments to gauge selectivity occurred well before any theoretical input into array design could be developed, and the experimental findings led to an immediate understanding of the additional "dimensions" we could explore in the sensor array space [66].

The sensor array line of experimentation proved exciting in the early 1980s because we found both computational methods [66] as well as new sensors could improve analytical performance. In fact, one of the earliest attempts to improve identification capabilities entailed altering the computational system. Every sensor array had a library; that is, the array was calibrated in the laboratory by exposing it to known chemicals at known concentrations, and the array of signals was recorded. In this way, when an array encountered an unknown in the field, the unknown's pattern could be compared with, in the library; and, if a close match was made, the unknown could be identified as the matching library entry. However, the library generated when the sensor array was calibrated needed improvement because, as more and more patterns were added to it, the library contained more and more close "chemical" or "molecular" neighbors, i.e., similar patterns. In this regard, a surprising result came from a simple manipulation of the library. We presorted the library into groups defined by the generic nature of the sensor responses. For example, some compounds responded slowly or with negative signals in certain sensor output channels. When the library was presorted according to the "type" of the sensor responses (slow, fast, positive, negative), then the unknown could also be presorted and in effect, compared to just a subgroup of its own "type" in the library. If the unknown had a negative signal in one or more specific sensor response channels, it belonged to a subgroup of the library with "negative" responses in channel X. Applying the pattern recognition algorithm to a subgroup with fewer compounds resulted in many fewer errors in identification [66]. This example leads to the issue of how experimental evidence, not always theory, can lead the way to discover the "easily useable" dimensions of the sensor space (positive, negative, slow responses), and then use them in practical computations to statistically compare responses and obtain improved results within that space. The method elucidated by experiment, drew on knowledge about the compound's response character, experimentally demonstrating improved identification as evidenced by a reduction in misidentifications. The technique, first reported in 1986, showed how segregation of signals into groups as defined by the sensors' responses - that is, in the sensory space created by the specific array being used - could enhance performance. Additional methods can improve the results by experimentally disregarding noise, and it has been demonstrated that noise weighting also aids sensor array accuracy [47, 54, 62, 66].

It follows from this initial work that the specific sensors chosen for inclusion in the array will define the space that is able to be mapped, i.e., the chemical space dimensions. One must choose a class and type of sensor(s) that can provide response and separation of the target analyte(s) by virtue of the responses being able to define a large enough feature space. Experimental work is required in order to both define the appropriate set of sensors for a given task and then validate response analytical characteristics, e.g., if you wish to separate a polar compound from a nonpolar compound, choose sensors that respond differently to the property of molecular polarity. All sensor arrays are not created equal and cannot address all problems. And it has been shown that the most versatile arrays are heterogenous (i.e., contain more than one class of sensor and have a "high" information content). In the end, organization of this field has led to powerful theoretical methods to approach analytical problems with sensor arrays, but it is still best to use both theory and experiment to traverse the complex parameter space of sensor arrays and analysis.

1.6.3   The Use of Pattern Recognition

The computational methods of library sorting and weighting improve sensor array analytical performance (more correct identifications) for good reasons. First, the mathematics of a good pattern recognition method can utilize sufficient unique features that are included in the data base to sort and separate and thereby identify the constituents. Using a presorted library allows manipulating the sensor feature space characteristics experimentally and adds information content that is knowledge-based. Second, experimentally, the best data channels have the best signal-to-noise ratio, and making decisions based on the experimental production of the best data provides signal processing that is likely to improve the quality of input to any algorithm. In addition, the sorting of the library into subgroups that contain specific features and eliminating others from consideration permits the remaining feature space to spread out along stretched dimensions. Thus, to arrive at an improved pattern recognition algorithm, a developer needs to consider weightings for (1) dimensions that are uniquely expressed by the sensor array and are important to analyte classification, (2) sensory channels that have the best signal-to-noise ratios for decision making, and (3) libraries that divide the feature space into categories that the sensor array system preferentially expresses. In this way, we are able to add the knowledge of the lab analytical chemist to bias the pattern recognition system and improve results.

For the same set of sensors, different algorithm and library design affect the analytical outcome. An experiment performed in 1993 illustrates that choice of a specific pattern recognition method can improve the performance of a sensor array using the same sensor data: the same sensor array headspace vapor data set, in this case data used to identify bacterial activity in grain [67], was analyzed by an artificial neural network (ANN) and by a k-NN pattern recognition algorithm method. Although the k-NN and the ANN are virtually identical in their ability to identify an unknown signal pattern when trained on about half the data set, the ANN outperformed the k-NN when error was introduced (via random noise generation) into the array sensors' signals that were being used for identification [67]. This experiment illustrates an important aspect of pattern recognition techniques - just as we said that all sensor arrays are not created equal, above all pattern recognition techniques are not equally effective – the ANN outperformed k-NN and was analytically more robust.

The experiment also raised a question: What specific feature created by the sensory system did the ANN consider and use that the k-NN did not use? Answering this last question specifically would allow us to design sensor arrays with more robust, unique identifying chemical parameters and thereby improve the analytical performance of the system. A further observation is that an optimum "system" can be found with an experimental approach in which the system of pattern recognition method and sensor array is matched to provide optimum analytical capability for the specific problem to be addressed by the array. In any case, the choice of sensor array and pattern recognition together can be important to system performance. It is no accident that the mammalian nose sensor array and brain pair evolved as it did. The data is provided in temporal and spatial digital patterns and the brain is an expert pattern recognition storing and analyzing system for these data streams.

1.6.4   Selectivity and Time-Dependent Signals

An experimental method for exploring the origin of selectivity involves the use of time-dependent chemical sensor signals. Such approaches invariably extend the options for creating information, enlarging feature space, and enabling effective analytical data treatment. In an early work with heterogeneous sensor arrays and synergistic signal generation, the time-dependent modulation of the analyte concentration entering a sensor [68–72] was found to significantly increase information in the recorded signal. The information can be related to important chemical reaction parameters like the activation energy for the oxidation of the analyte. Thus, the kinetic characteristics of analyte reactions with the chemical sensor's sensing layer can influence the selectivity of the sensor and/or sensor array. Since the origin of selectivity in the chembio-sensor is either from the kinetics or the thermodynamics of the analyte reaction(s) with the sensor's reacting layer, any modulation of this interaction will allow the value of the thermodynamic and kinetic constants to influence the recorded sensor response and be reflected in the data collected. If we can understand the specific reaction chemistry in a sensor system, the responses can be understood in terms of fundamental experimentally determined reaction constants (e.g., Arrhenius rate constants, thermodynamic equilibrium constants). Such knowledge can be used in creating a sensing system that can be understood theoretically and then optimized from theoretical knowledge for specific applications. Data treatment [47, 49, 69, 73, 74] using Fourier transform is also possible with time-dependent data, thereby allowing additional mathematical methods to store and extract large quantities of the sensor-generated information content.

1.6.5 The Use of "Learning Surfaces"

One can use many known algorithms to display differences in data sets generated by a chemical sensor array. But the above examples begin to show how very important it is to add a "learning surface" concept to the data analysis process. To advance the experimental design of sensors or arrays, they must be examined from the perspective of what knowledge is created in each experimental module of the sensor array instrument and what knowledge is uniquely created by the collective system. The first use of a branched library reported in 1985 [66], illustrated that dividing a library into subgroups, based on the dimensions being produced by the sensory system, could significantly improve the identification performance of the array when using the same algorithm. As this grouping was derived from knowledge of the sensory data, and not from an analytical table, we could refer to it as a learned surface. Each sensor array produces a unique set of "features" that represent the knowledge related to the molecular nature of the sample. Finding this knowledge and using it to maximize analytical utility should be more common in sensor array development than it is today.

The converse of this situation has important experimental consequences; that is, to ask what is the fewest sensors that can be used to define a feature space that will solve a given problem? Using fewer sensors significantly reduces instrumentation systems problems, simplifies design and calibration, lowers instrument costs, and improves user features such as weight, size, and reliability. Recently, an experiment using an array of 20 sensors for experimental fire detection under different conditions demonstrated that a subset of just 3 of the 20 sensors in the array could solve the entire fire detection problem, indicating both the kind of electrical fire and its stage of progress [75]. Simplifying a fielded sensor array from 20 to 3 sensors offers enormous economic and operational benefits.

A simple way to visualize the feature space created by sensor arrays is as a tree with many branches. The branches are the chemical dimensions, created by the sensors and can be treated like filtering questions (see Fig. 1.11) we use. The answer to the question allows one to select a specific branch over another because it provides characteristic information about the target compounds or endpoint (which is the fruit at the end of the smallest branch). After we ask the appropriate questions experimentally, we can realize the appropriate answers and eventually narrow down the possibilities such that only one answer or endpoint is possible among the many starting possibilities. Figure 1.11 illustrates some of the types of questions that a chemical sensor array is capable of answering about a given sample. Sorbent-based sensors ask/answer the question as to whether the target analyte absorbs or not on a specific polar/nonpolar coating. Electrochemical sensors ask/answer the question about the presence or absence of specific functional groups that can be electrooxidized or electroreduced. Each class and type of sensor asks/answers specific chemical questions about the sampled material. After we record enough sensory responses of the appropriate type, we know enough about our endpoint analyte to differentiate it within our data and we have solved the analytical

• Examples of chemical information
obtainable by sensor arrays and methods

— Absorbs or not
  • Differentiates permanent gases
  from semi-volatiles
  – H_2, CH_4, CO, O_2 from gasoline vapors

— Electro-oxidizes: specifies functional groups
  • NO_2, NH_3, SO_2, OH, CO, ...

— Electro-reduces: specifies functional groups
  • NO_2 ozone, Cl_2, ...

— Rate of reaction – K_e, ΔE_{ACT}

— Has high or low thermal conductivity

— Time-dependent signal character
  • Mass transport
  • Kinetics
  • Thermodynamics

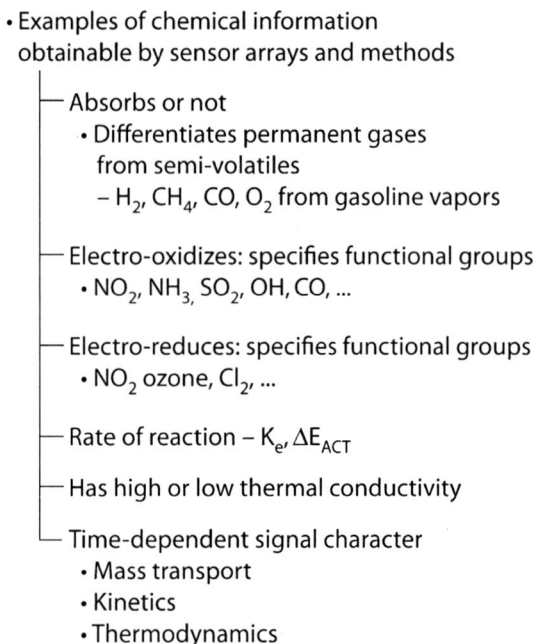

Fig. 1.11 All systems can be combined with a learning surface

problem. The key is to solve the problem with precision, accuracy, and validity. The exemplary Fig. 1.11 provides yet another way to express the idea that the array needs to be designed to solve the problem at hand. The challenge for the array experiment is to ask the most important distinguishing questions and with the minimum complexity (i.e., the fewest sensors and the fewest modes) so that low cost, speedy, and reliable information is produced. This "learning surface" approach provides an insightful way to examine biosensory systems, like the mammalian nose. In this way we can eventually understand how to mimic the brain's remarkable learning plasticity, as well as finally achieve the dog's legendary sensitivity and selectivity for analyte tracking.

1.7 Concluding Remarks: Future Directions

Despite the helpful analogy and experimental guidance of prior work and despite the proliferation of the types and styles of micro sensor arrays, it is apparent that acquiring the ability to discriminate a molecular analyte quantity and identity in a mixture or to determine a complex endpoint like quality (e.g., toxicity, hazard level, fire, or disease presence) with chemical sensor arrays often remains a difficult and unsolved experimental problem. The most successful application of sensor arrays has been in comparative analysis, i.e., wherein we have a standard material and want to determine if the sample is the same or different. Examples of successful applications can be found [76].

Information content in chemical sensor arrays can be created by a multitude of combinations of materials, structures and methods, and that variety is only now coming to be appreciated. Just as important, computational methods to arrange, store and analyze complex data sets are making significant impact on progress in this field of research and in applications. Experimental work in sensor arrays has shed light on the richness of this field of research.

The future will see the biological, physical, and chemical disciplines merge. New sensing performance based on new materials and structures will reach ever improving resolution of the "gedanken" experiment to completely analyze a single cc of gas. The new sensors will be able to gain discerning power because their signals represent fundamental thermodynamic and kinetic aspects of the molecules and molecule-sensor interactions. The merging of biological mechanisms and microsystems will produce cybernetic sensors of the type found in mammalian sensory organs. Improved sensor sensitivity, as well as selectivity will result. Applications will reach every fabric of society as sensory data will be generated and shared wirelessly. Hopefully, all this will be used for the common good to improve lifestyle, health, and environmental quality (Fig. 1.12).

Combinations of sensors of all types, computational theories, and experimental protocols will produce exciting discoveries. In fact, after 30 years of research, we are just beginning. The author's vision for the future of sensor array research is exemplified by the following quotations:

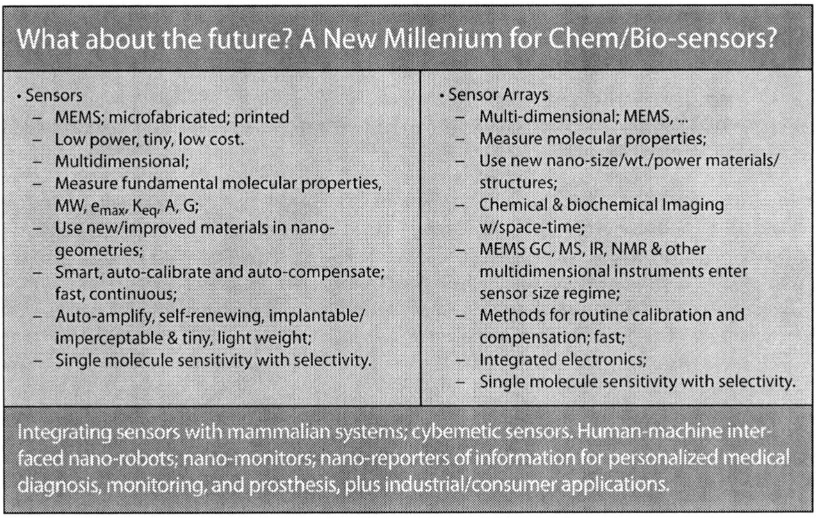

What about the future? A New Millenium for Chem/Bio-sensors?

• Sensors
- MEMS; microfabricated; printed
- Low power, tiny, low cost.
- Multidimensional;
- Measure fundamental molecular properties, MW, e_{max}, K_{eq}, A, G;
- Use new/improved materials in nano-geometries;
- Smart, auto-calibrate and auto-compensate; fast, continuous;
- Auto-amplify, self-renewing, implantable/imperceptible & tiny, light weight;
- Single molecule sensitivity with selectivity.

• Sensor Arrays
- Multi-dimensional; MEMS, ...
- Measure molecular properties;
- Use new nano-size/wt./power materials/structures;
- Chemical & biochemical Imaging w/space-time;
- MEMS GC, MS, IR, NMR & other multidimensional instruments enter sensor size regime;
- Methods for routine calibration and compensation; fast;
- Integrated electronics;
- Single molecule sensitivity with selectivity.

Integrating sensors with mammalian systems; cybernetic sensors. Human-machine interfaced nano-robots; nano-monitors; nano-reporters of information for personalized medical diagnosis, monitoring, and prosthesis, plus industrial/consumer applications.

Fig. 1.12 Future directions for Chem/Bio-sensors

"The significant problems we face cannot be solved at the same level of thinking we were at when we created them."

– Albert Einstein

"There will come a time when you believe everything is finished. That will be the beginning."

– Louis L'Amour

Acknowledgments I would like to thank all of my colleagues for their tremendously stimulating work and discussions that helped me remain dedicated, inspired, and diligent in my pursuit of the understanding of sensors and arrays and their analytical utility and application for the common good. Also, special appreciation to Susan Creamer, Lee Gerans, and editors at SRI for their help with the organization and presentation of this work.

References

1. Buck, L., Unraveling the Sense of Smell, Nobel Lecture, Stockholm, Sweden, Dec. 8, 2004
2. Rolfe, B. M., Toward nanometer-scale sensing systems: Natural and artificial noses as models for ultra-small, ultra-dense sensing systems, *Adv. Comput.* **2007**, 71
3. Schiffman, S. S.; Pearce, T. C., Introduction to olfaction: Perception, anatomy, physiology, and molecular biology, In Handbook of Machine Olfaction: Electronic Nose Technology; Pearch, T. C.; Schiffman, S. S., et al., Eds.; Wiley, New York, **2006**, 1–26
4. Crasto, C.; Marenco, L.; Miller, P. L.; Shepherd, G. S., Olfactory receptor database: A metadata-driven automated population from sources of gene and protein sequences, *Nucleic Acids Res.* **2002**, 1, 354–360
5. Crasto, C.; Marenco, L.; Skoufos, E.; Healy, M. D.; Singer, M. S.; Nadkarni, P. M.; Miller, P. L.; Shepherd, G. S., The Olfactory Receptor Database. Available at http://senselab.med. yale.edu/senselab/ORDB
6. Kurosawa, S.; Park, J. W.; Aizawa, H.; Wakida, S.; Tao, H.; Ishihara, K., Quartz crystal microbalance immunosensors for environmental monitoring, *Biosens. Bioelectron.* **2006**, 22, 473–481
7. Ricci, F.; Volpe, G.; Micheli, L.; Palleschi, G., A review on novel developments and applications of immunosensors in food analysis, *Anal. Chim. Acta.* **2007**, 605, 111–129; and Mendoza, A., et al., Scalable fabrication of immunosensors based on CNT polymer composites, *Nanotechnology* **2008**, 19, 075102
8. Kauer, J. S.; White, J., Representation of odor information in the olfactory system: From biology to an artificial nose, In Sensors and Sensing in Biology and Engineering; Barth, F. G.; Humphrey, J. A. C.; Secomb, T. W., Eds.; Springer Verlag, Berlin, **2002**
9. Craven, B.; Neuberg, T.; Paterson, E. G.; Webb, A. G.; Josephson, E. M.; Morrison, E. E.; Settles, G. S., Reconstruction and morphometric analysis of the nasal airway of the dog (*Canis familiaris*) and implications regarding olfactory airflow, *Anat. Rec.* **2007**, 209, 1325–1340
10. Stetter, J. R.; Penrose, W. R., Understanding Chemical Sensors and Chemical Sensor Arrays (Electronic Noses): Past, Present, and Future, Chapter 2.3 in Sensors Update, Wiley-VCH, Weinheim, Germany, **2002**, Vol. 10, 189–229
11. Stetter, J. R.; Zaromb, S.; Penrose, W. R.; Otagawa, T.; Sinclair, J.; Stull, J., Portable instrument for the detection and identification of air pollutants, J US Environ Prot Agency, Res. Dev., (Rep.) EPA (1984) Number: EPA/600/9–84/019, *Proc. Natl. Symp. Recent Adv. Pollut. Monit. Ambient Air Stationary Sources* **1984**, 73–81
12. Lavner, G. I.; Bloch, Y. G.; Azulai, O.; Goldblatt, A.; Terkel, J., A simple system for the remote detection and analysis of sniffing in explosives detection dogs, *Behav. Res. Methods Instrum. Comput.* **2003**, 35, 82–89. Also see http://k9.fgcu.edu/articles/gazit1.pdf)

13. Pauling, L.; Robinson, A. B.; Terashi, R., et al., Quantitative analysis of urine vapor and breath by gas-liquid partition chromatography, *Proc. Natl Acad. Sci. USA* **1971**, 68, 2374–2376
14. Phillips, M., Method for the collection and assay of volatile organic compounds in breath, Thesis intro chapter, *Anal. Biochem.* **1997**, 247, 272–278
15. Göpel, W., Chemical imaging: I. Concepts and visions for electronic and bioelectronic noses, *Sensors Actuat. B* **1998**, 52, 125–142
16. Stetter, J. R.; Zaromb, S.; Penrose, W. R.; Findlay, M. W.; Otagawa, T.; Sincali, A. J., Portable device for detecting and identifying hazardous vapors, In Journal: Hazardous Materials Spills Conference Proceedings, Prevention Behaviour Control Cleanup Spills Waste Sites, Paper 116; Ludwigson, J., Ed.; Government Institutes Inc., Rockville, MD, **1984**, 183–190
17. Stetter, J. R.; Jurs, P. C.; Rose, S. L., Detection of hazardous gases and vapors: Pattern recognition analysis of data from an electrochemical sensor arrays, *Anal. Chem.* **1986**, 58, 860–866
18. Stetter, J. R.; Otagawa, T., A chemical concentration – Modulation sensor for the selective detection of airborne chemicals, *Sensors Actuat.* **1987**, 11, 251–264
19. Janata, J., Modern topics in chemical sensing, *Chem. Rev.* **2008**, 108, 327–844
20. Vaihinger, S.; Göpel, W.; Stetter, J. R., Detection of halogenated and other hydrocarbons in air: response functions of catalyst/electrochemical sensor systems, *Sensors Actuat.* **1991**, B4, 337–343
21. Stratheropoulos, M.; Agapiou, A.; Spiliopoulou, C.; Pallis, G. C.; Sianos, E., Environmental aspects of VOCs evolved in the early stages of human decomposition, *Sci. Total Environ.* **2007**, 385, 221–227
22. Stratheropoulos, M.; Agapiou, A.; Spiliopoulou, C., A study of VOCs evolved from the decaying human body, *Forensic Sci. Int.* **2005**, 153, 147–155
23. Bhushan, A.; Yemane, D.; Overton, E. B.; Goettert, J.; Murphy, M. C., Fabrication and preliminary results for LiGA fabricated nickel micro gas chromatograph columns, *J. Microelectromech. Syst.* **2007**, 16, 383–393
24. Iqbal, J.; Overton, E. B.; Gisclair, D., Polycyclic aromatic hydrocarbons in Louisiana rivers and coastal environments: Source fingerprinting and forensic analysis, *Environ. Forensics* **2008**, 9, 63–74
25. Machado, R. F.; Laskowski, D.; Deffenderfer, O.; Burch, T.; Zheng, S.; Mazzone, P. J.; Mekhail, T.; Jennings, C.; Stoller, J. K.; Pyle, J.; Duncan, J.; Dweik, R. A.; Erzurum, S. C., Detection of lung cancer by sensor array analyses of exhaled breath, *Am. J. Respir. Crit. Care Med.* **2005**, 171, 1286–1291
26. Mitrovics, J.; Ulmer, H.; Weimar, U.; Göpel, W., Modular sensor systems for gas sensing and odor monitoring: The MOSES concept, In ACS Symposium Series: Chemical Sensors and Interfacial Design, **1998**, Vol. 31, 307–315
27. Eckenrode, B. A.; Ramsey, S. A; Stockham, R. A.; Van Berkel, G. J.; Asano, K. G.; Wolf, D. A., Performance evaluation of the scent transfer unit™ (STU-100) for organic compound collection and release, *J. Forensic Sci.* **2006**, 51, 780–789, doi:10.1111/j.1556–4029.2006.00178
28. Skoog, D. A.; West, D. M.; Holler, F. J.; Crouch, S. R., Fundamentals of Analytical Chemistry, 7th edn.; Thomas Learning, Belmont, CA, **1996**, 870 pages, ISBN 0030059380 or Skoog, D. A.; Leary, J., Principles of Instrumental Analysis, Saunders, Fort Worth, TX, **1992**
29. Willard, H.; Merritt, L.; Dean, J.; Settle, F., Instrumental Methods of Analysis, 7th edn.; Wadsworth, Belmont, CA, **1998**
30. Strobel, H.; Heineman, W., Chemical Instrumentation: A Systematic Approach 3rd edn.; Wiley, New York, **1989**, Validity is discussed in Heineman
31. Barsan, N.; Stetter, J. R.; Findlay, M.; Göpel, W., High performance gas sensing of CO: Comparative tests for semiconducting (SnO_2-based) and for amperometric gas sensors, *Anal. Chem.* **1999**, 71, 2512–2517
32. Stetter, J. R., Amperometric electrochemical gas sensors: Description and applications, In NIST Workshop on Gas Sensors: Strategies for Future Technologies, Sept. 8–9, **1993**, 61–64 (NIST Pub. #865)

33. Cao, Z.; Buttner, W. J.; Stetter, J. R., The properties and applications of amperometric gas sensors, *Electroanalysis* **1992**, 4, 253–266
34. Chang, S. C.; Stetter, J. R.; Cha, C. S., Amperometric gas sensors, *Talanta* **1993**, 40, 461–467
35. Stetter, J. R.; Li, J., Modern topics in chemical sensing: Chapter 4, Amperometric gas sensors – A review, *Chem. Rev.* **2008**, 108, 352–366
36. Chao, Y.-T.; Buttner, W. J.; Gupta, K.; Penrose, W. R.; Stetter, J. R., Shared sensor technology user facility at IIT. Hydrogen amperometric gas sensor: Performance evaluation by SSTUF, In Chemical Sensors VI: Chemical and Biological Sensors and Analytical Methods Proceedings of the International Symposium (as part of the 206th Meeting of the Electrochemical Society); Bruckner-Lea, C.; Hunter, G.; Miura, N.; Vanysek, P.; Egashira, M.; Mizutani, F., Eds.; Honolulu, Hawaii, **2004**, 117–121, http://www.iit.edu/~stetter/SSTUFOverview.html
37. Zellers, E. T.; Park, J.; Hsu, T.; Groves, W. A., Establishing a limit of recognition for a vapor sensor array, *Anal. Chem.* **1998**, 70, 4191–4201
38. Hsieh, M-D.; Zellers, E. T., Adaptation and evaluation of a personal electronic nose for selective multivapor analysis, *J. Occup. Environ. Hygiene* **2004**, 1, 149–160
39. Willis, C. M.; Church, S. M.; Guest, C. M.; Cook, W. A.; McCarthy, N.; Bransbury, A. J.; Church, M. R. T.; Church, J. C. T., Olfactory detection of human bladder cancer by dogs: Proof of principle study, *Br Med J* **2004**, 329, 712
40. Stetter, J. R.; Penrose, W. R., Eds., Artificial Chemical Sensing: Proceedings of Eighth International Symposium on Olfaction and the Electronic Nose (ISOEN8 - ISOEN 2001), Washington, DC, March 20–24, **2001**, ECS Publishing, Pennington, NJ, 2001, 229 pages, ISBN 1-56677-321-0
41. Pardo, M.; Kwong, L. G.; Sberveglieri, G.; Brubaker, K.; Schneider, J. F.; Penrose, W. R.; Stetter, J. R., Data analysis for a hybrid sensor array, *Sensors Actuat. B* **2005**, 106, 136–143
42. Penrose, W. R.; Penrose, S. E., Designing Portable Computerized Instruments, TAB Books, Inc., Blue Ridge Summit, PA, **1987**, 262
43. Lorber, A.; Faber, K.; Kowalski, B. R., Net analyte signal calculation in multivariate calibration, *Anal. Chem.* **1997**, 69, 1620–1626, 10.1021/ac960862b S0003–2700(96)00862–1
44. Stetter, J. R., Instrumentation to monitor chemical exposure in the synfuel industry, *Ann. Am. Conf. Govern. Ind. Hygienists* **1984**, 11, 225–269
45. Stetter, J. R.; Zaromb, S.; Findlay, M. W., Jr., Monitoring of electrochemically inactive compounds by amperometric gas sensors, *Sensors Actuat.* **1985**, 6, 269
46. Stetter, J. R., New toxic-gas detector could save lives, prevent disasters, logos, In C&ENews, Argonne National Lab, Argonne, IL, **1984**, Vol. 3, 2; Sensor array for toxic gas detection, US Patent 4670405, filing date 1984
47. Stetter, J. R.; Zaromb, S.; Penrose, W. R.; Otagawa, T.; Sinclair, J.; Stull, J., Portable instrument for the detection and identification of air pollutants, In Proceedings of Fourth National Symposium on Recent Advances in Pollutant Monitoring of Ambient Air and Stationary Sources, U.S. EPA, Research Triangle Park, NC, May **1984**
48. Stetter, J. R.; Zaromb, S.; Penrose, R.; Otagawa, T.; Sincali, A. J.; Stull, J. O., Selective monitoring of hazardous chemicals in emergency situations, In Proceedings of 1984 JANNAF Safety and Environmental Protection Annual Meeting, Las Cruces, NM, May **1984**
49. Vaihinger, S.; Stetter, J. R.; Göpel, W., Detection of Halogenated and Other Hydrocarbons in Air: Response Functions of Catalyst/Electrochem. Sensor Systems, Proc. of Eurosensors IV, Karlsruhe, Federal Republic of Germany, October 1–3, **1990**
50. Stetter, J. R.; Penrose, W. R.; Zaromb, S.; Stull, J. O., et al., A portable toxic vapor detector and analyzer using an electrochemical sensor array, In Analysis Instrumentation; Herbst, K. S., Ed.; ISA, RTP, NC 27709, **1985**, Vol. 21, 163–170
51. Stull, J. O.; Stetter, J. R.; Penrose, W. R.; Zaromb, S., Development and evaluation of a portable personal monitor for detection and identification of hazardous chemical vapors, In Proceedings of 1985 of Hazardous Materials Management Conference and Exhibit,

Long Beach Convention Center, CA, Dec. 3–5, **1985**, Tower Conf. Mgt. Co., Wheaton, IL, **1985**

52. Buttner, W. J.; Stetter, J. R., et al., A portable instrument for multiple compound detection and analysis, In Proceedings of Fourth Annual Technical Seminar on Chemical Spills, Feb 10–12, Toronto, ON, Sponsored by Environment Canada, **1987**

53. Buttner, W. J.; Stetter, J. R., et al., Portable Instrumentation for On-Site Analysis of Toxic Vapors, Technical Seminar on Chemical Spills, Toronto, ON, Canada, **1988**, 295–301

54. Stetter, J. R.; Penrose, W. R.; Zaromb, S.; Stull, J. O., et al., A portable toxic vapor detector and analyzer using an electrochemical sensor array, In Analysis Instrumentation; Herbst, K. S., Ed.; ISA, RTP, NC 27709, **1985**, Vol. 21, 163–170

55. Stetter, J. R.; Zaromb, S.; Penrose, W., [ARCH] Sensor array for toxic gas detection, US Patent 4,670,405: 6/2/87

56. Stetter, J. R.; Penrose, W. R.; Zaromb, S.; Christian, D.; Hampton, D. M.; Nolan, M.; Billings, M. W.; Steinke, K.; Otagawa, T., Evaluating the effectiveness of chemical parameter spectrometry in analyzing vapors of industrial chemicals, In Proceedings of Second Annual Technical Seminar on Chemical Spills, Toronto, ON, **1985**

57. Vaihinger, S.; Stetter, J. R.; Göpel, W., Detection of halogenated and other hydrocarbons in air: Response functions of catalyst/electrochemical sensor systems, In Proceedings of Eurosensors IV, Karlsruhe, Federal Republic of Germany, October 1–3, **1990**

58. Stetter, J. R., Method and apparatus for identifying and quantifying simple and complex chemicals, US Patent 4,818,348: Apr. 4, **1989**

59. Stetter, J. R., Electrochemical gas sensors for identification of solid and liquid compounds, Transducers '87, In Proceedings of the Fourth International Conference on Solid-State Sensors and Actuators, Institute of EE of Japan, Tokyo, Japan, June 2–5, **1987**, 557–560

60. Strathmann, S.; Penrose, W. R.; Stetter, J. R.; Göpel, W., Detection of TNT with chemical sensors, In Proceedings of International Symposium on Olfaction and the Electronic Nose, (ISOEN 99), Tuebingen, Germany, September 20–22, **1999**

61. Stetter, J. R.; Findlay, M. W.; Maclay, G. J.; Zhang, J.; Vaihinger, S.; Göpel, W., Sensor array and catalytic filament for chemical analysis of vapors and mixtures, *Sensors Actuat.* **1990**, B1, 43–47

62. Stetter, J. R.; Penrose, W. R.; Strathmann, S.; Göpel, W., Approaches to a more versatile electronic nose, In Proceedings of International Symposium on Olfaction and the Electronic Nose (ISOEN 99), Tuebingen, Germany, September 20–22, **1999**

63. Stetter, J. R.; Jurs, P. C.; Rose, S. L.; Stull, J. O., Enhancement of the toxic vapor identification capability of portable sensors using pattern recognition techniques, In Proceedings of 1985 JANNAF Safety and Environmental Protection Subcommittee, Naval Postgraduate School, Monterey, CA, CPIA, John Hopkins Univ. APL, Baltimore, MD, Nov. 4–8, **1985**

64. Weimar, U.; Göpel, W., Chemical imaging: II. Trends in practical multiparameter sensor systems, *Sensors Actuat. B* **1998**, 52, 143–161

65. Stetter, J. R., Electrochemical sensors, sensor arrays, and computer algorithms, In Fundamentals and Applications of Chemical Sensors; Schuetzle, D.; Hammerle, R.; Butler, J., Eds.; ACS Symposium Series, No. 309, **1986**, 299–308

66. Zaromb, S.; Battin, R.; Penrose, W. R.; Stetter, J. R.; Stamoudis, V. C.; Stull, J. O., Extending the capabilities of the portable chemical parameter spectrometer to the identification of up to 100 compounds, In Proceedings of the Second International Meeting on Chemical Sensors, Bordeaux, France, July 7–10; Aucouturier, J. L., et al., Eds.; **1986**, 739–742

67. Findlay, M. W.; Penrose, W. R.; Stetter, J. R., Quality classification of grain using a sensor array and pattern recognition, *Anal. Chim. Acta* **1993**, 284, 1–11

68. Stetter, J. R.; Otagawa, T., Selective detection of chemicals using energy modulated signals, In Proceedings of the Third International Conference on Solid-State Sensors and Actuators, Philadelphia, PA, June 10–14, IEEE, Piscataway, NJ 08854 (IEEE Cat. #85CH2127–9 and LC #84–62799), **1985**, 77–81

69. Stetter, J. R.; Otagawa, T., Selective detection of chemicals using energy modulated signals, In Proceedings of the Third International Conference on Solid-State Sensors and Actuators, Philadelphia, PA, June 10–14, IEEE, Piscataway, NJ 08854 (IEEE Cat. #85CH2127–9 and LC #84–62799), **1985**, 77–81

70. Stetter, J. R.; Otagawa, T., Chemical Detection by Energy Modulation of Sensors. USP# 5047352; 9/10/91

71. Maclay, G. J.; Stetter, J. R., Use of time-dependent chemical sensor signals for selective identification, Transducers '87, In Proceedings of the Fourth International Conference on Solid-State Sensors and Actuators, Institute of EE of Japan, Tokyo, Japan, June 2–5, **1987**, Vol. 28, 557–560

72. Stetter, J. R.; Maclay, G. J.; Christesen, S., Time dependent signals, In Proceedings of 1987 US Army Conference on Scientific Defense Research, APG MD, Nov. 16–21, **1987**

73. Stetter, J. R., Neural network approach to identify signatures from electrochemical sensor arrays, In Proceedings of NATO Advanced Research Workshop on Sensors and Sensory Systems for an Electronic Nose, University of Iceland, Reykjavik, Iceland, August 5–9, **1991**

74. Stetter, J. R., Chemical sensor arrays: Practical insights and examples, In Sensors and Sensory Systems for an Electronic Nose; Gardner, J.; Bartlett, P. N., Eds.; Kluwer, Dordrecht, **1992**, 273–301

75. Ni, M.; Stetter, J. R.; Buttner, W. J., Orthogonal gas sensor arrays with intelligent algorithms for early warning of electrical fires, *Sensors Actuat. B* **2008**, 130, 889–899

76. Röck, F.; Barsan, N.; Weimar, U., Electronic nose: Current status and future trends, *Chem. Rev.* **2008**, 108, 705–725

Chapter 2
Electromechanical and Chemical Sensing at the Nanoscale: DFT and Transport Modeling

Amitesh Maiti

Abstract Of the many nanoelectronic applications proposed for near to medium-term commercial deployment, sensors based on carbon nanotubes (CNT) and metal-oxide nanowires are receiving significant attention from researchers. Such devices typically operate on the basis of the changes of electrical response characteristics of the active component (CNT or nanowire) when subjected to an externally applied mechanical stress or the adsorption of a chemical or bio-molecule. Practical development of such technologies can greatly benefit from quantum chemical modeling based on density functional theory (DFT), and from electronic transport modeling based on non-equilibrium Green's function (NEGF). DFT can compute useful quantities like possible bond-rearrangements, binding energy, charge transfer, and changes to the electronic structure, while NEGF can predict changes in electronic transport behavior and contact resistance. Effects of surrounding medium and intrinsic structural defects can also be taken into account. In this work we review some recent DFT and transport investigations on (1) CNT-based nano-electromechanical sensors (NEMS) and (2) gas-sensing properties of CNTs and metal-oxide nanowires. We also briefly discuss our current understanding of CNT–metal contacts which, depending upon the metal, the deposition technique, and the masking method can have a significant effect on device performance.

2.1 Carbon Nanotube Basics

A carbon nanotube (CNT) is geometrically equivalent to a single sheet of graphite sheet rolled into a seamless cylinder in which a graphene lattice point (n_1, n_2) coincides with the origin $(0, 0)$. Thus, if a_1 and a_2 are the two lattice vectors of

A. Maiti
Lawrence Livermore National Laboratory, Livermore, CA 94551, USA
e-mail: amaiti@llnl.gov

M.A. Ryan et al. (eds.), *Computational Methods for Sensor Material Selection*,
Integrated Analytical Systems,
DOI 10.1007/978-0-387-73715-7_2, © Springer Science+Business Media, LLC 2009

graphene, the CNT circumference is equal to the length of the vector $(n_1\boldsymbol{a_1} + n_2\boldsymbol{a_2})$, while the CNT chiral angle θ is defined as the angle between vectors $(n_1\boldsymbol{a_1} + n_2\boldsymbol{a_2})$ and $\boldsymbol{a_1}$. With the choice of lattice vectors as in Fig. 2.1a, the chiral angle and diameter of a CNT are given respectively by the following formulas:

$$\theta = \tan^{-1}\left[\sqrt{3}n_2/(2n_1 + n_2)\right] \tag{2.1}$$

and

$$d = a\sqrt{\left(n_1^2 + n_1 n_2 + n_2^2\right)}/\pi, \tag{2.2}$$

where $a = |\boldsymbol{a_1}| = |\boldsymbol{a_2}| \sim 2.45$ Å is the lattice constant of graphene. The CNT diameter and chirality, and therefore its atomic geometry, are completely specified by the two integers (n_1, n_2), which are referred to as the chiral indices of the CNT. Because of the symmetry of the graphene lattice, a nanotube of any arbitrary chirality can be defined in the range $n_1 \geq n_2 \geq 0$ and $n_1 > 0$, which implies that the chiral angle θ for all CNTs lies between 0 and 30°. CNTs with the extreme chiral angles of 0 and 30° have special names: a CNT with $\theta = 0$ (i.e., $n_2 = 0$) is called *zigzag*, while a CNT with $\theta = 30°$ ($n_1 = n_2$) is called *armchair*. The names armchair and zigzag simply reflect the shape of the open edges of these CNTs (Fig. 2.1b, c). CNTs with any other chiral angles (i.e., $0 < \theta < 30°$) are called *chiral*.

As a CNT is just a rolled-up graphene sheet, one can obtain a good approximation to the CNT electronic structure simply by applying an appropriate boundary condition to the electronic structure of a graphene sheet, with a small perturbation due to the finite cylindrical *curvature* of the CNT surface. The boundary condition for a CNT with chiral indices (n_1, n_2) corresponds to the coincidence of the (n_1, n_2) lattice point of graphene with the origin $(0, 0)$. Taking into account small effects due to curvature, such boundary conditions lead to the following important result [1–6]: all *armchair* tubes are metallic; CNTs with $n_1-n_2 = 3n$ (n = any positive integer), which include the $(3n, 0)$ *zigzag* tubes as a special class, are quasimetallic (small bandgap $\sim$10 meV or less, arising from curvature effects); and CNTs with $n_1-n_2 \neq 3n$ are semiconducting, with a bandgap decreasing as $1/d$ as a function of tube diameter d (thereby converging to the zero bandgap of graphite in the limit $d \rightarrow \infty$). The presence of contact resistance and thermal effects often makes it difficult to experimentally distinguish between metallic and quasimetallic tubes. Thus for simplicity, experimentalists often classify CNTs as either metallic or semiconducting, and we follow the same convention in the discussion below. In spite of significant efforts, researchers have so far been unable to control the chiral indices of the synthesized CNTs (except, perhaps some control on the diameter). Therefore, given the preceding condition for metallic and semiconducting tubes, one could expect a random mix with roughly one-third metallic and two-third semiconducting CNTs. In our discussion so far it has been implied that the CNTs consist of only a single graphitic layer. Interestingly, such tubes, commonly called

(a) Graphite Sheet   (b) Armchair CNT   (c) Zigzag CNT

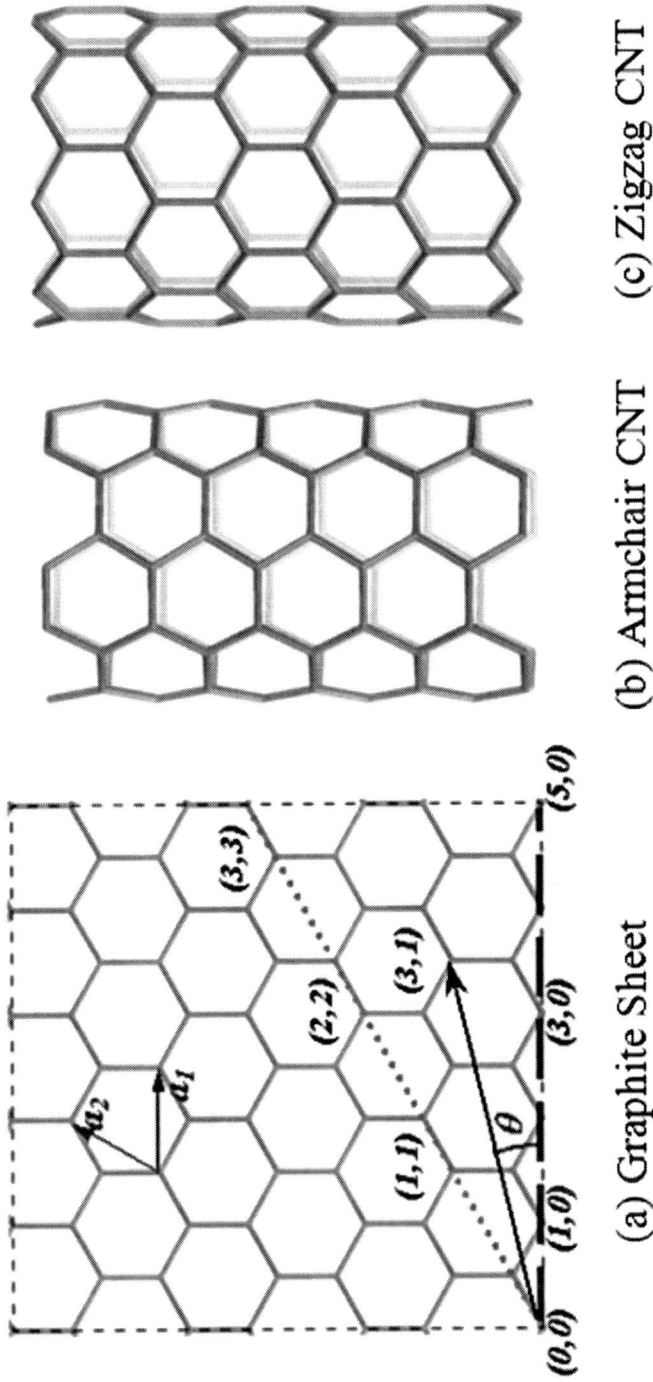

Fig. 2.1 Carbon nanotube (CNT) basics. (a) A graphene sheet with lattice vectors a_1, a_2. A few lattice points are indicated, as is the chiral angle θ for a (3, 1) CNT. *Dotted* and *dashed lines* are drawn along circumferences of *armchair* and *zigzag* tubes, respectively; (b) a (5, 5) *armchair* tube; (c) a (9, 0) *zigzag* tube

single-walled nanotubes (SWNTs), were discovered 2 years after the initial discovery of CNTs that consisted of several concentric layers. Such objects, now called multiwalled nanotubes (MWNTs), usually exhibit metallic characteristics.

More than 15 years after its initial discovery, CNTs continue to be one of the hottest research areas in all of science and engineering. The interest is driven by the possibility of several commercial applications [7–11], including field emission-based flat panel displays, transistors, quantum dots, hydrogen storage devices, structural reinforcement agents, chemical and electromechanical sensors, nanoscale manipulators, probes, and tweezers. At the same time, the highly regular atomic structure of CNTs and the large degree of structural purity make them accessible to accurate computer modeling using a variety of theoretical techniques. In fact, ever since the discovery of the CNT it has provided a fertile ground for theoretical simulations and analysis. The prediction of the dependence of CNT's electronic structure on its chirality [1–3] came within a year of the initial experimental discovery [12]. Since then, there have been a significant number of theoretical investigations [13–18] of growth mechanisms, structure and energetics of topological defects, mechanical and electrical response to various kinds physical perturbation, field-emission from tips of metallic CNTs, electronic effects of doping and gas adsorption, chemical reactivity, interaction with polymers, capillary effects, CNT–metal contacts, H- and Li-storage, thermal conductivity, encapsulation of organic and inorganic material, optical properties, as well as intrinsic quantum effects like quantized conductance, Coulomb Blockade, Aharonov–Bohm effect, Kondo effect, and so on. Computational approaches used in the above work include solving diffusion equations, quantum-mechanical (QM) simulations (DFT, tight-binding, and semiempirical methods), classical molecular dynamics, kinetic Monte Carlo, Genetic algorithms, and Green's-function-based electronic transport theory.

Focusing on electronics applications of CNTs, the areas that have received the most attention have been displays, transistors, and sensors. Displays require metallic CNTs, and naturally involve MWNTs. Although single-SWNT-based transistors have been demonstrated for a few years now, there are a number of serious challenges to be overcome for CNT-based integrated chips to become practical. Much of the recent work has therefore focused on sensor applications. Two types of sensors that have received the most attention are the electromechanical sensors and gas/chemical/bio sensors. Both of these operate on the basis of changes in electrical conductance upon either an external mechanical stimulus or the adsorption of an analyte. The electrical conductance changes can be directly related to either changes in the electronic band structure due to mechanical perturbation, or partial electron transfer between the analyte and the nanotube. Such studies can be carried out efficiently by the present-day DFT codes (in addition to semiempirical methods, which have also been employed in situations where parameters exist). As transport through nanotubes is essentially ballistic, one could use electronic transport calculations based on nonequilibrium Green's function (NEGF) to compute intrinsic conductance of the device (assuming "ideal" contacts).

2.1.1 SWNTs as Nanoelectromechanical Sensors

In a pioneering experiment Tombler et al. [19] demonstrated that when the middle part of a segment of a metallic SWNT suspended over a trench is pushed with an atomic force microscope (AFM) tip, the electrical conductance of the tube dropped by more than two orders of magnitude beyond a deformation angle of $\sim14°$. The effect was found to be completely reversible; i.e., through repeated cycles of AFM-deformation and tip removal, the electrical conductance displayed a cyclical variation with constant amplitude.

The drop in conductance in the AFM-deformed tube was much higher than the computationally predicted values for tubes bent under mechanical duress. Both tight-binding [20] and semiempirical extended-Hückel type calculations [21] concluded that even under large bending angles the reduction in electrical conductance was less than an order of magnitude. For AFM-deformed nanotubes, in contrast, O(N) tight-binding calculations [22] show that beyond a critical deformation several C-atoms close to the AFM tip become sp^3-coordinated. The sp^3 coordination ties up delocalized π-electrons into localized σ-states. This would naturally explain the large drop in electrical conductivity, as verified by explicit transport calculations.

Realizing that an AFM-deformed tube also undergoes tensile stretching, and a stretched tube belonging to certain chirality class can undergo significant changes in electrical conductance upon stretching, we carried out independent calculations to check the above sp^3 coordination idea. The smallest models of CNTs necessary in such simulations typically involve a few thousand atoms, which makes first-principles quantum mechanics simulations unfeasible. Therefore, as described below, we carried out a combination of first-principles DFT [23–25] and classical molecular mechanics [26, 27] to investigate structural changes in a CNT under AFM-deformation. Bond reconstruction, if any, is likely to occur only in the highly deformed, nonstraight part of the tube close to the AFM-tip. For such atoms, we used a DFT-based quantum mechanical description (~150 atoms including AFM-tip atoms), while the long and essentially straight part away from the middle was described accurately using the universal force field (UFF) [28]. For DFT calculations we employed the code DMol^3 [3, 29–32] distributed by Accelrys, Inc. The electronic wave functions were expanded in a double-numeric polarized (DNP) basis set with a real-space cutoff of 4.0 Å. Such a cutoff reduces computational requirements without significantly sacrificing accuracy, as has been explicitly verified in this and many other numerical experiments. An all-electron calculation was carried out on a "medium" integration grid using a gradient-corrected exchange-correlation functional due to Perdew, Burke, and Ernzerhof [33]. For calculations with periodic supercells (as described in Sects. 2.1.2, 2.1.3, and 2.2.1) we performed accurate Brillouin zone integration by a careful sampling of k-points [34]. Also, in order to estimate charge transfer to adatoms, the partial charge on each atom was computed using the Mulliken population analysis [35].

Because of known differences in electronic responses of zigzag and armchair tubes to mechanical deformation, we studied a (12, 0) zigzag and a (6, 6) armchair

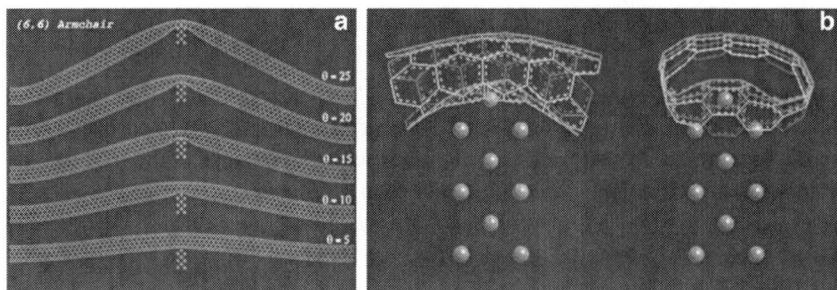

Fig. 2.2 (a) AFM deformation of a (6, 6) tube by a Li-needle. Respective deformation angles are indicated. (b) QM clusters at 25° of deformation showing no signs of sp^3 coordination

tube, each consisting of 2,400 atoms. The AFM tip was modeled by a 6-layer deep 15-atom Li-needle normal to the (100) direction, terminating in an atomically sharp tip (see Fig. 2.2) [36]. To simulate AFM-tip-deformation, the Li-needle was initially aimed at the center of a hexagon on the bottom-side of the middle part of the tube. The Li-needle tip was then displaced by an amount δ toward the tube along the needle-axis, resulting in a deformation angle $\theta = \tan^{-1}(2\delta/L)$, L being the nonstretched length of the tube. At each end of the tube, a contact region was defined by a unit cell plus one atomic ring (a total of 36 and 60 atoms for the armchair and the zigzag tube, respectively). The whole tube was then relaxed by UFF keeping the needle atoms and the end contact regions of the tube fixed. The contact region atoms were fixed in order to simulate an ideal nondeformed semi-infinite CNT lead, and to ensure that all possible contact modes were coupled to the deformed part of the tube. Following the UFF relaxation, a cluster of 132 C-atoms for the (6, 6) tube, and a cluster of 144 C-atoms for the (12, 0) tube were cut out from the middle of the tubes. These clusters (plus the AFM-tip atoms), referred to below as the QM clusters, were further relaxed with the DFT-code DMol^3 [3, 29] with the end atoms of the cluster plus the Li-tip atoms fixed at their respective classical positions.

Figure 2.2b displays the tip-deformed QM-cluster for the (6, 6) tubes at the highest deformation angle of 25° considered in these simulations. Even under such large deformations, there is no indication of sp^3 bonding [the same is true for a (12, 0) tube], and the structure is very similar to what was previously observed for a (5, 5) tube [36]. The absence of sp^3 coordination is inferred on the basis of an analysis of nearest-neighbor distances of the atoms with the highest displacements, i.e., the ones closest to the Li-tip. Although for each of these atoms the three nearest neighbor C-C bonds are stretched to a length between 1.45 and 1.75 Å, the distance of the fourth neighbor, required to induce sp^3 coordination is greater than 2.2 Å for all tubes in our simulations. The electronic charge density in the region between a C-atom and its fourth nearest neighbor is negligibly small, and none of the C-C-C angles between bonded atoms in the vicinity of the tip deviates by more than a few degrees from 120°, suggesting that the C-atoms near the AFM-tip essentially remain sp^2-coordinated. In order to test any possible dependence on the choice of our tip, we also performed limited calculations with a close-capped (5, 5) CNT as the AFM tip, and the results were very similar to that obtained with a Li-tip.

2.1.1.1 Transport Calculations

Following structural relaxation of the CNTs as described above, we computed the transmission and conductance through the deformed CNT using NEGF formalism. A brief description of the method is provided below.

Electron transport through molecular wires like SWNTs is essentially ballistic, i.e., highly coherent with little scattering and very long mean free paths. Ohm's law completely breaks down in this regime. Such transport is best described by an energy-dependent transmission function $T(E)$, which strongly depends on the (discrete) electronic levels of the molecular wire (in our case, a nanotube), the levels in the (usually metallic) leads or electrodes, and broadening of the electronic levels in the wire because of chemical coupling to the electrodes [37–42]. Such physics is most conveniently described by the NEGF formalism. The starting point is the Green's function of an *isolated* system at an energy E, which is defined by the following equation:

$$(E \cdot S_{ij} - H_{ij})G^{R,jk} = \delta_i^k, \qquad (2.3)$$

where δ_i^k is the Kronecker delta, and $S_{ij} = <i|j>$ and $H_{ij} = <i|H|j>$ are the overlap and Hamiltonian matrix elements between electronic states i and j, respectively. However, we are interested in systems in which a nanoscale region is coupled with two semi-infinite electrodes at the two ends (the so called *two-probe* system). In such a system, the coupling to the electrodes (mathematically expressed in terms of the so-called *self-energy* matrices Σ) modifies (2.1) to the form

$$(E \cdot S_{ij} - H_{ij} - \Sigma_{L,ij} - \Sigma_{R,ij})G^{R,jk} = \delta_i^k. \qquad (2.4)$$

In the above equation $\Sigma_{L,R}$ are the retarded self-energies of the left and the right semi-infinite contacts. The transmission at energy E is then found from the following equation:

$$T(E) = G^{R,ij}\Gamma_{L,jk}G^{A,kl}\Gamma_{R,li}, \qquad (2.5)$$

where $\Gamma_{L,R} = i\left(\Sigma_{L,R}^R - \Sigma_{L,R}^A\right)$ are the couplings to the left and right leads and the superscripts R and A represent retarded and advanced quantities, respectively. Finally, the total conductance of the tube is computed using the Landauer-Büttiker formula [43, 44]:

$$G = \frac{2e^2}{h} \int\limits_{-\infty}^{\infty} T(E)\left(-\frac{\partial f_0}{\partial E}\right) dE, \qquad (2.6)$$

$f_0(E)$ being the Fermi-Dirac function. The electronic states in the system, and more specifically the various matrix elements can be obtained either from first-principles or semi-empirical (e.g., tight-binding) quantum mechanical treatments. Also, if the interest is to investigate changes to the *intrinsic* electrical conductance of the nanotube, a common trick is to define "ideal" semi-infinite contacts on the basis of defect-free unstrained pieces of the pristine CNT. Such a procedure bypasses the necessity to model explicit metallic contacts, which is likely to involve additional chemical complexities on the top of extra computational burden.

The electrons in our model were described using a nearest-neighbor sp^3-tight-binding Hamiltonian in a nonorthogonal basis. The parameterization scheme explicitly accounts for effects of strain in the system through a bond-length-dependence of the Hamiltonian and the overlap matrices H_{ij} and S_{ij} [45].

Our results indicate that the conductance remains essentially constant for the (6, 6) armchair tube up to deformation as large as 25°. However, for the (12, 0) tube the conductance drops by a factor of $\sim$0.3 at 15°, two orders of magnitude at 20°, and four orders of magnitude at $\theta = 25°$. As sp^3 coordination could be ruled out, the only logical explanation of the observed behavior could be due to stretching. We verified that by computing conductance changes due to pure tensile stretching and comparing the results with that of AFM-deformed tubes (Fig. 2.3). It should also be noted that the (12, 0) tube displays only a 70% drop in electrical conductance at $\theta = 15°$, while the experimental tube in Tombler et al. [19] underwent more than two orders of magnitude drop. This can be explained by the fact that the (12, 0) tube has a diameter of only $\sim$1 nm, while the experimental tube was of diameter $\sim$3 nm. A (36, 0) CNT, with diameter similar to the experimental tube, indeed displays a much higher drop in electrical conductance (Fig. 2.3).

In order to explain the differences in conductance drops of the armchair (6, 6) and the zigzag (12, 0) tubes as a function of strain, we turn to the literature where a considerable amount of theoretical work already exists [46–51]. An important result [50] is that the rate of change of bandgap as a function of strain depends on the CNT chiral angle θ, more precisely as proportional to $\cos(3\theta)$. Thus, stretched armchair tubes ($\theta = 30°$) do not open any bandgap, and always remain metallic. On the other hand, a metallic (3n, 0) zigzag tube ($\theta = 0$) can open a bandgap of $\sim$100 meV when stretched by only 1%. This bandgap increases linearly with strain, thus transforming the CNT into a semiconductor at a strain of only a few percent. In general, all metallic tubes with $n_1 - n_2 = 3n$ ($n > 0$) will undergo the above metal-to-semiconductor transition, the effect being the most pronounced in metallic zigzag tubes. An experiment as in Tombler et al. [19] is therefore expected to show a decrease in conductance upon AFM-deformation for all nanotubes except the armchair tubes. Researchers are also beginning to explore electrical response of squashed CNTs [52–54] where sp^3 coordination is a possibility.

In addition to the above results for metallic CNTs, theory also predicts that [50] for semiconducting tubes ($n_1 - n_2 \neq 3n$), the bandgap can either increase (for $n_1 - n_2 = 3n-2$) or decrease (for $n_1 - n_2 = 3n-1$) with strain. These results have prompted more detailed experiments on a set of metallic and semiconducting CNTs deformed with an AFM-tip [55], as well as on CNTs under experimental tensile

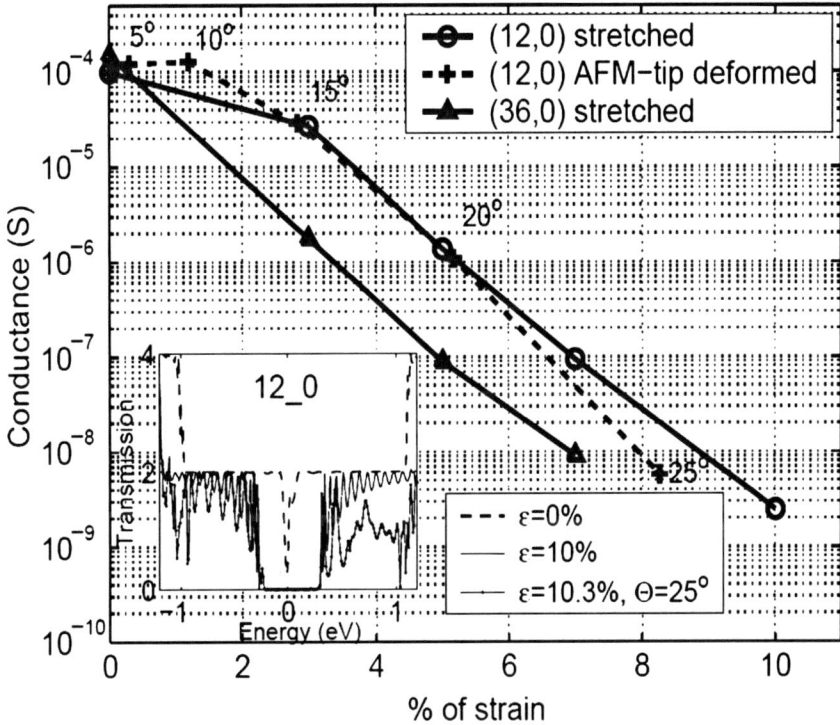

Fig. 2.3 Computed electrical conductance for (12, 0) CNT – comparison between AFM-deformed and uniformly stretched tubes. *Inset* displays density of states plot for the (12, 0) tube at the largest deformations, showing opening of a bandgap at the Fermi energy. The figure also displays the conductance of a (36, 0) tube subjected to a uniform stretch

stretch [56]. Commercial applications from such work could lead to novel pressure sensors, transducers, amplifiers, and logic devices [57].

Finally, progress is also being made in the application of CNTs as actuators [58, 59], where an externally applied small electric voltage, heat, or optical signal is transduced in the form of mechanical deformation of oscillations. Significant insight for the working of such devices can be provided through a theoretical understanding of the mode of electromechanical coupling in a CNT. A few such attempts using both DFT and tight-binding have appeared in the theoretical literature [60].

2.1.2   CNT–Metal Contacts

Conductance through a nanodevice depends strongly on the contact resistance of the metal electrodes, and CNT-based electronic devices are no exception. Besides,

CNTs interacting with metal nanoparticles are gaining considerable interest as sensing materials, catalysts, in the synthesis of metallic nanowires, as well as in nanoelectronics applications as field-effect-transistor (FET) devices. A systematic study of electron-beam-evaporation-coating of suspended CNT with various metals reveals that the nature of the coating can vary significantly depending upon the metal [61]. Thus, Ti, Ni, and Pd form continuous and quasicontinuous coating, while Au, Al, and Fe form only discrete particles on the CNT surface. In fact, Pd is a unique metal in that it consistently yields ohmic contacts to metallic nanotubes [62] and zero or even negative Schottky barrier at junctions [63] with semiconducting CNTs for FET applications. The Schottky barrier (for p-channel conductance) could, in principle, be made even lower if a higher work function metal, e.g., Pt is used. Unfortunately, Pt appears to form nonohmic contacts to both metallic and semiconducting CNTs with lower p-channel conductance than Pd-contacted junctions.

The computed interaction energy of a single metal atom on a CNT [64] follows the trend $E_b(\text{Ti}) >> E_b(\text{Pt}) > E_b(\text{Pd}) > E_b(\text{Au})$. These trends would suggest that Ti sticks the best to the CNT and Au the worst, in good agreement with experiment. However, it does not explain why Pt consistently makes worse contacts than Pd, and why Ti, in spite of its good wetting of a CNT surface, yields Ohmic contacts only rarely [62]. A detailed investigation of the metal–CNT contact at full atomistic detail is a significant undertaking, and is likely beyond the realm of today's first-principles quantum mechanics codes. Nevertheless, as a first attempt, it is instructive to look into the interactions of CNTs with metallic entities beyond single atoms.

We carried out binding energy calculations of metallic monolayers, multilayers, and 13-atom clusters with a sheet of graphene, which is a representative of wide-diameter CNTs. In addition, the interaction of a semiconducting (8, 0) tube with flat metallic surfaces was also studied. Three metals were considered for concreteness – Au, Pt, and Pd. Calculations were performed with the DFT code DMol^3 [3, 29] Details are given in Ref. [65]. We only summarize the main results below:

1. For isolated Au, Pd, and Pt atoms on a sheet of graphene, the respective binding energies are 0.30, 0.94, and 1.65 eV, respectively, i.e., in the same order as previous computed values on a (8, 0) CNT [64]. The binding sites are also quite similar, although the binding energies to graphene are ~40% smaller than that to the CNT, whose finite curvature imparts higher reactivity.

2. For monolayer or multilayer of metal atoms on graphene, most of the metal binding arises from metal–metal interaction rather than metal–graphene interaction. This is due to high cohesive energies of the metals in the bulk crystalline state. If only the metal–graphene part of the interaction is considered, the binding for Pt falls rapidly with layer thickness, and is less than that of Pd for three-layer films, perhaps indicating possible instability of Pt films beyond a certain thickness. This is likely due to much higher cohesive energy of Pt as compared to that of Pd. For Au, isolated atoms as well as films interact very weakly with graphene, in agreement with experimentally observed poor wetting properties.

3. A 13-atom Pd cluster binds more strongly to the graphene surface than a 13-atom Pt cluster. The Pd cluster, in particular, gets significantly distorted from its ideal icosahedral geometry (see Fig. 2.4). Spin might play an important role in such binding calculations, and requires a more careful analysis.

4. We predict a critical cluster size of a few tens of metal atoms below which the metal should wet a graphene (and therefore CNT) surface uniformly, and above which nonuniform clustering is likely. Using a simple model we show that the critical cluster size for Pt is $\sim$23, which is smaller than that for Pd ($\sim$30), implying higher propensity of Pt to form a nonuniform coating unless it is deposited in the form of ultrafine nanoparticles.

5. Finally, CNTs placed on flat Pt or Pd surface can form direct covalent bonds to the metal, which, along with the resulting deformation in tube cross-section might alter its electronic properties and impact performance of electronic devices based on such geometry. Interaction with an Au surface is weak, and the CNT cross-section remains circular.

Our emphasis has been on the structure and chemistry of metal–CNT contacts. In order to characterize the contact resistance one needs to investigate electronic transport across the contact. We would like to mention a few contributions that shed important light into various aspects of this problem: effect of contact geometry and oxide thickness on Schottky barrier [66], effect of tube diameter and gate dielectric constant on current–voltage (I–V) characteristics [67], dependence of

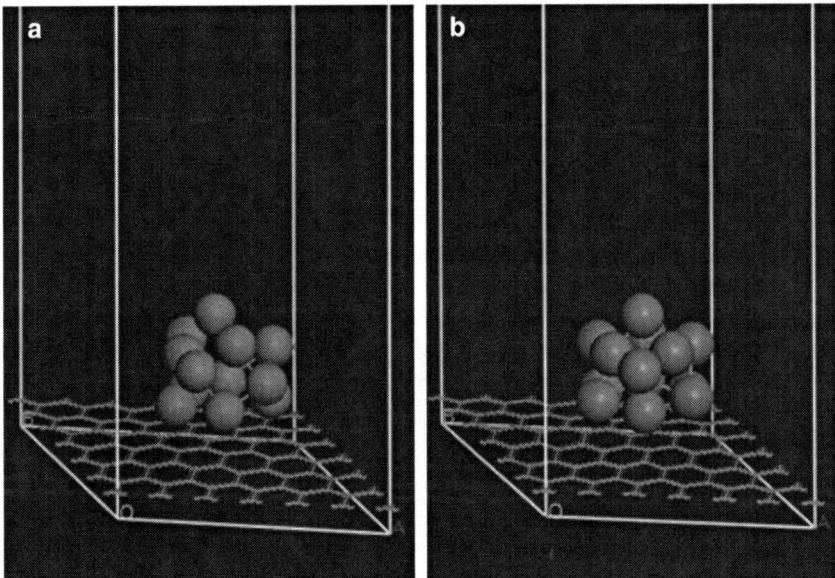

Fig. 2.4 Relaxed structures of (**a**) a 13-atom Pd cluster; and (**b**) a 13-atom Pt cluster on a (6 × 6) graphene surface. The binding energies (for the whole cluster) are 1.2 eV and 0.7 eV, respectively. Smaller Pd–Pd bond strength leads to a more open structure of the Pd-13 cluster and a higher binding energy to the graphene surface

Schottky barrier for various metals and surfaces [68], dependence of contact quality on charge distribution across the contact [69], and controlled lowering of Schottky barrier with chemical treatment [70]. For other theoretical aspects, including a discussion on electron–phonon scattering and effect of disorder on transport, we refer the reader to a recent review by Charlier et al. [71].

2.1.3   SWNTs as Chemical Sensors

Of the projected electronics-based application areas of CNTs, chemical/gas sensors appear to show a lot of commercial promise. Detection of gas molecules such as Ar, NO_2, O_2, NH_3, N_2, CO_2, H_2, CH_4, CO, or even water is important for monitoring environmental, medical, or industrial conditions. It has been reported [72–75] that the measured electrical conductance of semiconducting SWNTs at room temperature can increase or decrease upon the adsorption of different gas molecules. In particular, it was found that [72] the electrical conductance of a p-doped SWNT increases by three orders of magnitude upon exposure to 0.2% of NO_2, and decreases by two orders of magnitude upon exposure to 1% of NH_3. Amine containing molecules have been observed to significantly alter the doping type and the I–V characteristics of CNT-based transistors [76]. Exposure to O_2 also appears to consistently increase the electrical conductance, although the effect is not as dramatic as for NO_2. In addition to the above mentioned changes in electrical conductance, physisorption of molecules on CNTs has led to measurable modulations in capacitance [77] and thermopower [78]. Also, the selectivity and sensitivity of detection can be increased significantly by using appropriate functionalization, e.g., nanotubes coated by polymer [79] and single-stranded DNA [80]. Researchers have also extended nanotube-based sensors to the detection and immobilization of biomolecules and other biorelated applications [81–85].

Measured I–V characteristics of the CNT used in the experiment of Kong et al. [72] identified it to be effectively p-type (i.e. holes were the majority carriers). Thus, the observed behavior of conductance changes can be rationalized through a simple charge transfer model in which NO_2 molecules accept electrons from the CNT, thereby increasing the hole population, while the NH_3 molecules donate electrons, thereby depleting the hole population and the conductance. For NO_2 the computed binding energy has varied between 0.4 and 0.9 kcal mol^{-1} [81], and could be enhanced even more through the formation of NO_3 groups [86–88]. However, such a conclusion for NH_3 is inconsistent with theoretical results [89, 90] showing that NH_3 molecules interact very weakly with pure CNTs, which would make charge transfer very difficult to explain.

A possible explanation could be the presence of topological defects on the CNT incorporated either thermally or during high-temperature growth conditions. Evidence of the importance of defects has appeared in the experimental literature [91, 92]. Also, the nanotubes are not in isolation, and surrounding environment (oxygen, water vapor), or the substrate, or metal contacts at the ends might directly

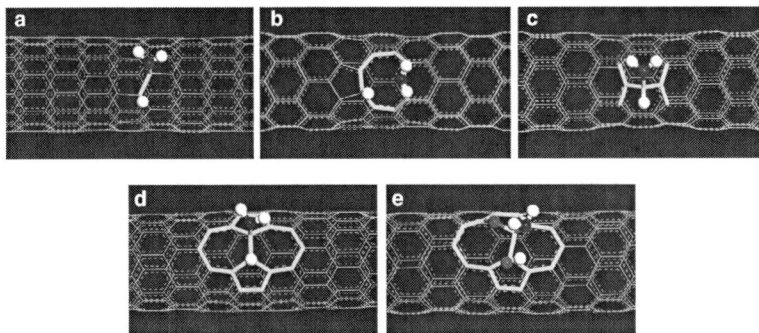

Fig. 2.5 NH$_3$ dissociated at various defects on a (8, 0) CNT: (**a**) defect-free tube; (**b**) vacancy; (**c**) interstitial; (**d**) a Stone–Wales (SW) defect; and (**e**) an O$_2$ molecule predissociated at a SW defect (SW_O_O). Dissociated NH$_2$ and H fragments are shown in *ball* representation. In (**e**), the second O breaks an O–C bond and creates a OH group single-bonded to the other C atom

Table 2.1 NH$_3$ dissociation (into NH$_2$ and H) on a (8, 0) CNT: reaction energetics (ΔE_{reac}), activation barrier (ΔE_{act}), and net electron transfer (Δq) from NH$_2$ and H groups to the CNT

Substrate	Resulting bonding configuration on CNT surface	ΔE_{reac} (eV)	ΔE_{act} (eV)	Δq from NH$_3$ (el)
CNT (defect-free)	C$_3$–NH$_2$ + C$_3$–H (Fig. 2.5a)	+0.77	2.38	0.025
V	C$_2$–NH$_2$ + C$_2$–H (Fig. 2.5b)	−2.49	0.35	0.063
I	NH$_2$–C$_{2, bridge}$–H (Fig. 2.5c)	−2.26	1.13	0.036
SW	C$_{3, 577}$–NH$_2$ + C$_{3, 577}$–H (Fig. 2.5d)	−0.17	1.50	0.044
SW_O_O	C$_{3, 577, O}$–NH$_2$ + C$_{3,577}$–O–H (Fig. 2.5e)	−2.77	0.25	0.176

Negative values of (ΔE_{reac}) denote an exothermic process
C$_3$ = Regular threefold coordinated sp^2 carbon on a defect-free CNT; C$_2$ = C$_3$ atom with a missing C neighbor; C$_{3, 577}$ = sp^2 carbon at a SW site shared by two heptagons and a pentagon; C$_{3, 577, O}$ = C$_{3, 577}$ atom with a bridging O separating it from one of its C neighbors

or indirectly provide a mechanism of binding of the gas molecules [66, 93]. Recently we performed DMol^3 [3] calculations on the chemisorption of a NH$_3$ molecule on structural defects on a semiconducting SWNT [94]. Figure 2.5 displays dissociated NH$_2$ and H fragments chemisorbed on a (8, 0) CNT containing various types of defects: (a) pristine CNT; (b) a vacancy (V); (c) an interstitial (I); (d) a Stone-Wales (SW) defect [95]; and (e) an O$_2$ molecule pre-dissociated into a SW defect (SW_O_O). Table 2.1 displays the reaction energetics (ΔE_{reac}), activation barrier (ΔE_{act}), and net electron transfer (Δq) from NH$_2$ and H groups to the CNT for the five structures described by Fig. 2.5a-e. We follow the convention that $\Delta E_{reac} < 0$ for an exothermic process. The important results can be summarized as follows:

1. Chemisorption to a defect-free CNT is an endothermic process with a large activation barrier, and therefore highly unlikely even at elevated temperatures.

2. At both V and I the dissociation becomes exothermic with energy gains of 2.49 and 2.26 eV, respectively. The activation barrier for dissociation is rather low at V, and dissociation should happen readily at room temperature, while that at I is also possible, although at a slower rate.
3. At SW the dissociation is marginally exothermic, and the activation barrier is lower than that for a defect-free tube, although a bit high for chemisorption to happen readily at room temperatures. However, the presence of pre-dissociated O (SW_O_O) significantly enhances the stability of chemisorbed NH_3 and makes the NH_3 dissociation process nearly spontaneous.
4. In all cases there is net electron transfer from the chemisorbed NH_3 to the CNT. As compared to the defect-free tube, the amount of charge transfer to V increases almost two-and-a-half fold, while that to SW_O_O is enhanced nearly seven-fold. It is also to be noted here that multiple O_2 molecules could potentially dissociate on the same SW defect, thus providing dissociation sites for multiple NH_3 molecules, leading to much higher net charge transfer to the SWNT.
5. Computed infrared (IR) spectrum of some of the defect structures of Fig. 2.5 provides proper interpretation of experimental FTIR data [96]. See Andzelm et al. [94] for more details.

Large charge transfer should qualitatively explain the observed drops in electrical conductance, although a quantitative comparison would require explicit electronic transport calculations [97]. An important part of that problem is an analysis of the electronic density of states (DOS), which has been carried out for several different adsorbates on nanotubes, both physisorbed and chemisorbed [94, 98, 99]. Most important to transport are the changes in DOS that occur close to the Fermi surface. In addition, most of the theoretical analyses have been performed on molecules adsorbing on the external surface of single isolated nanotubes. In reality, the sensor design often involves nanotube networks, bundles, and films. For such systems, one needs to analyze the effect of binding and diffusion in the intertube space within bundles, which could lead to more delayed desorption [100] than for the isolated tubes.

2.2 Metal-Oxide Nanowires

In spite of tremendous advances in carbon nanotube research, there remain some practical difficulties that hinder many applications. Cheap mass-production remains one of the biggest hurdles. Other technological challenges involve controlling CNT diameter, chirality, and doping levels, isolating/separating CNTs from bundles, alignment in nanocomposites, and so on. Chemical inertness of the CNT often poses big problems in sensor applications and adhesion to structural materials, although some of it is being overcome through chemical functionalization. The above deficiencies prompted researchers to explore other types of one-dimensional nanostructures, and led to the synthesis of *nanowires* and *nanoribbons*. Nanowires

[101]

10-30 nm

(10$\bar{1}$)

(010)

80-120 nm

5 – 100 μm

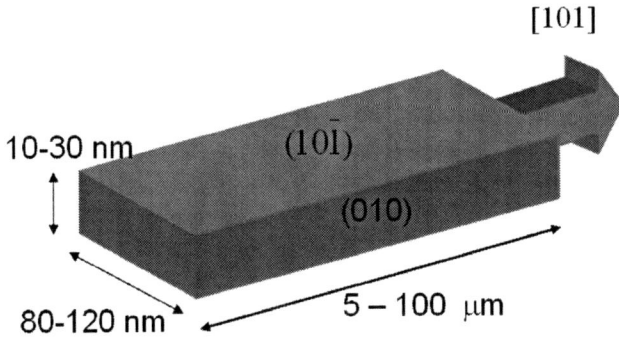

Fig. 2.6 Schematic diagram of a SnO₂ nanoribbon, showing typical dimensions, exposed planes, and growth direction

are typically solid (i.e., not hollow) cylindrical objects with a nearly uniform diameter of a few tens of nanometers or less. Most nanowires [101] have so far been synthesized from standard semiconductors: Si, Ge, GaAs, GaP, GaN, InAs, InP, ZnS, ZnSe, CdS, CdSe, and mixed compounds. Semiconducting nanowires have great potential in electronic and optoelectronic applications at the nanoscale. In addition, conducting nanowires made of transition and noble metals, silicides (ErSi₂), and polymeric materials have also been investigated in connection with interconnect applications.

Nanoribbons are a special type of nanowires. As the name suggests, they possess a uniform rectangular cross-section with well-defined crystal structure, exposed planes, and growth direction (see Fig. 2.6). So far, nanoribbons have primarily been synthesized from the oxides of metals and semiconductors. In particular, SnO₂ and ZnO nanoribbons [102–104] have been material systems of great current interest because of potential applications as catalysts, in optoelectronic devices, and as chemical sensors for pollutant gas species and biomolecules [105–107]. Although they grow to tens of microns long, the nanoribbons are remarkably single-crystalline and essentially free of dislocations. Thus they provide an ideal model for the systematic study of electrical, thermal, optical, and transport processes in one-dimensional semiconducting nanostructures, and their response to various external process conditions.

2.2.1 SnO₂ Nanoribbons as Gas Sensor

SnO₂ nanoribbons are usually synthesized by evaporating SnO or SnO₂ powder at high temperature, followed by deposition on an alumina substrate in the down-stream of an Ar-gas flow [102]. Field-emission scanning electron microscopy (FE-SEM) and transmission electron microscopy revealed that the ribbons (1) possess a highly crystalline rutile structure; (2) grow tens of microns long in the <1 0 1> direction; (3) display a uniform quasirectangular cross-section perpendicular to the growth direction; and (4) present the $(1\ 0\ \bar{1})$ and $(0\ 1\ 0)$ rutile planes as surface facets along the growth axis, with dimensions ranging from 80 to 120 nm by 10 to 30 nm (Fig. 2.6). Rutile SnO_2 is a wide-bandgap (3.6 eV) n-doped semiconductor, with the intrinsic carrier density determined by the deviation from stoichiometry, primarily in the form of oxygen vacancies [108]. Experiments with SnO_2 nanoribbons [109] indicate that these are highly effective in detecting even very small amounts of harmful gases like NO_2. Upon adsorption of these gases, the electrical conductance of the sample decreases by several orders of magnitude. More interestingly, it is possible to get rid of the adsorbates by shining UV light, and the electrical conductance is completely restored to its original value. Such single-crystalline sensing elements have several advantages over conventional thin-film oxide sensors: low operating temperatures, no ill-defined coarse grain boundaries, and high active surface-to-volume ratio.

Electron withdrawing groups like NO_2 and O_2 are expected to deplete the conduction electron population in the nanoribbon, thereby leading to a decrease in electrical conductance. To investigate this, we performed $DMol^3$ calculations of the adsorption process of NO_2, O_2 and CO on the exposed $(1\ 0\ \bar{1})$ and $(0\ 1\ 0)$ surfaces, as well as the edge atoms of a SnO_2 nanoribbon [110]. The nanoribbon surfaces were represented in periodic supercells (Fig. 2.7).

In bulk rutile SnO_2, the Sn atoms are octahedrally coordinated with six O neighbors, while each O atom is a threefold bridge between neighboring Sn centers. At both $(1\ 0\ \bar{1})$ and $(0\ 1\ 0)$ surfaces the Sn atoms lose an O neighbor, thereby becoming fivefold coordinated (Fig. 2.7a, b). The surface O atoms become twofold-coordinated bridges connecting neighboring surface Sn atoms (Fig. 2.7a, b). Both surfaces were represented by three layers of Sn, each layer being sandwiched between two O layers. The bottom SnO_2 layer was fixed in order to simulate the presence of several bulk-like layers in the actual sample. In order to reduce interaction with periodic images, the surface unit cell was doubled in the direction

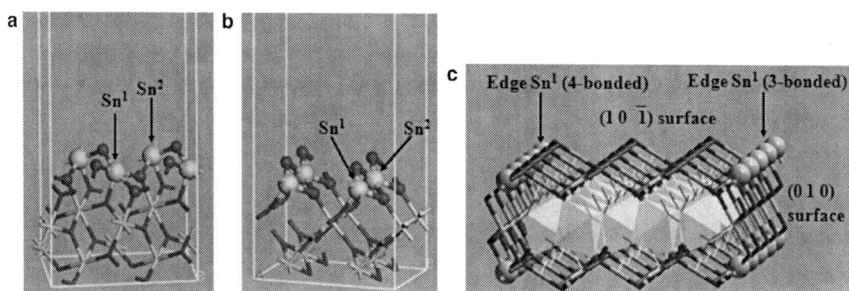

Fig. 2.7 Simulation supercells representing exposed surfaces of a SnO_2 nanoribbon: (**a**) $(1\ 0\ \bar{1})$ surface; (**b**) $(0\ 1\ 0)$ surface; and (**c**) nanoribbon edge. For clarity, the periodic cell is not shown in (**c**), and the interior atoms are represented by *polyhedra*. Surface atoms in (**a**) and (**b**), and edge atoms in (**c**) are shown in *ball* representation. *Larger balls* and *smaller balls* represent Sn and O atoms, respectively. Sn^1 and Sn^2 are neighboring Sn atoms connected with a bridging O. Reprinted with permission from Ref. [110]. Copyright (2005) from the American Chemical Society

of the smaller surface lattice constant, and a vacuum of 15 Å was placed normal to the surface. To simulate nanoribbon edges [i.e., lines of intersection of $(1\ 0\ \bar{1})$ and $(0\ 1\ 0)$ planes], a structure as in Fig. 2.7c was embedded in a periodic supercell with the smallest repeat period (5.71 Å) along the length of the ribbon, and a vacuum of 15 Å normal to both the $(1\ 0\ \bar{1})$ surface (y-axis) and the $(0\ 1\ 0)$ surface (x-axis). At the nanoribbon edges the Sn atoms can be either threefold- or fourfold-coordinated (Fig. 2.7c).

Details of the results on binding energy and charge transfer are discussed elsewhere [110]. The important results are summarized below.

All adsorbate structures involve one or more bonds to surface Sn atoms. The binding energy on different surfaces and edges increases in the sequence $(0\ 1\ 0) <$ $(1\ 0\ \bar{1}) <$ threefold edge $<$ fourfold edge.

NO_2 adsorption displays a very rich chemistry because it can either form a single bond to a surface Sn, or can adsorb in the bidented form through two single bonds to neighboring Sn atoms. The doubly bonded NO_2 is 2–3 kcal mol^{-1} more stable than the single-bonded NO_2, and the binding energies are in general 4–5 kcal mol^{-1} higher on the $(1\ 0\ \bar{1})$ surface than on the $(0\ 1\ 0)$. Activation barrier between the doubly-bonded and single-bonded structures is expected to be low, which should make the NO_2 species mobile on the exposed faces by performing a series of random walk steps along well-defined rows of Sn atoms on the surface.

When two NO_2 molecules meet on the surface, either through random walk as described above, or through the incidence of a second NO_2 from gas phase in the vicinity of an already chemisorbed NO_2, there is a transfer of an O atom from one NO_2 to the other, thus converting it to a surface NO_3 species. The net disproportionation reaction $NO_2 + NO_2 \rightarrow NO_3 + NO$ is well known in chemistry. The bidented NO_3 group has a substantially higher binding energy, especially on the $(1\ 0\ \bar{1})$ surface, and should not therefore be mobile. The resulting NO species is only weakly bound to the surface and should desorb easily. Synchrotron measurements using X-ray absorption near-edge spectroscopy (XANES) confirmed the abundance of NO_3 species on the nanoribbon surface following NO_2 adsorption.

On a defect-free surface (i.e., surface with no O-vacancies), the O_2 molecules can only weakly physisorb. In this configuration, there is no charge transfer to the O_2, and therefore, a nanoribbon surface without surface O-vacancies should be insensitive to atmospheric oxygen. However, at O-vacancy sites, the O_2 molecule has a strongly bound chemisorbed structure in the form of a peroxide bridge.

Both NO_3 groups and chemisorbed O_2 (at O-vacancy sites) accept significant amount of electronic charge from the surface. Therefore, such adsorbates should lead to the lowering of electrical conductance of the effectively n-doped sample. CO, on the other hand, donates a moderate amount of electrons to the surface, and is therefore expected to increase the electrical conductance. All these results are consistent with direct experimental measurements of sample conductance [105, 109, 111]. Charge transfer between molecular species (donor or acceptor alike) and the nanoribbon surface could thus serve as a general mechanism for ultrasensitive chemical and biological sensing using single crystalline semiconductor nanowires.

A CO likes to adsorb in the following manner: the C forms two single bonds to the surface – one with a surface Sn and another with a bridging O, while the O of the CO forms a double bond to the C and sticks out of the surface. This way, the C atoms attains its preferred 4-valency and the O has its bivalency satisfied.

To end this section, we would like to note that pioneering work by Lieber and his network of collaborators has led to the synthesis and exploration of many different configurations of nanowires, including core-shell structures, cross-bar architecture, heterostructures, and so on. Versatility in materials choice and architectural flexibility are enabling a vast array of potential applications including optical waveguides, nanophotonics, nanomedicine, NEMS, biosensing, biomedicine, flexible electronics, and smart materials. For a recent review see Ref. [112]. Some of the previous work has been reviewed by Law et al. [113].

2.3   Conclusions

Through a few application examples we have illustrated the use of a variety of theoretical techniques spanning a wide spectrum of length and time-scales. As better synthesis and experimental manipulation methods emerge, and more complex and newer applications are proposed, they open up exciting opportunities for theory, modeling, and simulations. However, several challenges, both experimental and theoretical, remain as a roadblock to successful commercial deployment of most technologies. As with any nanosystem, contact remains a critical issue. Small atomic-level changes in the structure of the contact can have a significant impact on the contact resistance, and very little characterization data exists on most experimental contacts. Besides, even though DFT-based NEGF codes are becoming faster and more accurate, they are still too limited in realistically representing metal–CNT contacts. Standard DFT or tight-binding treatments, as described here, also do not take into account complex many-body electron-correlation effects that may arise at low temperatures, or Coulomb blockade effects when electrons in CNTs get highly localized because of large mechanical deformation or highly resistive contacts.

Acknowledgment The author would like to acknowledge collaborations with M. P. Anantram, A. Svizhenko, and A. Ricca (NASA, Ames), J. Andzelm, N. Govind, and P. Kung (Accelrys), J. Rodriguez (Brookhaven National Lab), and Prof. P.Yang (UC, Berkeley). Stimulating discussions with Prof. H. Dai and Dr. A. Javey (Stanford) are also greatly appreciated. The work was performed under the auspices of the U.S. Department of Energy by the UC LLNL under Contract W-7405-Eng-48.

References

1. Mintmire, J. W.; Dunlap, B. I.; White, C. T., Are fullerene tubules metallic, *Phys. Rev. Lett.* **1992**, 68, 631–634

2. Hamada, N.; Sawada, S.; Oshiyama, A., New one-dimensional conductors: Graphitic micro-tubules, *Phys. Rev. Lett.* **1992**, 68, 1579–1582

3. Saito, R.; Fujita, M.; Dresselhaus, G.; Dresselhaus, M. S., Electronic structure of chiral graphene tubules, *Appl. Phys. Lett.* **1992**, 60, 2204–2206

4. White, C. T.; Robertson, D. H.; Mintmire, J. W., Helical and rotational symmetries of nanoscale graphitic tubules, *Phys. Rev. B* **1993**, 47, 5485–5488

5. Jishi, R. A.; Inomata, D.; Nakao, K.; Dresselhaus, M. S.; Dresselhaus, G., Electronic and lattice properties of carbon nanotubes, *J. Phys. Soc. Jpn* **1994**, 63, 2252–2260

6. White, C. T.; Mintmire, J. W.; Mowrey, R. C.; Brenner, D. W.; Robertson, D. H.; Harrison, J. A.; Dunlap, B. I., In Buckminsterfullerenes; Billups, W. E.; Ciufolini, M. A., Eds.; VCH Publishers, Deerfield Beach, FL, 1993

7. Articles in *Phys. World* **2000**, 13, 29–53

8. Terrones, M., Science and technology of the twenty-first century: Synthesis, properties, and applications of carbon nanotubes, *Annu. Rev. Mater. Res.* **2003**, 33, 419–501

9. Baughman, R. H.; Zakhidov, A. A.; de Heer, W. A., Carbon nanotubes - The route toward applications, *Science* **2002**, 297, 787–792

10. Ajayan, P. M.; Zhou, O., in Carbon Nanotubes Synthesis, Structure, Properties and Applications; Dresselhaus, M. S.; Dresselhaus, G.; Avouris, P., Eds.; Springer, Berlin, **2001**, 391–425

11. Meyyappan, M., Ed., Carbon Nanotubes – Science and Applications, CRC, Boca Raton, FL, **2004**

12. Iijima, S., Helical microtubules of graphitic carbon, *Nature* **1991**, 354, 56–58

13. Articles in NSTI Technical Proceedings, NSTI Publications, Cambridge, MA, Vol. 2, **2001**

14. Articles in NSTI Technical Proceedings, NSTI Publications, Cambridge, MA, Vol. 2, **2002**

15. Articles in NSTI Technical Proceedings, NSTI Publications, Cambridge, MA, Vol. 3, **2003**

16. Bernholc, J.; Brenner, D.; Nardelli, M. B.; Meunier, V.; Roland, C., Mechanical and electrical properties of nanotubes, *Annu. Rev. Mater. Res.* **2002**, 32, 347–375

17. Tománek, D.; Enbody, R., Eds., Science and Applications of Nanotubes, Kluwer, Netherlands, **2000**

18. Articles in *Phys. B: Condensed Matter* **2002**, 323, No. 1–4

19. Tombler, T. W.; Zhou, C.; Alexseyev, L.; Kong, J.; Dai, H.; Liu, L.; Jayanthi, C. S.; Tang, M.; Wu, S. Y., Reversible electromechanical characteristics of carbon nanotubes under local-probe manipulation, *Nature* **2000**, 405, 769–772

20. Nardelli, M.; Bernholc, J., Mechanical deformations and coherent transport in carbon nanotubes, *Phys. Rev. B* **1999**, 60, R16338–R16341

21. Rochefort, A.; Avouris, P.; Lesage, F.; Salahub, D., Electrical and mechanical properties of distorted carbon nanotubes, *Phys. Rev. B* **1999**, 60, 13824–13830

22. Liu, L.; Jayanthi, C. S; Dai, H., Controllable reversibility of an sp^2 to sp^3 transition of a single wall nanotube under the manipulation of an AFM tip: A nanoscale electromechanical switch, *Phys. Rev. Lett.* **2000**, 84, 4950–4953

23. Parr, R. G.; Yang, W., Density Functional Theory of Atoms and Molecules, Oxford University Press, Oxford, **1989**

24. Hohenberg, P.; Kohn, W., Inhomogeneous electron gas, *Phys. Rev.* **1964**, 136, B864–B871

25. Kohn, W.; Sham, L. J., Self-consistent equations including exchange and correlation effects, *Phys. Rev.* **1965**, 140, A1133–A1138

26. Jensen, F., Introduction to Computational Chemistry, Wiley, New York, **1999**

27. Hill, J.-R.; Subramanian, L.; Maiti, A., Molecular Modeling Techniques in Material Sciences, CRC/Taylor & Francis, Boca Raton, FL/London, **2005**

28. Rappe, A. K.; Casewit, C. J.; Colwell, K. S.; Goddard, W. A.; Skiff, W. M., UFF: A full periodic table force field for molecular mechanics and molecular dynamics simulations, *J. Am. Chem. Soc.* **1992**, 114, 10024–10039

29. Delley, B., An all-electron numerical method for solving the local density functional for polyatomic molecules, *J. Chem. Phys.* **1990**, 92, 508–517

30. Delley, B., Fast calculation of electrostatics in crystals and large molecules, *J. Phys. Chem.* **1996**, 100, 6107–6110
31. Delley, B., A scattering theoretic approach to scalar relativistic corrections on bonding, *Int. J. Quantum Chem.* **1998**, 69, 423–433
32. Delley B., From molecules to solids with the DMol³ approach, *J. Chem. Phys.* **2000**, 113, 7756–7764
33. Perdew, J. P.; Burke, K.; Ernzerhof, M., Generalized gradient approximation made simple, *Phys. Rev. Lett.* **1996**, 77, 3865–3868
34. Monkhorst, H. J.; Pack, J. D., Special points for Brillouin-zone integrations, *Phys. Rev. B* **1976**, 13, 5188–5192
35. Szabo, A.; Ostlund, N. S., Modern Quantum Chemistry, Dover, New York, **1996**
36. Maiti, A., Application of carbon nanotubes as electromechanical sensors – Results from First-Principles simulations, *Phys. Stat. Sol. B* **2001**, 226, 87–93
37. Datta, S., Electronic Transport in Mesoscopic Systems, Cambridge University Press, Cambridge, **1997**
38. Datta, S., Quantum Transport: Atom to Transistor, Cambridge University Press, Cambridge, **2005**
39. Imry, Y., Introduction to Mesoscopic Physics, Oxford University Press, Oxford, **1997**
40. Brandbyge, M.; Mozos, J.; Ordejón, P.; Taylor, J.; Stokbro, K., Density-functional method for nonequilibrium electron transport, *Phys. Rev. B* **2002**, 65, 165401.1–165401.17
41. Ferry, D. K.; Goodnick, S. M., Transport in Nanostructures, Cambridge University Press, Cambridge, **1997**
42. Beenakker, C. W. J., Random-matrix theory of quantum transport, *Rev. Mod. Phys.* **1997**, 69, 731–808
43. Büttiker, M., Four-terminal phase-coherent conductance, *Phys. Rev. Lett.* **1986**, 57, 1761–1764
44. Landauer, R., Conductance determined by transmission: Probes and quantised constriction resistance, *J. Phys.: Condens: Matter* **1989**, 1, 8099–8110
45. Papaconstantopoulos, D. A.; Mehl, M. J.; Erwin, S. C.; Pederson, M. R., Tight-binding approach to computational materials science, In MRS Proceedings 491; Turchi, P.E.A.; Gonis, A.; Colombo, L., Eds.; Materials Research Society, Warrendale, PA, **1998**
46. Maiti, A.; Svizhenko, A.; Anantram, M. P., Electronic transport through carbon nanotubes: Effects of structural deformation and tube chirality, *Phys. Rev. Lett.* **2002**, 88, 126805.1–126805.4
47. Kane, C. L.; Mele, E. J., Size, shape, and low energy electronic structure of carbon nanotubes, *Phys. Rev. Lett.* **1997**, 78, 1932–1935
48. Heyd, R.; Charlier, A.; McRae, E., Uniaxial-stress effects on the electronic properties of carbon nanotubes, *Phys. Rev. B* **1997**, 55, 6820–6824
49. Yang, L.; Anantram, M. P.; Han, J.; Lu, J. P., Band-gap change of carbon nanotubes: Effect of small uniaxial and torsional strain, *Phys. Rev. B* **1999**, 60, 13874–13878
50. Yang, L.; Han, J., Electronic structure of deformed carbon nanotubes, *Phys. Rev. Lett.* **2000**, 85, 154–157
51. Kleiner, A.; Eggert, S., Band gaps of primary metallic carbon nanotubes, *Phys. Rev. B* **2001**, 63, 073408.1–073408.4
52. Lammert, P. E.; Zhang, P.; Crespi, V. H., Gapping by squashing: Metal–insulator and insulator-metal transitions in collapsed carbon nanotubes, *Phys. Rev. Lett.* **2000**, 84, 2453–2456
53. Lu, J-Q.; Wu, J.; Duan, W.; Liu, F.; Zhu, B. F.; Gu, B. L., Metal-to-semiconductor transition in squashed armchair carbon nanotubes, *Phys. Rev. Lett.* **2003**, 90, 156601.1–156601.4
54. Svizhenko, A.; Mehrez, H.; Anantram, M. P.; Maiti, A., Sensing mechanical deformation in carbon nanotubes by electrical response: A computational study, *Proc. SPIE* **2005**, 5593, 416–428
55. Minot, E. D.; Yaish, Y.; Sazonova, V.; Park, J-Y.; Brink, M.; McEuen, P. L., Tuning carbon nanotube band gaps with strain, *Phys. Rev. Lett.* **2003**, 90, 156401.1–156401.4

56. Cao, J.; Wang, Q.; Dai, H., Electromechanical properties of metallic, quasimetallic, and semiconducting carbon nanotubes under stretching, *Phys. Rev. Lett.* **2003**, 90, 157601.1–157601.4

57. Maiti, A., Carbon nanotubes: Band gap engineering with strain, *Nat. Mater. (London)* **2003**, 2, 440–442

58. Baughman, R. H., et al., Carbon nanotube actuators, *Science* **1999**, 284, 1340–1344

59. Sazonova, V.; Yaish, Y.; Ustunel, H.; Roundy, D.; Arias, T. A.; McEuen P. L., A tunable carbon nanotube electromechanical oscillator, *Nature* **2004**, 431, 284–287

60. Hartman, A. Z.; Jouzi, M.; Barnett, R. L.; Xu, J. M, Theoretical and experimental studies of carbon nanotube electromechanical coupling, *Phys. Rev. Lett.* **2004**, 92, 236804.1–236804.4

61. Zhang, Y.; Franklin, N. W.; Chen, R. J.; Dai, H., Metal coating on suspended carbon nanotubes and its implication to metal–tube interaction, *Chem. Phys. Lett.* **2000**, 331, 35–41

62. Mann, D.; Javey, A.; Kong, J.; Wang, Q.; Dai, H., Ballistic transport in metallic nanotubes with reliable Pd ohmic contacts, *Nano Lett.* **2003**, 3, 1541–1544

63. Javey, A.; Guo, J.; Wang, Q.; Lundstrom, M.; Dai, H., Ballistic carbon nanotube field-effect transistors, *Nature* **2003**, 424, 654–657

64. Durgun, E.; Dag, S.; Bagci, V. M. K.; Gülseren, O.; Yildirim, T.; Ciraci, S., Systematic study of adsorption of single atoms on a carbon nanotube, *Phys. Rev. B* **2003**, 67, 201401.1–201401.4

65. Maiti, A.; Ricca, A., Metal–nanotube interactions – binding energies and wetting properties, *Chem. Phys. Lett.* **2004**, 395, 7–11

66. Heinze, S.; Tersoff, J.; Martel, R.; Derkcke, V.; Appenzeller, J.; Avouris, P., Carbon nanotubes as Schottky barrier transistors, *Phys. Rev. Lett.* **2002**, 89, 106801.1–106801.4

67. Guo, J.; Datta, S.; Lundstrom, M., A numerical study of scaling issues for Schottky-barrier carbon nanotube transistors, *IEEE Trans. Electron Devices* **2004**, 51, 172–177

68. Shan, B.; Cho, K. J., Ab initio study of Schottky barriers at metal-nanotube contacts, *Phys. Rev. B* **2004**, 70, 233405.1–233405.4

69. Nemec, N.; Tomanek, D.; Cuniberti, G., Contact dependence of carrier injection in carbon nanotubes: An ab initio study, *Phys. Rev. Lett.* **2006**, 96, 076802.1–076802.4

70. Auvray, S. et al., Chemical optimization of self-assembled carbon nanotube transistors, *Nano Lett.* **2005**, 5, 451–455

71. Charlier, J. C.; Blasé, X.; Roche, S., Electronic and transport properties of nanotubes, *Rev. Mod. Phys.* **2007**, 79, 677–732

72. Kong, J.; Franklin, N.R; Zhou, C.; Chapline, M.G.; Peng, S.; Cho, K.; Dai, H., Nanotube molecular wires as chemical sensors, *Science* **2000**, 287, 622–625

73. Collins, P.G.; Bradley, K.; Ishigami, M.; Zettl, A., Extreme oxygen sensitivity of electronic properties of carbon nanotubes, *Science* **2000**, 287, 1801–1804

74. Valentini, L.; Armentano, I.; Kenny, J. M.; Cantalini, C.; Lozzi, L.; Santucci, S., Sensors for sub-ppm NO_2 gas detection based on carbon nanotube thin films, *Appl. Phys. Lett.* **2003**, 82, 961–963

75. Chopra, S.; McGuire, K.; Gothard, N.; Rao, A. M.; Pham, A., Selective gas detection using a carbon nanotube sensor, *Appl. Phys. Lett.* **2003**, 83, 2280–2282

76. Klinke, C.; Chen, J.; Afzali, A.; Avouris, P., Charge transfer induced polarity switching in carbon nanotube transistors, *Nano Lett.* **2005**, 5, 555–558

77. Snow, E. S.; Perkins, F. K.; Houser, E. J.; Badescu, S. C.; Reinecke, T. L., Chemical detection with a single-walled carbon nanotube capacitor, *Science* **2005**, 307, 1942–1945

78. Sumanasekera, G. U.; Pradhan, B. K.; Romero, H. E.; Adu, K. W.; Eklund, P. C., Giant thermopower effects from molecular physisorption on carbon nanotubes, *Phys. Rev. Lett.* **2002**, 89, 166801.1–166801.4

79. Qi, P.; Vermesh, O.; Grecu, M.; Javey, A.; Wang, Q.; Dai, H.; Peng, S.; Cho, K., Toward large arrays of multiplex functionalized carbon nanotube sensors for highly sensitive and selective molecular detection, *Nano Lett.* **2003**, 3, 347–351

80. Staii, C.; Johnson, A. T.; Chen, M.; Gelperin, A., DNA-decorated carbon nanotubes for chemical sensing, *Nano Lett.* **2005**, 5, 1774–1778

81. Dai, H., Carbon nanotubes: Synthesis, integration, and properties, *Acc. Chem. Res.* **2002**, 35, 1035–1044

82. Chen, R.; Zhang, Y.; Wang, D.; Dai, H., Noncovalent sidewall functionalization of single-walled carbon nanotubes for protein immobilization, *J. Am. Chem. Soc.* **2001**, 123, 3838–3839

83. Grüner, G., Carbon nanotube transistors for biosensing applications, *Anal. Bioanal. Chem.* **2006**, 384, 322–335

84. Heath, J. R. In Nanobiotechnology II; Mirkin, C.; Niemeyer, C. M., Eds.; Wiley, New York, 2007, Chap. 12, 213

85. Asuri, P.; Bale, S. S.; Pangule R. C.; Shah, D. A.; Kane, R. S.; Dordick, J. S., Structure, function, and stability of enzymes covalently attached to single-walled carbon nanotubes, *Langmuir* **2007**, 23, 12318–12321

86. Peng, S.; Cho, K. J.; Qi, P.; Dai, H., Ab initio study of CNT NO₂ gas sensor, *Chem. Phys. Lett.* **2004**, 387, 271–276

87. Rodriguez, J. A.; Jirsak, T.; Sambasiban, S.; Fischer, D.; Maiti, A., Chemistry of NO₂ on CeO₂ and MgO: Experimental and theoretical studies on the formation of NO₃, *J. Chem. Phys.* **2000**, 112, 9929–9939

88. Rodriguez, J. A.; Jirsak, T.; Liu, G.; Hrbek, J.; Dvorak, J.; Maiti, A., Chemistry of NO₂ on oxide surfaces: Formation of NO₃ on TiO₂(110) and NO₂:O vacancy interactions, *J. Am. Chem. Soc.* **2001**, 123, 9597–9605

89. Chang, H.; Lee, J. D.; Lee, S. M.; Lee, Y. H., Adsorption of NH₃ and NO₂ molecules on carbon nanotubes, *Appl. Phys. Lett.* **2001**, 79, 3863–3865

90. Zhao, J.; Buldum, A.; Han, J.; Lu, J. P., Gas molecule adsorption in carbon nanotubes and nanotube bundles, *Nanotechnology* **2002**, 13, 195–200

91. Valentini, L.; Mercuri, F.; Armentano, I.; Cantalini, C.; Picozzi, S.; Lozzi, L.; Santucci, S.; Sgamellotti, A.; Kenny, A., Role of defects on the gas sensing properties of carbon nanotubes thin films: Experiment and theory, *Chem. Phys. Lett.* **2004**, 387, 356–361

92. Robinson, J. A.; Snow, E. S.; Bǎdescu, S. C.; Reinecke, T. L.; Perkins, F. K., Role of defects in single-walled carbon nanotube chemical sensors, *Nano Lett.* **2006**, 6, 1747–1751

93. Yamada, T., Modeling of carbon nanotube Schottky barrier modulation under oxidizing conditions, *Phys. Rev. B* **2004**, 69, 125408.1–125408.8

94. Andzelm, J.; Govind, N.; Maiti, A., Carbon nanotubes as gas sensors – Role of structural defects, *Chem. Phys. Lett.* **2006**, 421, 58–62

95. Stone, A. J.; Wales, D. J., Theoretical studies of icosahedral C₆₀ and some related species, *Chem. Phys. Lett.* **1986**, 128, 501–503

96. Ellison, M. D.; Crotty, M. J.; Koh, D.; Spray, R. L.; Tate, K. E., Adsorption of NH₃ and NO₂ on single-walled carbon nanotubes, *J. Phys. Chem. B* **2004**, 108, 7938–7943

97. Maiti, A., Multiscale modeling with carbon nanotubes, *Microelectron. J.* **2008**, 39, 208–221

98. Latil, S.; Roche, S.; Charlier, J. C., Electronic transport in carbon nanotubes with random coverage of physisorbed molecules, *Nano Lett.* **2005**, 5, 2216–2219

99. Santucci, S.; Picozzi, S.; Di Gregorio; F., Lozzi; L., Cantalini, C., L'Aquila, C.; Valentini, L.; Kenny, J. M.; Delley, B., NO₂ and CO gas adsorption on carbon nanotubes: Experiment and theory, *J. Chem. Phys.* **2003**, 119, 10904–10910

100. Ulbricht, H.; Kriebel, J.; Moos, G.; Hertel, T., Desorption kinetics and interaction of Xe with single-wall carbon nanotube bundles, *Chem. Phys. Lett.* **2002**, 363, 252–260

101. Wang, Z. L., Ed., Nanowires and Nanobelts: Materials, Properties, and Devices, Kluwer, Netherlands, **2003**

102. Dai, Z. R.; Pan, Z. W.; Wang, Z. L., Ultra-long single crystalline nanoribbons of tin oxide, *Solid State Commun.* **2001**, 118, 351–354

103. Huang, M.; Wu, Y.; Feick, H.; Tran, N.; Weber, E.; Yang, P., Catalytic growth of zinc oxide nanowires by vapor transport, *Adv. Mater.* **2001**, 13, 113–116

104. Huang, M.; Mao, S.; Feick, H.; Yan, H.; Wu, Y.; Kind, H.; Weber, E.; Russo, R.; Yang, P., Room-temperature ultraviolet nanowire nanolasers, *Science* **2001**, 292, 1897–1899

105. Comini, E.; Faglia, G.; Sberveglieri, G.; Pan, Z.; Wang, Z. L., Stable and highly sensitive gas sensors based on semiconducting oxide nanobelts, *Appl. Phys. Lett.* **2002**, 81, 1869–1871

106. Cui, Y.; Wei, Q.; Park, H.; Lieber, C. M., Nanowire nanosensors for highly sensitive and selective detection of biological and chemical species, *Science* **2001**, 293, 1289–1292

107. Favier, F.; Walter, E. C.; Zach, M. P.; Benter, T.; R. M. Penner, Hydrogen sensors and switches from electrodeposited palladium mesowire arrays, *Science* **2001**, 293, 2227–2231

108. Founstadt, C. G.; Rediker, R. H., Electrical properties of high-quality stannic oxide crystals, *J. Appl. Phys.* **1971**, 42, 2911–2918

109. Law, M.; Kind, H.; Kim, F.; Messer, B.; Yang, P., Photochemical sensing of NO_2 with SnO_2 nanoribbon nanosensors at room temperature, *Angew. Chem. Int. Ed.* **2002**, 41, 2405–2408

110. Maiti, A.; Rodriguez, J.; Law, M.; Kung, P.; McKinney, J.; Yang, P., SnO_2 nanoribbons as NO_2 sensors: insights from First-Principles calculations, *Nano Lett.* **2003**, 3, 1025–1028

111. Kind, H.; Yan, H.; Law, M.; Messer, B.; Yang, P., Nanowire ultraviolet photodetectors and optical switches, *Adv. Mater.* **2002**, 14, 158–160

112. Lieber, C. M.; Wang, Z. L., Functional nanowires, *MRS Bull.* **2007**, 32, 99–104

113. Law, M.; Goldberger, J.; Yang, P., Semiconductor nanowires and nanotubes, *Annu. Rev. Mater. Res.* **2004**, 34, 83–122

Chapter 3
Quantum Mechanics and First-Principles Molecular Dynamics Selection of Polymer Sensing Materials

Mario Blanco, Abhijit V. Shevade, and Margaret A. Ryan

Abstract We present two first-principles methods, density functional theory (DFT) and a molecular dynamics (MD) computer simulation protocol, as computational means for the selection of polymer sensing materials. The DFT methods can yield binding energies of polymer moieties to specific vapor bound compounds, quantities that were found useful in materials selection for sensing of organic and inorganic compounds for designing sensors for the electronic nose (ENose) that flew on the International Space Station (ISS) in 2008–2009. Similarly, we present an MD protocol that offers high consistency in the estimation of Hildebrand and Hansen solubility parameters (HSP) for vapor bound compounds and amorphous polymers. HSP are useful for fitting measured polymer sensor responses with physically rooted analytical models. We apply the method to the JPL electronic nose (ENose), an array of sensors with conducting leads connected through thin film polymers loaded with carbon black. Detection relies on a change in electric resistivity of the polymer film as function of the amount of swelling caused by the presence of the analyte chemical compound. The amount of swelling depends upon the chemical composition of the polymer and the analyte molecule. The pattern is unique and it unambiguously identifies the compound. Experimentally determined changes in relative resistivity of fifteen polymer sensor materials upon exposure to ten vapors were modeled with the first-principles HSP model.

M. Blanco (✉)
Division of Chemistry and Chemical Engineering, California Institute of Technology, BI 139-74, Pasadena, CA 91125, USA
e-mail: mario@wag.Caltech.edu

A.V. Shevade and M.A. Ryan
Jet Propulsion Laboratory, California Institute of Technology, Pasadena, CA 91109, USA

M.A. Ryan et al. (eds.), *Computational Methods for Sensor Material Selection*, Integrated Analytical Systems,
DOI 10.1007/978-0-387-73715-7_3, © Springer Science+Business Media, LLC 2009

3.1 Introduction

In the context of sensor development work, the main objective of molecular modeling is to provide theoretical support for sensor material selection. Material selection for the JPL electronic nose (ENose) made use of computer modeling. The JPL (ENose) is an array of polymer–carbon composite sensors designed to monitor breathing air quality for chemical contaminants aboard space shuttles/space stations [1–6]. The JPL ENose is under development as an array–based sensing system, which can run continuously and monitor for the presence of toxic chemicals in the air in near real time. The sensing array in the JPL ENose is made from polymer–carbon composite sensing films. The previous two generations of the JPL ENose have successfully detected organic contaminants at parts-per-million (ppm) to sub ppm concentration levels. In general, the methods outlined here are applicable to amorphous materials principally made of individual monomer units, synthetic or of biological origin.

Methods that correlate polymer–carbon composite sensor response with molecular descriptors [4, 6] are quite helpful. These quantitative structure property relationships (QSPR) models require experimental measurements to create a predictive model. In addition, there are limitations regarding predictions outside the initial training set.

Ideally, a first principles approach can be used to predict the interactions between any vapor bound analyte and individual or collective organic moieties built into a polymer sensor, without the need to perform experiments. These predictions can then be fed back into the experimental effort to optimize the material selection process. This is the approach we take in this chapter.

Although direct computational estimates of the energetics of analyte/polymer interactions with simulated bulk samples of the full polymer environment could eventually become quite useful in this field [7], we have chosen to focus on a methodology that involves calculating binding energies for individual organics-analyte binary systems, including common classes of organic compounds such as alkanes, alkenes, aromatics, primary, secondary and tertiary amines, aldehydes, carboxylic acids, and esters. We have also found useful to model the physical properties of the polymer, particularly Hansen and Hildebrand solubility parameters (HSP) as critical factors in creating predictive algebraic models [8]. However, the direct estimates of the free energy of binding of analytes in simulated polymer bulk samples await the application of more advanced methods [9].

3.1.1 First Principles Modeling

First-principles modeling refers to the estimation of physical properties without the use of experimental information. Typically this involves the solution of the Schrödinger equation (quantum mechanics) or if a suitable force field (an algebraic representation of the energy of the molecules involved as a function of geometry) is available it might also involve solutions to Hamilton's equations of motion (Newtonian classical molecular dynamics). First principle properties that are now available for computation include electronic energies, binding or interaction energies and entropies, electronic charge distributions, dipole moments, dielectric properties, swelling, densities, free volume distributions, solubility parameters, surface tension, and diffusion coefficients.

Quantum interaction energies are typically calculated using quantum mechanics (QM) density functional theory (DFT) to solve the Schrödinger equation for all electrons in the system:

$$H\Psi = E\Psi \tag{3.1}$$

where H is the Hamiltonian of the electronic degrees of freedom (all atoms in the system).

$$H = -\sum_k \frac{h^2}{2M_k}\nabla^2 - \sum_i \frac{h^2}{2m_i}\nabla^2 + \sum_k \sum_{l>k} \frac{Z_k Z_l e^2}{R_{kl}} + \sum_i \sum_{j>i} \frac{e^2}{r_{ij}} - \sum_{kj} \frac{Z_k e^2}{r_{kj}}. \tag{3.2}$$

Here, k, l run over all the nuclei, and i, j over all electrons. The first two terms give the kinetic energy of the nuclei and electrons, respectively, and the third term yields the repulsion between the atomic nuclei, while the fourth and fifth terms are the most difficult terms to solve, representing the repulsion of all electrons and the attraction between electrons and nuclei, respectively.

We recommend B3LYP for routine calculations [10, 11]. For DFT methods that might offer advantages in estimating dispersion (van der Waals) interactions we refer the reader to the X3LYP method [12] or to the most current M06 suite of DFT functionals [13–16]. Often it is necessary to use basis superposition error (BSSE) corrections in these calculations. Superposition errors arise from the use of a finite basis set. The number of functions that represent the electron densities must be manageable, an unavoidable process in estimating numerical solutions of the Schrödinger equation. The polymer moiety and the analyte are represented with different basis sets; particularly the locations of the centers of these basis functions are different. To avoid errors coming from these differing basis sets we include in the calculation of the polymer moiety the basis functions (as ghost functions, not the electrons or nuclei, just the numerical basis functions) for the analyte and vice versa. Thus, both energy calculations use exactly the same basis set for the combined polymer/analyte calculation and the error is avoided.

The quantum calculations yield mainly three types of sensor useful information:

1. The relative energies of polymer/analyte together vs. polymer and analyte (separately), which measure the binding energy of the pair
2. The atomic charges (typically Mulliken population charges) for each atom
3. The energy as a function of polymer/analyte distance at various orientations

These results are useful in evaluating the potential use of a given material for detection of a particular analyte. In addition, these results are also of practical importance in creating an accurate "force field" that can be employed to represent the energies, forces, and molecular dynamics of analyte/polymer interactions using classical (Newtonian) dynamics. We refer to the process of using quantum mechanically derived information to build a suitable force field for molecular dynamics also as first principles modeling because no experimental data are required.

3.2 Identifying Chemical Functionalities in Polymers for Analyte Detection: SO$_2$ Polymer Sensor Example

The target compound list for the third generation JPL ENose, now undergoing deployment in the International Space Station (ISS), included a set of inorganic compounds (notably mercury, Hg, and sulfur dioxide, SO$_2$) in addition to a subset of previously detected organics (Table 3.1).

We use B3LYP DFT calculations [10, 11] to estimate binding energies of typical organic functionalities present in homo- and copolymers of interest (see Table 3.2) with these analytes. Subsequently the QM results were used to develop a first principles force field for use in the calculation of interaction energies of SO$_2$ molecules with various organic functional groups. As an example, the quantum binding energies for methylamine/SO$_2$ system at various scan distances, measured between the sulfur atom in SO$_2$ and the nitrogen atom in methylamine are shown in Table 3.1. Scan distances from 1.5 to 6 Å in increments of 0.1 Å were chosen in all cases. In most cases the molecular orientation was fixed around the most optimal approach found with a generic force field [17], using Mulliken population quantum charges, and a docking algorithm previously developed by one of the authors [18]. The QM binding energy for the methylamine/SO$_2$ system is $\sim$10.7 kcal mol^{-1}. In this case BSSE error corrections are quite small, on the order of 0.1 kcal mol^{-1}. Nonetheless, we included them in all energy calculations. The molecules are shown in the optimal, fixed, molecular approach orientation (Fig. 3.1).

Table 3.1 Calculated analyte properties for third generation JPL ENose

	Hansen Solubilities (cal cm^{-3})$^{1/2}$			Molecular volume (Å^3)
	Electrostatics	Dispersion	H-bond	
Acetone	4.79	7.53	0	65.7
Ammonia	6.66	3.83	8.31	22.2
Dichloromethane	2.31	8.75	0	57.8
Ethanol	5.72	6.82	4.33	53.2
Freon 218 (C$_3$F$_8$)	1.5	5.18	0	104.1
Mercury	0	20.11	0	15.5
Methanol	7.33	6.16	5.2	35.9
2-Propanol	4.42	7.14	3.53	71.9
Sulfur Dioxide	19.07	6.95	0	40.7
Toluene	0.99	8.92	0	98.98

Table 3.2 Polymer candidates for the Third Generation JPL ENose

Polymer	Cohesive energy components			Hansen solubilities		
	Electrostatic	Dispersion (cal cm^{-3})	H-Bond	Electrostatic	Dispersion (cal cm^{-3})$^{1/2}$	H-Bond
Poly(4-vinyl phenol)	−5.06	−46.78	−3.24	2.25	6.84	1.80
Methyl vinyl ether/maleic acid, 50/50	−22.05	−29.65	−6.88	4.70	5.45	2.62
Poly(styrene-co-maleic acid)	−7.62	−29.99	−2.65	2.76	5.48	1.63
Polyamide resin	−7.32	−45.51	−1.46	2.71	6.75	1.21
Poly(N-vinyl pyrrolidone)	−16.73	−41.36	0	4.09	6.43	0.00
Vinyl alcohol/vinyl butyral, 20/80	−15.39	−52.29	−2.54	3.92	7.23	1.59
Ethyl cellulose	−8.97	−27.05	−0.14	2.99	5.20	0.37
Poly(2,4,6-tribromostyrene), 66%	−0.55	−33.37	0	0.74	5.78	0.00
Poly(vinyl acetate)	−22.31	−29.17	0	4.72	5.40	0.00
Poly(caprolactone)	−20.92	−69.12	0	4.57	8.31	0.00
Soluble polyimide, Matrimid	−7.27	−54.89	0	2.70	7.41	0.00
Poly(epichlorohydrin-co-ethylene oxide)	−10.48	−56.7	0	3.24	7.53	0.00
Poly(vinylbenzyl chloride)	−1.81	−48.07	0	1.35	6.93	0.00
Styrene/isoprene, 14/86 ABA block polymer	−1.69	−39.08	0	1.30	6.25	0.00
Ethylene-propylene diene terpolymer	−0.57	−55.77	0	0.75	7.47	0.00
Polyethylene oxide, PEO100	−18.38	−72.61	0	4.29	8.52	0.00

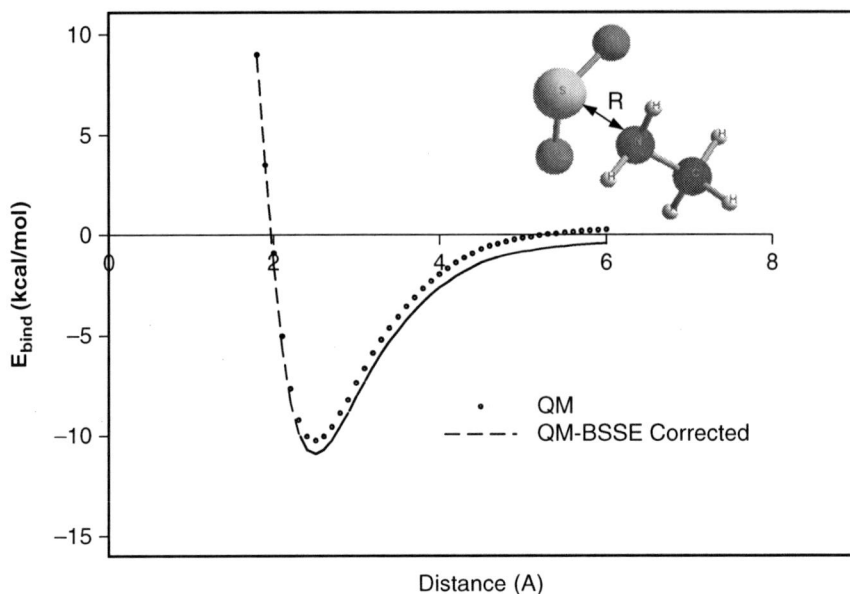

Fig. 3.1 Binding energy as a function of the distance between analyte (SO_2) and a potential polymer moiety (methylamine). *Insert* shows the varied distance (*white arrow*) between the nitrogen in methylamine and sulfur in SO_2

Figure 3.2 shows B3LYP DFT results for a variety of other moieties of interest and SO_2. Binding energies range between 0.0 and 15.0 kcal mol^{-1} (customarily we drop the sign when referring to these energies as "binding" energies). PL1 refers to an approach from above the plane of the molecule, parallel to the plane of symmetry, which appeared more stable than an in plane approach for both benzene and ethylene (not shown). An axial approach to ethylene was also tried and found even less favorable. We note that aliphatic moieties (ethane, ethylene) as well as aromatics (benzene) are nonbinding. These results exclude polymers that contain ethylene and styrene as their main constituents for the detection of SO_2. On the other hand, amine-containing polymers, particularly secondary and tertiary amines, are predicted to have good binding energies (ca. 10–15 kcal mol^{-1}), suitable for their reversible detection. Secondary cyclic amines, such as methylpyrrolidone, and carboxylate moieties, including formic acid, formaldehyde, and formamide, are predicted to be intermediate in binding (between 4 and 8 kcal mol^{-1}).

The fundamental understanding on the basis of the quantum mechanical molecular interactions between various organic moieties/SO_2 systems was useful to prioritize the selection of polymers sensor materials for the JPL ENose for SO_2 detection.

Two polymers were selected and made into polymer carbon black composite sensors [19]. These two polymers are both poly-4-vinyl pyridine derivatives with a quaternary and a primary amine. The polymers were designated EYN2 and EYN7; the structures are shown in Fig. 3.3. The polymers were synthesized from poly-4-vinyl pyridine and made into polymer-carbon composite sensing films using

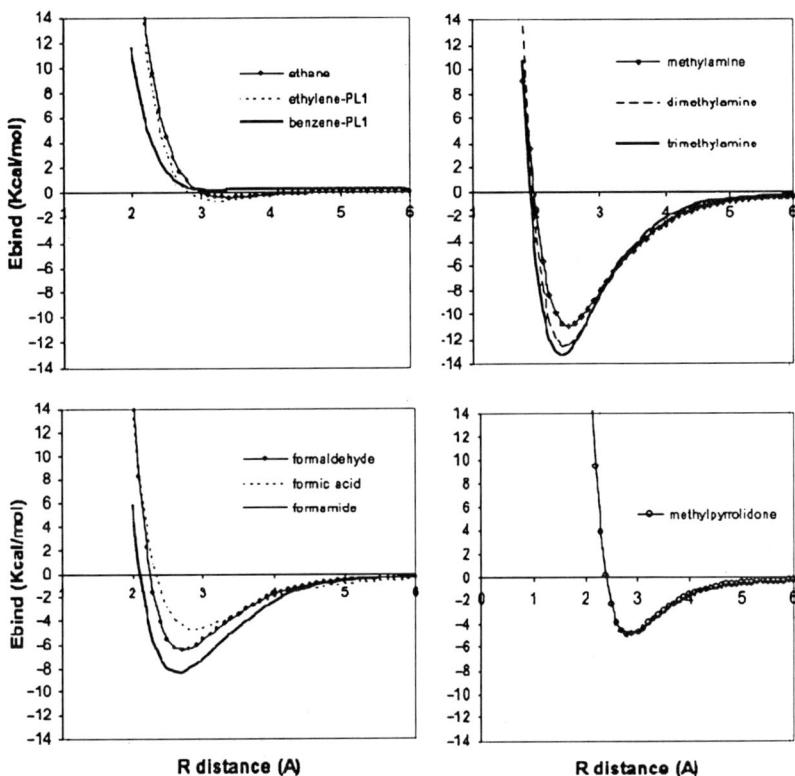

Fig. 3.2 Binding energy as a function of distance between analyte (SO_2) and a host of potential polymer moieties

protocols which have been previously described [1, 2]. These films were loaded with 8–10% carbon by weight and solution deposited onto micro-hotplate sensor substrates with a sensor area of 200 μm by 200 μm (4×10^{-8} cm^2). The baseline resistance of each sensor was ~10 kΩ.

Previously tested polymers, shown in Fig. 3.4, yielded less optimal changes in resistance, on the order of less than 2% at higher (e.g.,15 ppm) concentrations. The molecular level predictions (weak/strong binding) when compared to experimental sensor responses follow closely the same order:

Tertiary amine > primary amine > carboxylate > aliphatic polyimide > polyamide resin > polycaprolactone > ethylene–propylene.

New sensor responses are shown in Fig. 3.5. Primary and cyclic amines provide sufficient binding energy for the detection of SO_2 in the gas phase, as predicted. Changes in base line resistance are on the order of 6% at 9 ppm SO_2 concentrations. On the basis of the quantum binding energy predictions, even higher sensitivities would have been obtained if secondary and/or pure tertiary amines (without surrounding carbonyl moieties as is the case in polyimide) had been used in the polymer sensor synthesis.

Fig. 3.3 Monomer structures for the preparation of polymers and co-polymers selected for SO_2 detection

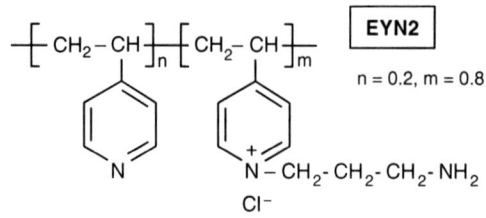

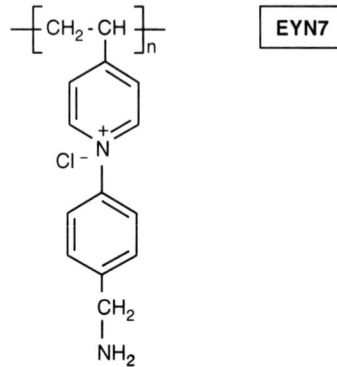

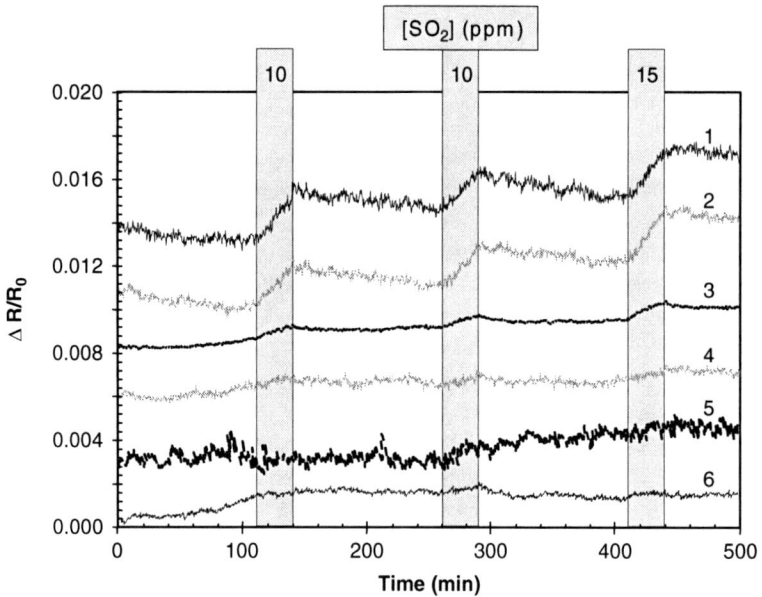

EYN2

$n = 0.2, m = 0.8$

EYN7

Fig. 3.4 Sulfur dioxide response of six polymer sensors: 1 polyimide, 2 polyamide resin, 3 polycaprolactone, 4 ethylene-propylene, 5 poly 4-vinylphenol, and 6 polyvinyl acetate

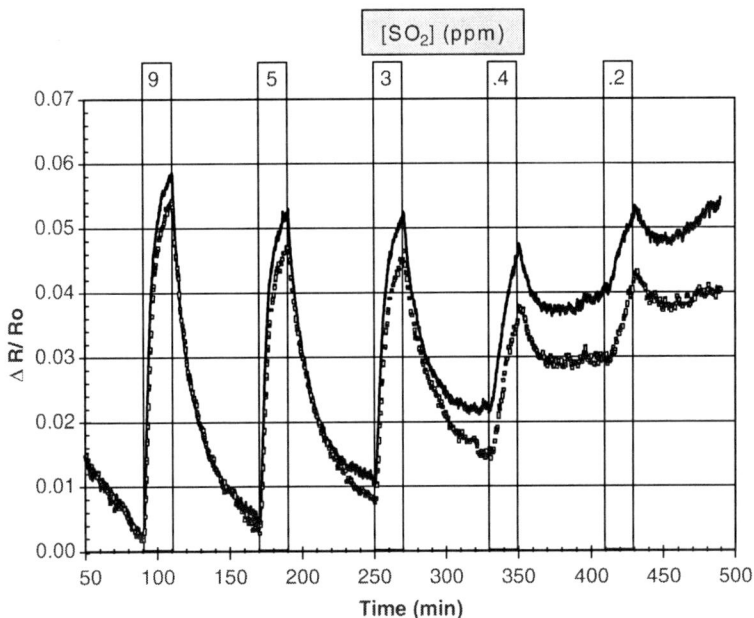

Fig. 3.5 Response of two sensors, both made from polymer EYN2, to 0.2–9 ppm SO_2 in air

3.3 Molecular Dynamics Principles for Modeling Polymer and Analyte Bulk Properties

Following the quantum mechanical calculations, which often yield enough data to make informed decisions for sensing materials, we pursue the modeling of the properties of the polymer in the bulk. We hope to achieve a first-principles understanding of the interactions that give rise to the full gamut of molecular interactions with the analyte. We use molecular dynamics (MD), solving the classical Hamilton's equations of motion. Beginning with the Lagrangian,

$$L = T - V(q). \tag{3.3}$$

$$H = \sum_j p_i \dot{q}_i - L, \quad \frac{\partial H}{\partial q_i} = -\dot{p}_i, \quad \frac{\partial H}{\partial p_i} = \dot{q}_i, \quad p_i = \frac{\partial L}{\partial \dot{q}_i}. \tag{3.4}$$

Here q represents the atomic coordinates p their momenta, and T the kinetic energy of all the atoms (here we no longer deal with individual electrons). $V(q)$ is the potential energy function, an algebraic expression often referred to as the "force field", giving the energy of the molecules for a given set of geometric positions of their constituent atoms. $V(q)$ should be on the basis of solutions to the Schrödinger equation, as much as possible. In practice a generic force field is used for valence terms (covalently bound atoms) while the nonbond interactions (electrostatics, hydrogen bonds, and van der Waals or dispersion interactions) are calculated from quantum mechanical scans as those shown in Sect. 3.2. If the state (coordinates and momenta) of the system is well known at time t, we can find the position and velocity of all atoms at time $t + dt$ by integration of the partial differential (3.4). For this we typically use the Verlet algorithm:

$$q(t + \Delta t) = 2q(t) - q(t - \Delta t) - \frac{\dot{p}(t)}{m} \Delta t^2. \tag{3.5}$$

It gives accurate results to $O\ (\Delta t^4)$. For a full description of MD the reader is referred to standard molecular dynamics textbooks [21, 22].

The experimentally determined changes in relative resistivity, $\Delta R/R$, of the polymer sensors upon exposure to vapors have been correlated with computer calculated Hansen solubility parameters [9].

$$\Delta R/R = R_0 \exp(-\gamma V_s) \exp\left[\sum_{i=1}^{3} \beta_i \left(\delta_i^s - \delta_i^p\right)\right]. \tag{3.6}$$

γV_s is the activation energy of diffusion of the solute in the polymer, proportional to the molar volume of the odorant, V_s. The exponential factor γ is a best-fit parameter. We base this relation on the experimental observation that the diffusion coefficient of various molecules is linearly related to the molar volume of the solute above the glass transition temperature (T_g) of the polymer [22]. δ_i^s are the Hansen solubility parameters (HSP) of the solvent s, where $i = 1, 2$, and 3 and refer to the electrostatic, dispersion, and hydrogen bond components, respectively. Similarly, δ_i^p is the ith HSP component of the polymer sensor p. The exponential coefficients β_i are treated as best fit parameters and so is the pre-exponential term R_0. It should be noted that we preserve the sign of the energy components in (3.6), which is usually lost in the definition of Hansen and Hildebrand parameters. This is important because such interactions can be attractive or repulsive, depending on the polymer/odorant mixture in question.

The Hildebrand solubility parameter for a pure liquid substance is defined as the square root of the cohesive energy density.

$$\delta = \left[(\Delta H_v - RT)/V_m\right]^{1/2}, \tag{3.7}$$

ΔH_v is the heat of vaporization and V_m the molar volume. RT is the ideal gas pV term and it is subtracted from the heat of vaporization to obtain the energy of vaporization. Typical units are as follows:

1 hildebrand = 1 cal$^{1/2}$ cm$^{-3/2}$ = 0.48888 $\times$ MPa$^{1/2}$ = 2.4542 $\times$ 10^{-2} (kcal mol^{-1})$^{1/2}$ A$^{-3/2}$

Hansen [23] proposed an extension of the Hildebrand parameter to estimate the relative miscibility of polar and hydrogen bonding systems:

$$\delta^2 = \delta_d^2 + \delta_p^2 + \delta_h^2, \tag{3.8}$$

where, δ_d, δ_p, and δ_h are the dispersion, electrostatic, and hydrogen bond components of δ, respectively. For molecules whose heats of vaporization can be measured, or calculated, one can easily determine the value of δ. The Hansen solubility parameters in (3.8) are typically determined empirically on the basis of multiple experimental solubility observations, including observed solubility or swelling with a series of solvents, NMR and IR signals, elution times in chromatographic columns, etc. The reported values in the literature can vary over a large range, however, owing to the multiple experiments used to determine these components. Instead, we rely on a MD protocol to estimate the HSPs.

3.4   First-Principles Force Field

The MD protocol is defined as a sequence of individual modeling tasks which begin with the proper selection of a force field. The energy of the polymers, $V(q)$ in (3.3), is represented as a sum of valence and non-bond terms as follows:

$$V(q) = \sum_{\text{bonds}} V_r + \sum_{\text{angles}} V_\theta + \sum_{\text{torsions}} V_\varphi + \sum_{\text{charges}} V_{\text{coul}}(R_{ij}; q_{ij})$$

$$+ \sum_{\text{charges}} V_{\text{coul}}(R_{ij}; q_{ij}) + \sum_{\text{O,N,S}} V_{\text{Hbond}}(R_{ij}). \tag{3.9}$$

Valence force field: we use a generic force field, Dreiding [17] to estimate valence terms. Bond stretch terms are given by

$$V_r = \frac{1}{2} K_r (R - R_0)^2. \tag{3.10}$$

Bond angle distortions are given by

$$V_\theta = \frac{1}{2} \frac{K_\theta}{\sin^2 \theta_0} (\cos\theta - \cos\theta_0)^2 \tag{3.11}$$

and dihedral angle torsions by

$$V_\varphi = \frac{1}{2} V (1 - d\cos3\phi). \tag{3.12}$$

We also approximate the noncovalent hydrogen bond term, which runs over heteroatoms such as O, N, and S, using the published Dreiding Lennard-Jones 12–10 potential form

$$V_{\text{Hbond}}(R_{ij}) = D_0 \left[\left(\frac{R_0}{R_{ij}}\right)^{12} - \left(\frac{R_0}{R_{ij}}\right)^{10} \right]. \tag{3.13}$$

We employ $D_0 = 3.2$ kcal mol^{-1} and $R_0 = 2.5$ Å, which give better agreement with the heat of vaporization of water (580 cal cm^{-3}) than the published Dreiding hydrogen bond values when MP2/6–31g** atomic charges are used for the isolated water molecule [$q(O) = -0.72866$, $q(H) = 0.36433$].

Electrostatics: all electrostatics interaction pairs were included in the calculation without the use of cutoffs or spline functions.

$$V_{coul}(R_{ij}) = c \sum_j \sum_{i>j} \frac{q_i q_j}{R_{ij}}. \tag{3.14}$$

Quantum charges, q_i, were calculated at the quantum optimized (minimum energy) geometry of each molecule (polymer segment and analytes) using polarized Mulliken charges at the minimum conformation. We use the Jaguar suite of programs [24], B3LYP DFT with a good (6–31g**) basis set. $\varepsilon = 1$ is the dielectric constant, q_i and q_j are the atomic charges in electron units, and R_{ij} is the distance in Å; $c = 322.0637$ converts the electrostatic energy to kcal mol^{-1}. For SO_2, the Mulliken population charges are $q(S) = 0.83914$ and $q(O_1, O_2) = -0.41957$ electrons.

The noncovalent terms in (3.9) play a crucial role in determining the structure of the polymer and the HSP values for both polymer and analyte. Thus, the force field is further refined for the dispersion (van der Waals) noncovalent interactions.

3.4.1 Refinement of van der Waal Interactions

The van der Waals interactions can be as important in determining the binding energetics of analyte/polymer sensor as much as the electrostatics, particularly when nonpolar chemical groups are involved at either end.

The binding energies for molecular pairs in Fig. 3.2, typical organic molecules interacting with SO_2, were calculated as follows:

$$E_{bind} = E_{AB} - (E_{AB}^* - E_{BA}^*). \tag{3.15}$$

Here E_{bind} is the calculated energy of the AB pair, and E_{AB*} and E_{BA*} are the energies of molecules A and B. These energies are calculated with the basis set of B and A present (B*, A*, respectively); the process described above is basis set superposition error correction (BSSE). As mentioned above we carried out a distance sweep between the organic molecule and SO_2 calculating the quantum energy of interaction between 1.5 and 6 Å in increments of 0.1 Å. In some cases, e.g., ethane, ethylene, and benzene we used various directions for the sulfur dioxide approach to the organic compounds, such as parallel (PL) or perpendicular (PR) to the molecular axis of symmetry. The binding energies exclude all covalent interactions (as the individual molecular energies have been subtracted) but include all nonbond terms, i.e., electrostatics, hydrogen bond, and dispersion. After subtraction of the electrostatic terms we fit the remaining binding energy with a Morse potential:

$$V(R_{ij}) = D_e(1 - e^{-\xi(R_{ij}-R_e)})^2 - D_e, \tag{3.16}$$

where D_e represents the van der Waals contribution to the total binding energy at the equilibrium distance R_e and ξ is related to the harmonic force constant k associated with the noncovalent binding mode by

$$\xi = \sqrt{\frac{k}{2D_e}}. \tag{3.17}$$

R_{ij} is the distance between the two atom centers. Because elements have different environments we need to estimate a set of van der Waals dispersion parameters for each atom type pair. Thus, an sp^2 carbon, such as a carbon in ethylene, will have its own specific set of values (D_e, R_e, and ξ). The process is somewhat involved, but for each geometry present in the distance sweep between the analyte and the organic moiety all the interatomic distances are recorded and used to fit the Morse parameters after subtraction of the electrostatic Coulomb energy. A full least square procedure on all Morse parameters combined is carried out.

The force field represents the calculated quantum binding energies quite well. We obtained the force-field energy curves shown in Fig. 3.6 for the secondary amine with SO_2. Small open circles are the quantum binding energies given by (3.15)

Tables 3.3a and 3.3b contain the set of (D_e, R_e, and ξ) for each type of atom pairs involving SO_2 and for the atom types present in the organic functionalities included

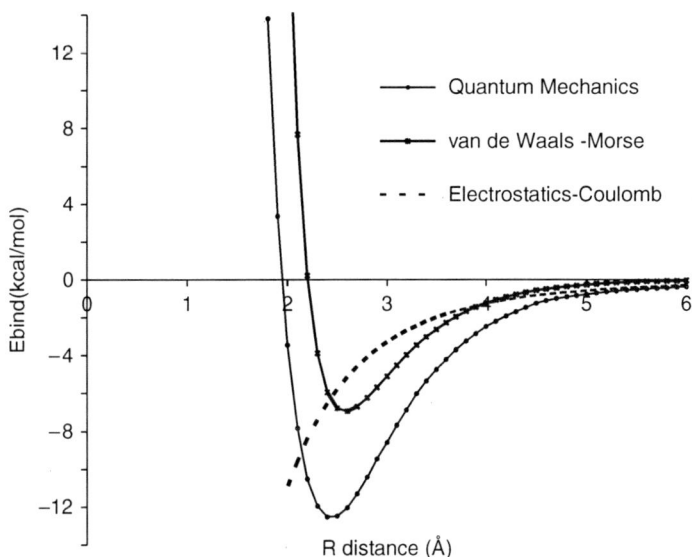

Fig. 3.6 Quantum binding energies and the force field van der Waals and Coulomb components for dimethylamine interacting with SO_2

Table 3.3a van der Waals force field parameters determined from DFT (B3LYP) binding energetics. Morse parameters for sulfur (S_3) Dreiding atom type in SO_2 with other atom types present in ten organic compounds

Atom types	Hybridization	R_e	D_e	ξ
C_3	sp^3	3.132617	0.749458	11.99488
C_2	sp^2	2.737065	1.340003	13.17681
C_R	sp^2 aromatic	5.344190	0.002262	14.72481
H_$^\$$	sp^3	3.973471	0.029141	12.00958
H_$^\$$	sp^2	2.961596	2.245861	13.49589
H_A[a]		3.499816	0	11.99997
H_A[b]		3.954983	0.000694	12.00027
H_R	sp^2 aromatic	3.077340	0.193491	13.22020
N_3	sp^3	2.230163	8.252832	11.92240
N_R	sp^2 aromatic	3.498603	0	11.91701
O_2	sp^2	3.459701	0.000602	11.81915
O_R	sp^2 aromatic	5.089213	0.167158	12.62331

[a]Attached to heteroatoms (e.g., O)
[b]Attached to N atom only

Table 3.3b Morse parameters for oxygen atoms of SO_2 (O_2) with common Dreiding atom types present in ten organic compounds

Atom types	Hybridization	R_e	D_e	ξ
C_3	sp^3	3.450184	0.000471	11.99898
C_2	sp^2	4.053726	0	11.07503
C_R	sp^2 aromatic	3.485616	0	12.00207
H_$^\$$	sp^3	3.900214	0.013218	12.03888
H_$^\$$	sp^2	2.822144	0	12.03854
H_A[a]		4.001108	0	13.15887
H_A[b]		3.969541	0.296846	11.84432
H_R	sp^2 aromatic	3.028448	0.427043	12.66140
N_3	sp^3	3.345748	0.846320	11.87265
N_R	sp^2 aromatic	3.499287	0	13.63104
O_2	sp^2	3.728343	0.106303	15.86427
O_R	sp^2 aromatic	2.999365	0	12.37445

[a]Attached to hetero-atoms (e.g., O)
[b]Attached to N atom only

in this study. A brief glance at this table shows that the sp^3 Nitrogen atom in the amine interacts the strongest ($D_e = 8.25$ kcal mol^{-1}) with the sulfur in sulfur dioxide.

3.5 Molecular Dynamics Protocol for Polymer Sensor Responses

As shown in the previous section quantum mechanics can provide sufficient information to guide material sensor selection successfully. A wealth of information is generated in the process, which includes the force field parameters in Tables 3.3a and 3.3b. These parameters can be used to estimate polymer and analyte material properties that can provide the basis for developing predictive models of sensor responses, such as the model given by (3.6). Thus, in this section we aim at estimating without experimental input Hansen solubility parameters. The most common problem in computer simulations of polymers is the long time required to obtain an equilibrated simulated sample. This is a common problem with amorphous condensed phases. We have developed a method that overcomes the common equilibration problems with condensed phase molecular dynamics, i.e., how to choose initial molecular configurations not far from equilibrium at normal densities. Significant amounts of simulation time are usually required to equilibrate the initially random packed molecules often generated with Monte Carlo methods. In particular, densely packed simulated polymers often lead to highly nonequilibrated dihedral populations. Thus, care must be taken to generate an ensemble of thermally accessible conformations not far from equilibrium. These two requirements, condensed phase densities and equilibrated molecular conformations, are satisfied through the following MD protocol:

1. A cubic periodic unit cell containing a given number of molecules is built at a low density, ρ_{low}, typically 50% of the target density. Generally four polymer chains are sufficient, although for very high molecular weights even one chain can be adequate. For solvents 16–64 solvent molecules are adequate. We find that for packing the structure, it is useful to scale van der Waals radii by a factor of 0.30 to get initial structures that will eventually lead to a good ensemble. In cases where the compounds are polymers, or a molecule with a large number of torsional degrees of freedom, we use the Amorphous Builder in Cerius^2 [25] to create the initial low-density sample. The initial polymer amorphous structures are constructed using the rotational isomeric state (RIS) table [26] and a suitable Monte Carlo procedure to achieve a correct distribution of conformational states in the low-density sample. The Amorphous Builder converts an existing model into an amorphous structure by manipulating the model's rotatable bonds. Each unique torsion can be defined using a Monte Carlo procedure with statistical weights given by a previously built rotational isomeric state table determined with well established molecular mechanics dihedral sampling procedures. Conformations are rejected if two or more atoms come closer than a van der Waals scale distance. In polymer calculations, the number of monomers in each chain is usually determined such that the total volume of the four chains is at least 6,000 Å [3]. Alternatively, a degree of polymerization of 30 suffices to give values comparable to those from experiment. In such polymer samples, the minimum number of atoms is at least 1,000. Larger samples are recommended whenever possible.

2. For convenience we used the experimental densities of the solvents and polymers as target values as these are commonly available in the literature. For liquid systems with unknown densities we typically run a preliminary MD calculation with a rough "trial" density, such as that predicted from group

additivity methods, to obtain a good initial estimate. The procedure below will increase the density to a maximum, ρ_{high}, typically 125% of the target density. The resulting amorphous structure is then relaxed, resulting in a predicted target density for the start of the definitive MD calculation.

3. The charges of the isolated solvent or polymer molecules are defined using those obtained from quantum mechanical calculations (Mulliken population charges) as previously explained.

4. The force field parameters are taken from a generic force field, such as the generic Dreiding force field with modifications of the van der Waals parameters for higher accuracy.

5. Minimization: The potential energy of the bulk system is minimized for M steps, typically $M = 5,000$ steps, or until the atom rms force converges to 0.10 kcal $(\text{mol Å})^{-1}$, whichever comes first.

6. Annealing dynamics to allow the structures to equilibrate typically 750 steps of MD (1 fs per step) at high temperature (typically between 400 and 800 K, with 700 K generally adequate) using canonical fixed volume dynamics (NVT) are carried out to anneal the sample.

7. Compression: The reduced cell coordinates are shrunk such that the density is increased by $(\rho_{\text{high}} - \rho_{\text{low}})/N$, where N is typically 5.

8. The atomic coordinates are minimized and dynamics run on the system with the previously described procedure holding the cell fixed (steps 5–6).

9. A total of N compression, minimization, and dynamics cycles are performed until the density reaches ρ_{high}, typically 125% of the target density, steps 5–8.

10. The cell parameters are then increased in N cycles of expansion, minimization, and dynamics, until the target density is reached

11. The sample is allowed to relax in M steps of minimization allowing both the cell and the atomic coordinates to relax.

12. Molecular dynamics are performed for a time to thermalize and then to measure properties. Typically we perform these for as few as 20 ps but longer times are recommended for high molecular weight compounds. The first 10 ps are used for thermalization of the sample at the desired temperature. The last 10 ps are used for averaging of cell volume and potential energy components: van der Waals (dispersion), electrostatic (polar), and hydrogen bonding.

13. The Hansen enthalpy components are calculated by subtracting the potential energy of the bulk system from the sum of the potential energies of the individual molecules in vacuum.

$$\delta_k^2 = \sum_{i=1}^{n} \langle E_i^k - E_c^k \rangle / N_0 \langle V_c / n \rangle. \tag{3.18}$$

14. Here $\langle \rangle$ indicates a time average over the duration of the dynamics, n the number of molecules, $k = 1,2,3$ for coulomb (polar), van der Waals (dispersion), and hydrogen bond components, and N_0 is Avogadro's number.

15. This process is repeated P times with different initial random conformations and packing. Typically $P = 10$ is adequate but higher values are recommended.

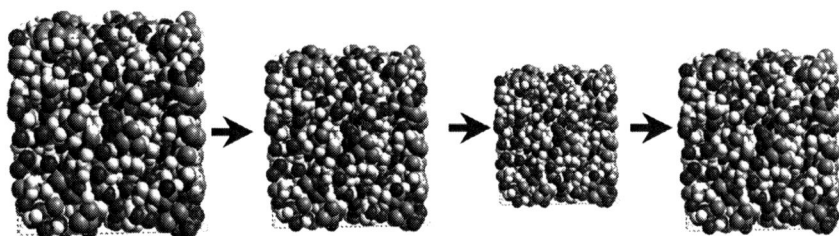

Fig. 3.7 A polymer or solvent sample is put through a series of compression and expansion steps until the proper density and packing are obtained. On the *left* the initial density is 40% of the target density. After compression, second step, the sample is overcompressed by 20%. Finally the sample is allowed to relax. Through NPT molecular dynamics a final prediction of the density and cohesive energy of the sample is obtained. The process is repeated for several samples and statistics are gathered

16. Hansen solubility parameters and molar volumes are computed as well as their standard deviations. We use the 95% confidence limit of an F statistical distribution test, with two standard deviations from the average value, to identify outliers. Typically a $P = 10$ sample run will have no outliers; more than two outliers are rare.

The overall procedure is schematically illustrated in Fig. 3.7. Tables 3.1 and 3.2 show the results of this procedure, the Hansen solubility parameters for analytes, and polymer candidates for the JPL ENose.

At this point we introduce experimental information to complete the model in (3.6). Once these free parameters in (3.6) are determined we can employ this expression to make sensor response predictions for new analytes, not initially included in the training set. Tables 3.1 and 3.2 contain the calculated (using the MD protocol outlined above) Hansen solubilities for analytes and polymers, respectively. Figure 3.8 shows the experimental vs. predicted sensor responses for some typical polymers. Figure 3.9 shows the model predictions vs. experimental data for some of the polymer sensors.

We next tested the predictive power of (3.6) by calculating the responses of the JPL ENose sensor array to an analyte not originally included in the training set. The results are shown in Fig. 3.10. There is good agreement between the model's prediction and the measured JPL ENose responses to Freon113. This is an important test because it shows that these model have true predictive power, beyond simple data regression or statistical correlations.

3.6   Discussion and Conclusions

We have presented a first-principles approach to materials selection for polymer sensors for specific analytes. The quantum (B3LYP/DFT) binding energies of analyte with a few representative organic moieties can yield a great deal

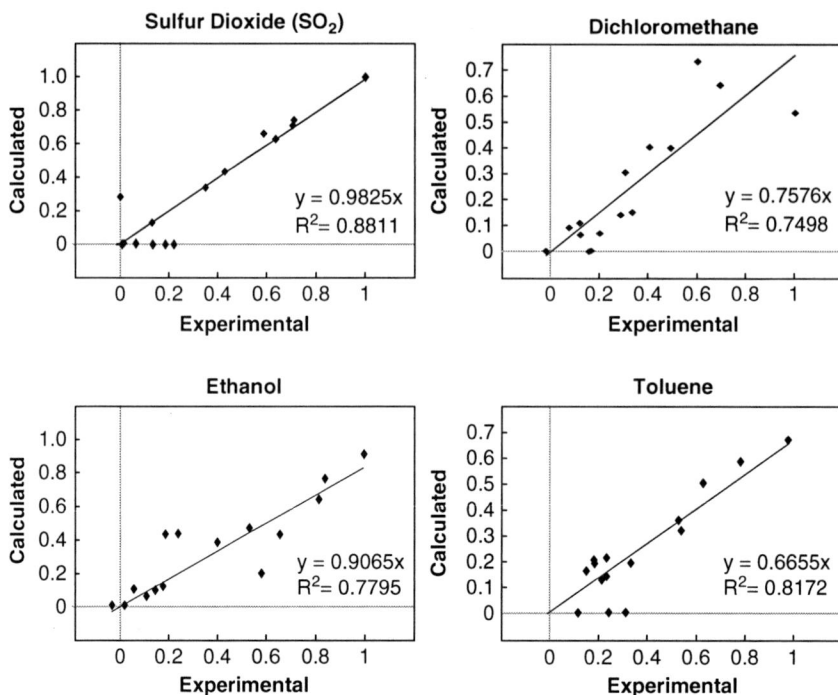

Fig. 3.8 Calculated (3.6) vs. experimental sensor responses to four analytes in the third generation JPL ENose compound list

of information. The quantum binding energies, as shown in the sulfur dioxide example, are strongly correlated with the measured responses, and therefore, the quantum calculated energy data can significantly aid the selection of sensing materials. Furthermore, the data can be employed to successfully model the self-interactions, in the form of Hansen solubility parameters, of analytes and polymers. These first-principles MD-derived solubilities can be used to model the responses of an array of polymer sensors to a wide variety of analytes and provide the basis for extrapolations outside the original training set.

Table 3.4 shows the values of the fitting parameters in (3.6), obtained using a least square fit of the experimental data and the calculated Hansen parameters in Tables 3.1 and 3.2 for each of analyte and polymer sensor pairs in the study. Over half of the polymer sensors follow (3.6), with a Pearson's correlation coefficient above 0.75. Clearly there are some sensors that deviate significantly from this equation, particularly copolymers such as methyl vinyl ether/maleic acid (50/50) and ethyl cellulose. Overall, the model represents the data with an average Pearson's correlation coefficient of $R^2 = 0.73$. In a previous study, the experimental responses of a different set of polymer sensors for the ENose had been fitted with the same expression and methodology [8]. Comparison between theory, (3.6), and experimental changes in resistivity of seven polymer sensors exposed to 24 solvents

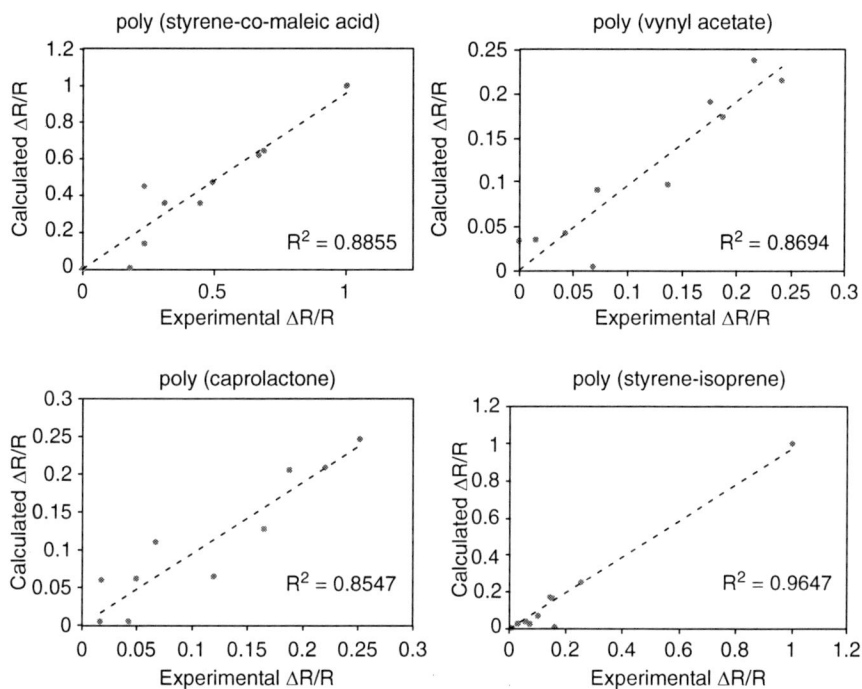

Fig. 3.9 Calculated (3.6) vs. experimental sensor responses of four polymer sensors to ten analytes in the third generation JPL ENose compound list

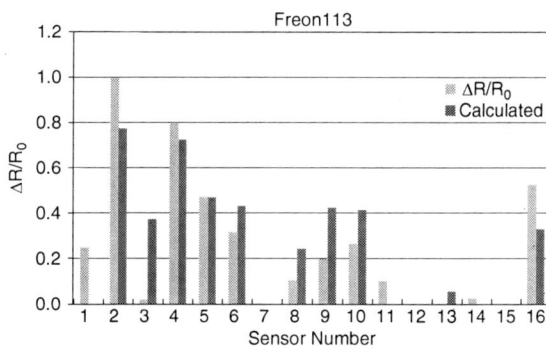

Fig. 3.10 Calculated (*dark*) vs. experimental (*light*) pattern of changes in resistivity by Freon113 on the JPL ENose array. Hansen solubility parameters for Freon113 are 2.82, 8.54, and 0 (cal per cm^3)$^{1/2}$ for electrostatic, dispersion, and hydrogen bonding respectively. The molar volume is 110.91 Å^3

gave a correlation coefficient $R^2 = 0.89$. The model was used to predict the response of Freon 113, an analyte not present in the training set. The predicted pattern of changes in resistivity agrees quite well with the measured polymer responses.

Table 3.4 Fitting model parameters, (3.6), for the JPL Enose polymer sensors.

	Polymer[a]	Ro	γ	βe	βd	βh	Pearson's R^2
1	Poly(4-vinyl phenol)	9.4311×10^{2}	9.2379×10^{-2}	1.4550×10^{0}	2.4467×10^{0}	-1.1473×10^{0}	0.75
2	Methyl vinyl ether/maleic acid, 50/50	2.0412×10^{-1}	-6.4933×10^{-3}	4.3703×10^{-2}	2.7624×10^{-3}	-2.7005×10^{-1}	-1.10
3	Poly(styrene-co-maleic acid)	1.4201×10^{1}	1.6597×10^{-2}	5.7604×10^{-1}	5.6279×10^{-1}	2.0015×10^{-2}	0.89
4	Polyamide resin	3.6871×10^{-2}	-2.9980×10^{-2}	-5.9101×10^{-2}	3.2918×10^{-1}	-2.0119×10^{-1}	0.57
5	Vinyl alcohol/vinyl butyral, 20/80	1.9976×10^{-2}	-3.0485×10^{-2}	-1.1009×10^{-3}	1.0789×10^{0}	-6.9556×10^{-1}	0.68
6	Ethyl cellulose	2.7476×10^{-2}	-3.8149×10^{-2}	9.4130×10^{-1}	-2.5777×10^{-1}	-5.3688×10^{-1}	0.48
7	Poly(2,4,6-tribromostyrene), 66%	9.2849×10^{0}	2.1263×10^{-2}	7.4222×10^{-2}	4.0145×10^{-1}	1.7336×10^{-1}	0.58
8	Poly(vinyl acetate)	2.6723×10^{-2}	-2.7333×10^{-2}	2.0736×10^{-1}	-4.3585×10^{-2}	2.4584×10^{-3}	0.87
9	Poly(caprolactone)	1.3872×10^{-2}	-3.3264×10^{-2}	1.4970×10^{-1}	9.2264×10^{-2}	-1.7486×10^{-1}	0.85
10	Soluble polyimide, Matrimid	6.3407×10^{0}	1.3727×10^{-1}	-4.9892×10^{-1}	1.0439×10^{1}	-2.3117×10^{0}	0.68
11	Poly(epichlorohydrin-co-ethylene oxide)	2.1315×10^{2}	1.1886×10^{-1}	8.3737×10^{-2}	5.1148×10^{-1}	1.6265×10^{-1}	0.64
12	Poly(vinylbenzyl chloride)	1.1729×10^{3}	3.1200×10^{-2}	1.5584×10^{0}	2.5691×10^{0}	5.7381×10^{-1}	0.54
13	Styrene/isoprene, 14/86 ABA block polymer	9.7786×10^{-6}	-9.6770×10^{-1}	-4.2760×10^{-1}	1.7732×10^{-3}	-4.1881×10^{-1}	0.96
14	Ethylene-propylene diene terpolymer	4.6806×10^{-6}	-1.2221×10^{-1}	-4.1192×10^{-1}	1.3660×10^{0}	3.2794×10^{-2}	0.92
15	Polyethylene oxide, PEO100	4.4420×10^{-2}	-1.6183×10^{-1}	-1.4877×10^{-1}	3.6562×10^{-2}	-2.9365×10^{-1}	0.77
	Average[b]						0.73

[a]Poly(N-vinyl pyrrolidone) data were not available. Sensor was dropped from experimental consideration

[b]Excludes methyl vinyl ether/maleic acid

[c]Pearson's correlation coefficient = 1 for a perfect correlation

Ideally, one might wish to predict directly the swelling of polymer sensors, loaded with carbon black, and transform this into a change in resistivity. This is however a great challenge for various reasons, from the ability to estimate the free energy of mixing analytes and polymers, to good knowledge of the true chemical composition of carbon black and how adsorbed molecules change its intrinsic conductivity. Nonetheless, we have shown that a combination of quantum mechanics with first-principles MD can afford a great deal of information that it is useful in designing and selecting materials for specific analytes. Future work involves the use of a newly developed MD method for the direct estimation of free energies [9].

Acknowledgments This work was supported in part by the Materials and Process Simulation Center, Beckman Institute at the California Institute of Technology and by a grant from NASA.

References

1. Ryan, M. A.; Shevade, A. V.; Zhou, H.; Homer, M. L., Polymer-carbon black composite sensors in an electronic nose for air-quality monitoring, *Mrs Bull.* **2004**, 29, 714–719
2. Ryan, M. A.; Zhou, H. Y.; Buehler, M. G.; Manatt, K. S.; Mowrey, V. S.; Jackson, S. R.; Kisor, A. K.; Shevade, A. V.; Homer, M. L., Monitoring space shuttle air quality using the jet propulsion laboratory electronic nose, *IEEE Sensors J.* **2004**, 4, 337–347
3. Zhou, H. Y.; Homer, M. L.; Shevade, A. V.; Ryan, M. A., Nonlinear least-squares based method for identifying and quantifying single and mixed contaminants in air with an electronic nose, *Sensors* **2006**, 6, 1–18
4. Shevade, A. V.; Homer, M. L.; Taylor, C. J.; Zhou, H. Y.; Jewell, A. D.; Manatt, K. S.; Kisor, A. K.; Yen, S. P. S.; Ryan, M. A., Correlating polymer–carbon composite sensor response with molecular descriptors, *J. Electrochem. Soc.* **2006**, 153, H209–H216
5. Shevade, A. V.; Ryan, M. A.; Homer, M. L.; Kisor, A. K.; Manatt, K. S.; Lin, B.; Fleurial, J. P.; Manfreda, A. M.; Yen, S. P. S., Calorimetric measurements of heat of sorption in polymer films: A molecular modeling and experimental study, *Anal. Chim. Acta* **2005**, 543, 242–248
6. Shevade, A. V., Developing sensor activity relationships for the JPL electronic nose sensors using molecular modeling and QSAR techniques, *2005 IEEE Sensors (IEEE Cat. No.05CH37665C)* **2005**, 4 pp.
7. Cozmuta, I.; Blanco, M.; Goddard, W. A., Gas sorption and barrier properties of polymeric membranes from molecular dynamics and Monte Carlo simulations, *J. Phys. Chem. B* **2007**, 111, 3151–3166
8. Belmares, M.; Blanco, M.; Goddard, W. A.; Ross, R. B.; Caldwell, G.; Chou, S. H.; Pham, J.; Olofson, P. M.; Thomas, C., Hildebrand and Hansen solubility parameters from molecular dynamics with applications to electronic nose polymer sensors, *J. Comput. Chem.* **2004**, 25, 1814–1826
9. Lin, S. T.; Blanco, M.; Goddard, W. A., The two-phase model for calculating thermodynamic properties of liquids from molecular dynamics: Validation for the phase diagram of Lennard-Jones fluids, *J. Chem. Phys.* **2003**, 119, 11792–11805
10. Becke, A. D., Density-functional thermochemistry. 3. The role of exact exchange, *J. Chem. Phys.* **1993**, 98, 5648–5652
11. Lee, C. T.; Yang, W. T.; Parr, R. G., Development of the Colle–Salvetti correlation-energy formula into a functional of the electron-density, *Phys. Rev. B* **1988**, 37, 785–789
12. Xu, X.; Goddard, W. A., The X3LYP extended density functional for accurate descriptions of nonbond interactions, spin states, and thermochemical properties, *Proc. Natl Acad. Sci. USA* **2004**, 101, 2673–2677

13. Zhao, Y., Development and assessment of a new hybrid density functional model for thermo-chemical kinetics, *J. Phys. Chem. A* **2004**, 108, 2715–2719

14. Zhao, Y., A density functional that accounts for medium-range correlation energies in organic chemistry, *Org. Lett.* **2006**, 8, 5753–5755

15. Zhao, Y., Comparative DFT study of van der Waals complexes: Rare-gas dimers, alkaline-earth dimers, zinc dimer, and zinc-rare-gas dimers, *J. Phys. Chem. A* **2006**, 110, 5121–5129

16. Zhao, Y., Density functionals with broad applicability in chemistry. *Acc. Chem. Res.* **2008**, 41, 157–167

17. Mayo, S. L.; Olafson, B. D.; Goddard, W. A., Dreiding – A generic force-field for molecular simulations. *J. Phys. Chem.* **1990**, 94, 8897–8909

18. Blanco, M., Molecular silverware.1. General-solutions to excluded volume constrained problems. *J. Comput. Chem.* **1991**, 12, 237–247

19. Ryan, M. A.; Shevade, A. V.; Taylor, C. J.; Homer, M. L.; Jewell, A. D.; Kisor, A. K.; Manatt, K. S.; Yen, S. P. S.; Blanco, M.; Goddard III, W. A. In expanding the capabilities of the JPL electronic nose for an international space station technology demonstration, In Proceedings of 36th International Conference on Environmental Systems **2006**, 2006–01–2179, Norfolk, Virginia, USA

20. Allen, M. P.; Tildesley, D. J., Computer Simulations of Liquids, Oxford University Press, Oxford, **1987**

21. Frenkel, D., Computer Simulation in Chemical Physics, Kjeuver, New York, **1993**

22. van Krevelen, D. W., Properties of Polymers: Their Correlation with Chemical Structure; their Numerical Estimation and Prediction from Group Contributions, Elsevier Science, New York, **1990**

23. Hansen, C. M., The three dimensional solubility parameter – Key to paint component affinities I. – Solvents, plasticizers, polymers, and resins. *J. Paint Technol.* **1967**, 39, 104–117

24. Jaguar, 7.207; Schrodinger, LLC, NY, **2007**

25. Accelrys, I. Cerius^2, 4.01, Accelrys, Inc.: San Diego, CA, **2005**

26. Lin, S. T.; Blanco, M. Rotational Isomeric State Table Algorithm, California Institute of Technology, Pasadena, CA, **2003**

Chapter 4
Prediction of Quartz Crystal Microbalance Gas Sensor Responses Using Grand Canonical Monte Carlo Method

Takamichi Nakamoto

Abstract Our group has studied an odor sensing system using an array of Quartz Crystal Microbalance (QCM) gas sensors and neural-network pattern recognition. In this odor sensing system, it is important to know the properties of sensing films coated on Quartz Crystal Microbalance electrodes. These sensing films have not been experimentally characterized well enough to predict the sensor response. We have investigated the predictions of sensor responses using a computational chemistry method, Grand Canonical Monte Carlo (GCMC) simulations. We have successfully predicted the amount of sorption using this method. The GCMC method requires no empirical parameters, unlike many other prediction methods used for QCM based sensor response modeling. In this chapter, the Grand Canonical Monte Carlo method is reviewed to predict the response of QCM gas sensor, and the modeling results are compared with experiments.

4.1 Introduction

An odor sensing system is required in fields such as food, beverage, cosmetics, environmental testing, and medical diagnostics, to evaluate smells objectively. A variety of odor sensing systems that use non-specific sensors, often called electronic noses, have been studied [1, 2]. In our study, we have used multiple sensors with partially overlapping specificities and a pattern recognition technique similar to that of the olfactory system. The sensing devices used in this work are Quartz Crystal Microbalance (QCM) gas sensors [3]. A QCM gas sensor consists of a resonator coated with a sensing film. We must select the optimum combinations of sensing films to discriminate among target odors. A statistical method based upon

T. Nakamoto
Graduate school of Science and Engineering, Tokyo Institute of Technology, Japan
e-mail: nakamoto@mn.ee.titech.ac.jp

M.A. Ryan et al. (eds.), *Computational Methods for Sensor Material Selection*,   93
Integrated Analytical Systems,
DOI 10.1007/978-0-387-73715-7_4, © Springer Science+Business Media, LLC 2009

discrimination analysis has been used to determine the combination of sensors [4]; this method requires experimental data from all the sensors for all the target odors, prior to selection. As a great deal of effort is required to collect the large amount of sensor data, prediction of the sensor response without experiment is essential.

Although several methods have been proposed to predict the response of an acoustic wave gas sensor, these methods require empirical parameters and therefore, experiments to obtain them. We report here, a method based on computational chemistry [5] to predict sensor responses without the use of an experimentally obtained database. As the method enables the calculation of a number of sensor responses, sensing film characterization may proceed quickly. Moreover, it may be easier to design the sensor array in an odor sensing system using this approach. Another advantage of this computational chemistry approach is that we can evaluate which forces are dominant in the sorption process. The contributions of electrostatic and van der Waals forces can also be evaluated in the sorption simulation.

In this chapter, we first discuss the principle of a quartz crystal microbalance gas sensor, including sorption characteristics, and then derive an expression for the relationship between the partition coefficient and the sensor response. This is followed by a review of previously proposed methods of sensor response prediction for a Quartz Crystal Microbalance sensor and SAW (Surface Acoustic Wave) gas sensors. Although the predictions are achieved using these prior methods, experiments are required to develop the model for prediction. Following this discussion, the simulation principles and methods using the Grand Canonical Monte Carlo (GCMC) approach are explained. The methodology involves using optimized chemical structures of gas and sensing film and performing sorption simulations. Measured sensor responses to evaluate the simulation result are reported at the end of the chapter. The simulation results agree with the corresponding experiments. During the GCMC simulations, we evaluated the electrostatic force contributions, which cannot be measured during experiments. A summary of the work is provided at the end.

4.2   Principle of Quartz Crystal Microbalance Gas Sensors

A Quartz Crystal Microbalance is a device sensitive to mass changes at its surface. This mass change causes a shift in the resonance frequency of the quartz resonator. The amount of the frequency shift is proportional to the mass change and this phenomenon is called mass loading effect [6]. The QCM mass loading effect had been initially utilized to monitor the metal thickness in an evaporator and was later applied to chemical sensors both in the gas and liquid phases [3, 7–9]. The QCM behavior can be analyzed using a one-dimensional model based on an equivalent circuit. The simplified model consists of a series of LCR (an inductor, a capacitor and a resistor) and a parallel parasitic capacitor between the electrodes. The acoustic load, such as mass loading and viscous loading, can be expressed as a circuit parameter, with mass loading expressed as an inductance change in the equivalent circuit [10].

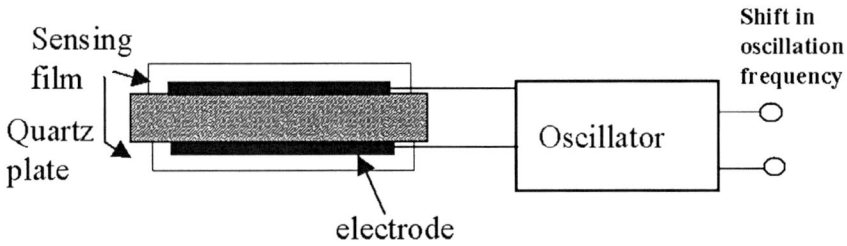

Fig. 4.1 Structure of Quartz Crystal Microbalance gas sensor

The structure of a Quartz Crystal Microbalance gas sensor is shown in Fig. 4.1. The quartz resonator consists of a quartz plate and electrodes deposited on both sides. The electrodes are deposited by evaporating or sputtering. The quartz resonator itself is typically used as an oscillator with its stable frequency. It works as a gas sensor when a sensing film is coated over the electrodes.

The quartz resonator is connected to an oscillator circuit, and the shift in oscillation frequency is measured using a frequency counter. When vapor is sorbed in the sensing film, the oscillation frequency decreases because of mass loading effects. When the vapor is replaced by air, the gas desorbs, and the oscillation frequency returns to the original frequency. Thus, the sensor can be used repetitively. The methods to coat sensing films are casting, painting, spin coating, spray coating, dip coating, plasma [11], Langmuir Blodgett [12] and atomizer [13]. In most methods, the sensing film material is dissolved in an organic solvent that evaporates quickly, such as chloroform. After solution deposition and evaporation of the solvent, only the sensing film remains on the electrode surface. The amount of coating is evaluated from the shift in resonance frequency due to deposited film. Although the sensitivity is governed by the film thickness, Q (Quality factor) of the resonator [14] decreases due to viscoelastic effect when the film is too thick. The deterioration of the Q causes the instability of the oscillation frequency. In the worst case, the oscillation stops. The Q factor is typically monitored by an impedance analyzer.

The sensing films can be characterized using a parameter called partition coefficient K, which is defined as:

$$K = \frac{C_s}{C_v}, \qquad (4.1)$$

where C_s (g/ml) is the gas concentration in the film and C_v (g/ml) is that in air in the equilibrium. Here the natural logarithm of K, $\ln K$, is used as an index of the amount of gas molecules sorbed into a sensing film.

The frequency shift Δf due to mass loading ΔM is

$$\Delta f = -f_r \frac{\Delta M}{M}, \qquad (4.2)$$

where M and f_r are the mass of the quartz plate and the resonant frequency, respectively. If m_s is the mass of coating, the frequency shift due to film coating Δf_s is

$$\Delta f_s = -f_r \frac{m_s}{M}. \tag{4.3}$$

The frequency shift due to sorption Δf_v is:

$$\Delta f_v = -f_r \frac{m_v}{M} = -\frac{m_v}{m_s} \Delta f_s, \tag{4.4}$$

where m_v is the amount of the gas sorption. Using the volume of the sensing film V and the film density ρ_s,

$$m_s = \rho_s V, \tag{4.5}$$

and

$$m_v = C_s V = KC_v V. \tag{4.6}$$

Using (4.4)–(4.6),

$$\Delta f_v = \frac{m_v}{m_s} \Delta f_s = \frac{KC_v V}{\rho_s V} \Delta f_s = \frac{KC_v}{\rho_s} \Delta f_s. \tag{4.7}$$

Thus,

$$K = \frac{\Delta f_v}{\Delta f_s} \frac{\rho_s}{C_v}. \tag{4.8}$$

In this work, ln K, the natural logarithm of K, was calculated using the Grand Canonical Monte Carlo (GCMC) method and it was compared with the experimentally obtained ln K.

4.3 Previously Proposed Prediction Methods for Sensor Responses

Although several researchers have reported methods for predicting sensor responses, they require empirical parameters determined by experimentation. Grate et al. proposed Linear Solvation Energy Relationships (LSER) to characterize the sorption characteristics of a sensing film and applied it to the prediction of SAW gas sensor response [15, 16]. In LSER the natural logarithm of K is expressed as a linear combination of the effective hydrogen-bond acidity, the effective hydrogen-bond basicity, dipolarity/polarization parameters and dispersion parameter. Although this relationship is used by many researchers, the coefficients of those parameters must be obtained experimentally, prior to the prediction. This method is described in a chapter by Grate in this book volume.

A method similar to LSER is to use the Hansh-Fujita equation [17]. The parameters in this equation are hydrophobicity, electronic property and steric effects; this method is known as SQAR (Structure-Activity Relationship) and is often used for developing medicine. Ohnishi applied this method to predict SAW gas sensor response [18].

A method for utilizing Gas Chromatograph (GC) database was proposed [19]. McReynolds has published a GC retention-volume database [20]; these data are useful for sensing films of GC stationary phases. The partition coefficient of the sensing film can be calculated from the retention data at high temperature after they are converted to room temperature. Moreover, a database of parameters of gas/liquid equilibrium, the Wilson parameter [21], is available to predict sensor response if the sensing film is the liquid phase [22]. However, experiments are necessary if there are no data of specified gas-sensing film combination in the database. Okahata proposed a prediction method using the parameter of slenderness of a vapor molecule [23], with the claim that the slenderness in addition to hydrophobicity parameter was effective in characterizing QCM sensor response. Kurosawa et al. also proposed a method using the Small number [24, 25].

All the above proposed methods have used extracted parameters to explain the sensor response. Even if the extracted parameters are different from method to method, they all require a coefficient for each parameter, which must be empirically determined. Thus, there has been no method for predicting sensor responses without experimentation. The method of computational chemistry is promising from this view point. However, the reports of sensor response prediction using computational chemistry are few. Hehl et al. evaluated the interaction between the sensing film and nitrobenzene-derivative vapor using the semi-empirical molecular orbital method, AM1 [26]. Fujimoto et al. did the simulation of the surface reaction of SnO_2 gas sensor to aminic and carboxylic vapors using molecular orbital calculation (MOPAC 97) [27]. Those researches are at a preliminary stage of sensor response prediction.

Sorption simulation was reported in 1987, where the sorption of methane gas into zeolite was calculated using Molecular Dynamics (MD) and GCMC methods. In the case of zeolite, the simulation might be easy because of its rigid and regular structure [28].

It is possible to predict the QCM gas sensor response without experiment. The first report of the application of sorption simulation to sensor response prediction using computational chemistry was in 1999 [29]. The following sections explain the method of sorption simulation using the Grand Canonical Monte Carlo method (GCMC) and its results in detail.

4.4 Grand Canonical Monte Carlo Method

4.4.1 Monte Carlo Simulation

There are two methods for evaluating sorption. One is Molecular Dynamics (MD) and the other is the Monte Carlo (MC) method. In case of MD calculation, an equation of motion of N molecular system is directly solved. It can be used to describe the transient phenomenon dependent upon time. In the Monte Carlo method, the configuration of the molecules is generated using random numbers. Only the quantities at equilibrium are obtained but temporal information is lost.

4.4.2 Sensing Film Model

The model of a sensing film was prepared according to the procedure below:

(a) Sketching a sensing-film molecule
(b) Preliminary molecular structure optimization
(c) Molecular structure optimization by Molecular Mechanics (MM) calculations [30]
(d) Random placement of film molecules in the unit cell
(e) Periodical and three-dimensional replication of a unit cell throughout space to form the model of a bulk sensing film
(f) Optimization of the sensing film model by MD and MM calculation

Preliminary molecular structure optimization in (b) is a simplified MM calculation to approach the convergence point rapidly. We assume that the gas molecules are not adsorbed at the sensing film surface but are in the bulk to start the simulation. Since the number of molecules placed in the simulation is limited by the computer performance, the three-dimensional periodic boundary condition was used in the step (e) above. Since these methods are computationally intensive, reduction in the amount of calculation is important, as discussed below.

Figure 4.2 shows the periodic boundary condition. Although a three-dimensional periodic boundary condition is actually used, the two-dimensional condition is shown for simplicity. The box, a unit cell is replicated throughout space to form an infinite lattice. As a molecule moves in the original box in the simulation, its periodic image in each of the neighbouring boxes moves exactly in the same manner. There are no walls at the boundary of the central box, and no surface molecules.

The computation time may be reduced by potential truncation. As the calculation of the potential between the two atoms needs a long time, it is impossible to calculate the interactions of one molecule with all other molecules. Thus,

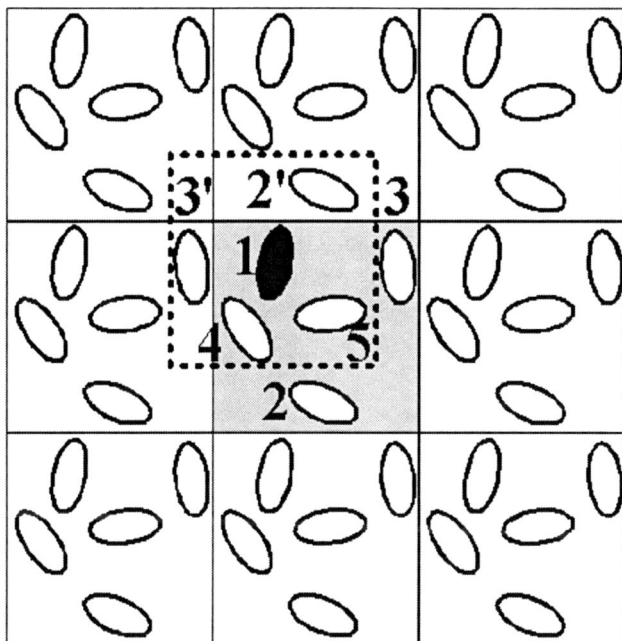

Fig. 4.2 Periodic boundary condition and cutoff distance

a spherical cutoff for short-range forces is normally applied. It means that the pair potential $v(r)$ is set to zero for $r > r_c$, where r_c is the cutoff distance. In Fig. 4.2, the interaction forces acting as molecule 1 are effective only within the dashed box. In case of molecule 1, only interactions with molecules 4, 5 and the periodic images $2'$, $3'$ are considered because they are located within the distance of r, where r_c is typically set to a half length of a unit-cell side.

There is a tradeoff between the number of sensing-film molecules in the unit cell and the calculation time. First we investigated the minimum number of those molecules accompanied with the tolerable error. In our simulation, five molecules were put in each unit cell and the sensing-film density was set to be 0.3 g/ml. The initial position and orientation of these molecules were determined by random number.

Then, the structure optimization was performed to reduce the energy of the system and to make the structure of the system more stable. In this simulation, we used Dreiding2.21 force field, used extensively in modeling organic molecules. In this step, bond stretching, bond angle bending, dihedral angle torsion, inversion, van der Waals, coulomb and hydrogen bond energy terms were considered. After the optimization, we confirmed that the molecules were uniformly placed in the unit cell by plotting its radial distribution function.

4.4.3   Gas Molecule Model

The model for gas molecule was prepared according to the following procedure:

(a) Sketching the gas molecule
(b) Preliminary molecular structure optimization
(c) Molecular structure optimization by MM calculation

Preliminary molecular structure optimization in (b) was done in the same manner as that in Sect. 4.4.2. After the optimization, the conformation of the gas molecule was set to be all trans. We used this model in the sorption simulation. Since sensors were used in air, nitrogen and oxygen molecules were also taken into account.

4.4.4   Sorption Simulation

The Grand Canonical Monte Carlo (GCMC) method was used for the sorption simulation. The simulation uses the grand canonical statistical ensemble. In this ensemble, the number of the molecules in the system is not fixed, whereas the chemical potential of each species in the gas phase, and the volume and the temperature of the system are fixed. This ensemble agrees with the environment of the static measurement system described in the next section. The probability of a certain microscopic state given by the grand canonical ensemble is

$$\rho(\boldsymbol{p},\boldsymbol{r},N) = \frac{\exp\left(-\frac{H(\boldsymbol{p},\boldsymbol{r})-\mu N}{k_{\mathrm{B}}T}\right)}{N!h^{3N}\Xi},\qquad(4.9)$$

where N is the number of molecules in the system, and $\boldsymbol{r}$ and $\boldsymbol{p}$ are the positions and momenta of the N molecules. $H(\boldsymbol{p},\boldsymbol{r})$ is the Hamiltonian and represents the total energy of the system. k_{B} is the Boltzmann constant, μ the chemical potential and T the temperature. Ξ is the Grand Partition Function and is expressed as

$$\Xi = \sum_{N}\frac{1}{N!h^{3N}}\exp\left(\frac{\mu N}{k_{\mathrm{B}}T}\right)\iint \exp\left(-\frac{H(\boldsymbol{p},\boldsymbol{r})}{K_{B}T}\right),\qquad(4.10)$$

where h is Planck's constant.

The Monte Carlo simulation in the film model is performed according to the following procedure as illustrated in Fig. 4.3.

(a) Create a gas molecule
(b) Destroy a gas molecule
(c) Translate a gas molecule
(d) Rotate a gas molecule

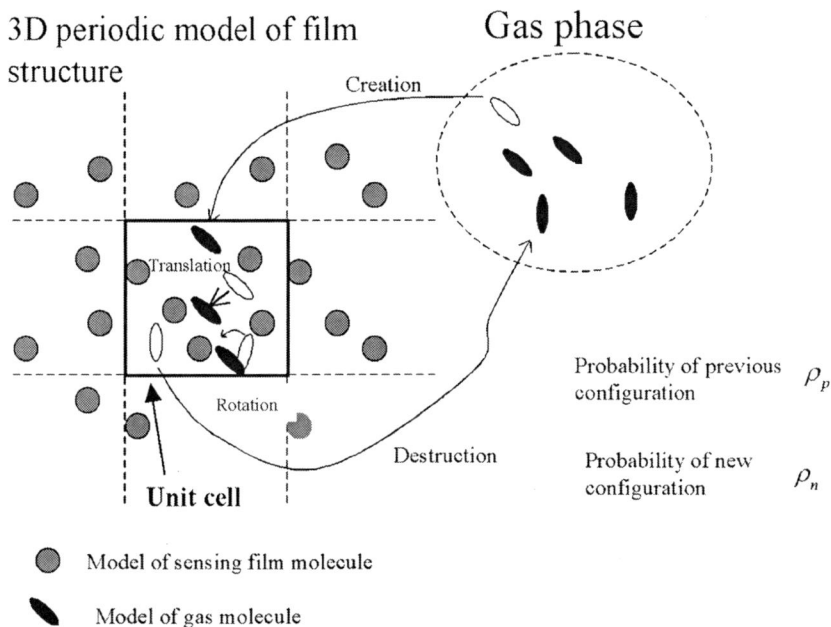

Fig. 4.3 Sorption simulation procedure

The acceptance or rejection of the configuration is determined by the probability of a certain microscopic state given by the grand canonical ensemble in (4.9). Let ρ_p and ρ_n be the probabilities of the previous and the new configurations, respectively. If ρ_n/ρ_p is greater than 1, the new configuration is accepted. When ρ_n/ρ_p is less than 1, it is compared with a random number, and the new configuration is accepted only if it is greater than the random number. Equilibrium of gas sorption is achieved when the chemical potential of the gas inside the bulk model of the sensing film is equal to that of the free gas outside the film.

We used Dreiding 2.21 force field for the sorption simulation. The number of parameters is much smaller than that of configurations of gas and sensing film molecules, and we can determine sorption equilibrium of those molecules, only using these parameters. Moreover, we considered only the intermolecular interactions, van der Waals, hydrogen bonding and coulomb for elements such as C, H and O. We used approximately ten parameters for energy terms in all. For example, van der Waals energy term is expressed by the Lennard Jones potential function,

$$E(R) = D_0 \left\{ \left(\frac{R_0}{R} \right)^6 - \left(\frac{R_0}{R} \right)^{12} \right\}, \tag{4.11}$$

and the parameters in this function are D_0 and R_0. All the parameters were fixed during the entire simulation and default values were generally used.

The sorption simulation of the ternary mixture of gas, nitrogen, oxygen and the sample gas was also performed. In this simulation, we assumed that the interactions between gas and film molecules were too weak to change the structure of gas molecule. The partial pressures of nitrogen, oxygen and the sample gas were 80, 20 and 0.1 kPa, respectively, and the temperature was 300K. The partial pressures of almost all sample gases are very low, compared with these saturated pressures.

4.5   Experiment on Gas Sensor Measurement

4.5.1   Measurement System

The measurement system in this study is a static measurement system for obtaining the steady-state sensor responses of QCM gas sensors. A schematic diagram of the system is shown in Fig. 4.4. A small amount of the sample liquid is injected into the closed chamber and the sensor steady-state response is obtained after the complete evaporation of the sample liquid. The sample concentration can be accurately determined because the amount of the injected sample and the chamber volume are fixed. Moreover, in this system the temperatures of the gas and the sensor are kept constant, in contrast to a flow measurement system, where the temperature of the sensor is sometimes different from that of the gas.

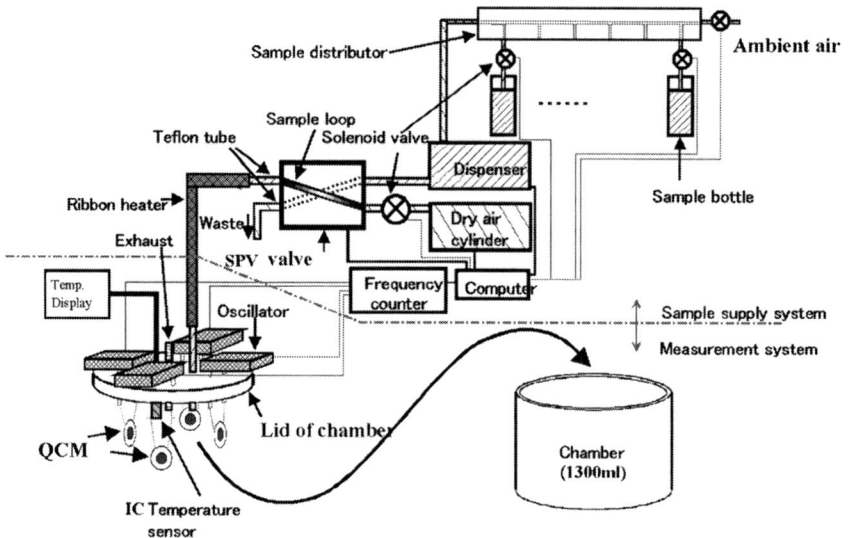

Fig. 4.4 Schematic diagram of static measurement system

A chamber made of Teflon with a volume of 1,300 ml was placed in a thermal bath. Eight QCM gas sensors and oscillation circuits are attached on the inner and outer sides of the chamber lid, respectively. The thermal bath with the chamber is placed in a heat-insulated box (not shown in Fig. 4.4) to keep the temperature at 27°C. The oscillation frequency shift due to gas sorption is measured using a frequency counter and data transferred to a computer.

Samples are automatically injected into the chamber through a sample dispenser (Bunchu-kun, BLSQ) and an SPV automatic valve (SPV-N-6A, GL Science), both controlled by a single computer. The injected volume of the liquid sample is determined from the volume of the sample loop (7 μl) attached to the SPV automatic valve. After the sample is injected and is evaporated in the chamber, the frequency shift is measured at the equilibrium point. Subsequent samples are injected in the same way. The dead volume of the sample path should be minimized in the automatic system using the SPV automatic valve; there is little dead volume in a manual injection.

As it takes a long time for the liquid sample in the chamber to evaporate, a teflon tube connecting the SPV automatic valve and the chamber is heated by a ribbon heater to force rapid evaporation. The liquid samples are sucked into the dispenser through the sample distributor as shown in Fig. 4.4. We can select the sample by opening the appropriate solenoid valve under the distributor. Since these processes are controlled by computer, we can measure seven samples automatically and continuously.

4.5.2 QCM Gas Sensors

We used quartz crystals (20 MHz, AT-CUT) with silver electrodes. The sensing films were squalane and polyethyleneglycol-400 (PEG400). The molecular structures of the sensing films are shown in Fig. 4.5. The molecular weights of squalane and PEG400 are 423 and 415, respectively. It was difficult to use molecules heavier than 1,000 as sensing film material in the simulation of the current computational environment. These materials were dissolved in chloroform and deposited by spin coating. Squalane and PEG400 are typical GC stationary phase materials without and with polarity, respectively. They are not solid but liquid phases at room temperature. In addition, a solid sensing film of PEG1000 was used.

Gas samples of alcohols, aromatics, ketones, esters, alkanes and perfumes were tested. The samples in each group have the same functional group but differ in the number of carbon atoms. When we measure a low-volatility sample, it takes a long time for the response to reach equilibrium even if it is heated using the ribbon heater. Therefore, we diluted the liquid samples with diethyl ether. We confirmed in advance that the sensor responses to those solvents were negligible.

The actual simulation was performed using commercially available software (Cerius^2, BIOSYSM/MSI) on an Indigo2 computer (Silicon Graphics).

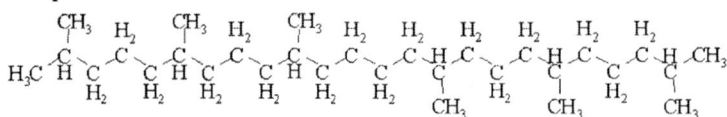

<PEG400>

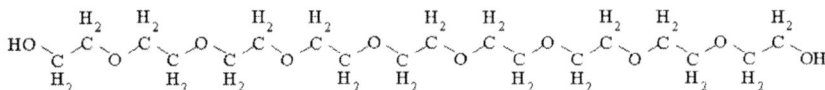

Fig. 4.5 Molecular structure of sensing film

4.6 Results

4.6.1 Evaluation of Sensing Film Model

The sensing film structure is optimized after initial placement of the sensing-film molecules according to step (f) in the procedure in the Sect. 4.4.2. Since the placement of the sensing film molecules is complicated, molecular mechanics calculation is insufficient due to possible trapping at local minima. Thus, MD calculation is performed to do the annealing in the sensing film simulation. As the temperature of the sensing film increases, the structure of the sensing film model is relaxed. Each sensing film molecule can move freely at high temperature. After an increase in temperature, it is decreased to room temperature. The sensing film is expected to be more stable after the annealing process. In the annealing simulations, the temperature is raised from 300 to 700°C in steps of 50°C and then decreased to 300°C with the same temperature step. In case of squalane, the energy in the unit cell was decreased from 233.1 (kcal/cell) to 57.5 (kcal/cell) due to the annealing process in MD calculation.

The uniformity of the sensing-film structure was evaluated using the radial distribution function $g(r)$. The function $g(r)$ gives the probability of finding a pair of atoms a distance r apart, relative to the probability expected for a completely random distribution at the same density. The value of $g(r)$ approaches 1 when r is large, a case of uniform distribution. Moreover, it is preferable for $g(r)$ to converge to 1 at the point within the cutoff distance. Figure 4.6 shows the radial distribution function of squalane. As the cutoff distance of the squalane model was 11.35 Å, that requirement was sufficiently satisfied. A uniform structure of the sensing film model was obtained in this simulation.

4.6.2 Evaluation of Sorption Simulation

In the sorption simulations, the value of the number of sorbed gas molecules in the unit cell was obtained and converted into the partition coefficient only if

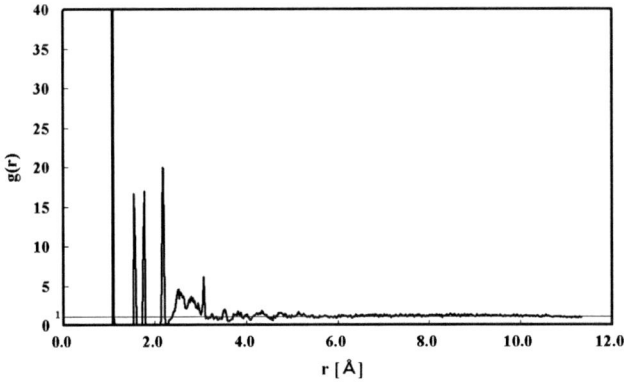

Fig. 4.6 Radial distribution function of squalane

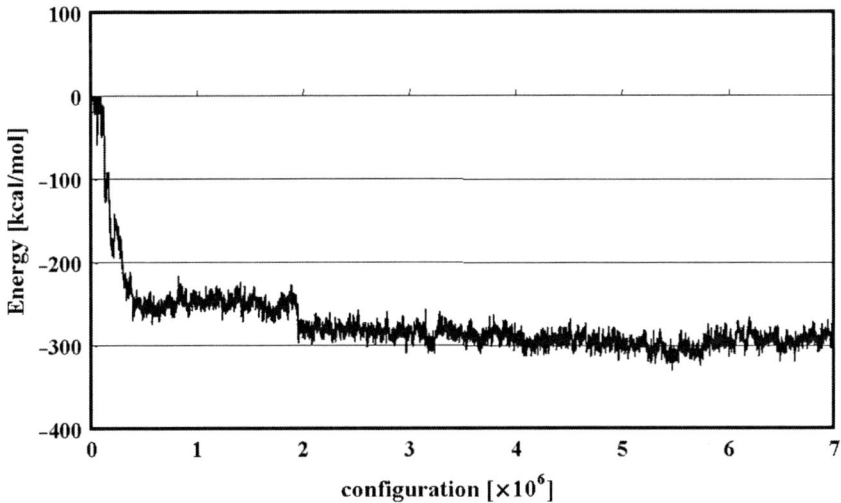

Fig. 4.7 Relationship between number of configurations and total energy of gas molecules in film (sensing film: squalane, gas: geranial)

convergence was achieved. One example of convergence characteristics is shown in Fig. 4.7. The relationship between the number of configurations and the total energy of the gas molecules in the film structure is depicted when the film material is squalane and the gas molecule is geranial.

As the number of configurations increases, that energy decreases and converges to a constant value around two million configurations. The number of sorbed gas molecules in the unit cell also converges around two million configurations. The convergences were achieved in case of other gases.

As a result of sorption simulation, a distribution of gas molecules is obtained. The distribution in the unit cell is shown in Fig. 4.8. The stick structure represents the squalane molecule and the cloud represents the sorbed gas molecules. Higher cloud density represents more sorbed gas molecules. This analysis method might

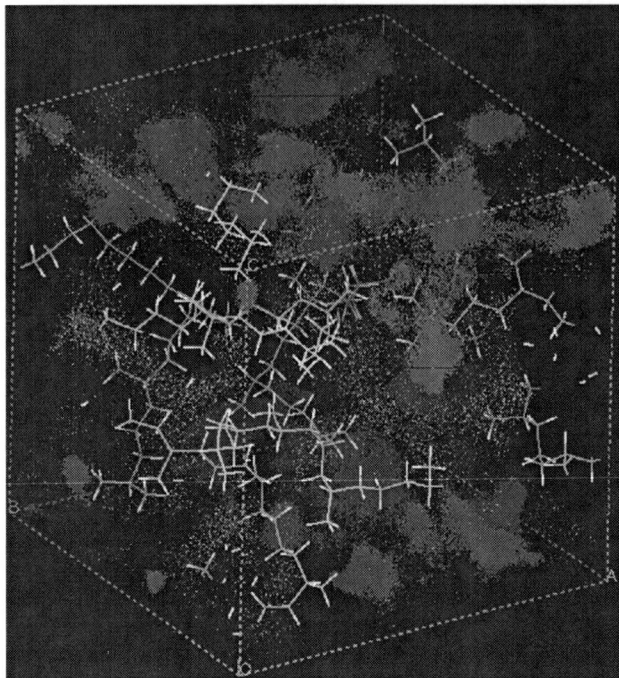

Fig. 4.8 Distribution of sorbed gas molecules in a unit cell

provide us with information about the interaction between the gas molecule and the sensing film molecule, e.g., the functional group dominant in the sorption.

4.6.3 Comparison of Predicted Results with Experiment

The calculated ln K values of ketones and perfumes for squalane and PEG400 are compared with experimental values, as in Fig. 4.9a, b. If the data plot falls on the solid diagonal line, the simulation results agree with the experimental one. As can be seen in Fig. 4.9a, b, the simulation data are in close agreement with the experimental values.

It is important to consider parameters which reflect the difference between squalane and PEG400 sensing films. The effect of parameter values in the force field is related to the contributions of each energy term. Although we have not rigorously studied contributions from van der Waals forces that might influence the results, Coulomb and hydrogen bond energy also influence the results since the difference between squalane and PEG400 films is also due to different film polarity.

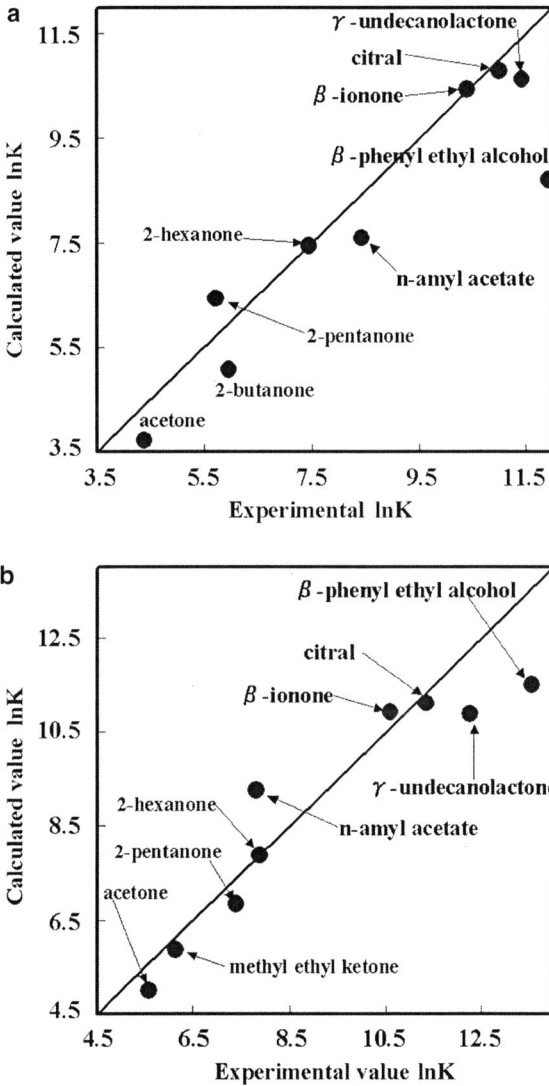

Fig. 4.9 Comparison of experimental and calculated ln K values for (**a**) squalane and (**b**) for PEG400. Reprint with permission from ref. 31. Copyright 2000 Elsevier Science

In Fig. 4.9a, b, it can be seen that the simulation results are in good agreement with the experimental results over a wide range of odor intensities, from acetone to citral. PEG1000 was also used to do both simulation and measurement. The prediction of the sensor response over a wide range of odor intensity could be achieved in the same manner as those of squalane and PEG400 although its data are shown elsewhere [31]. This simulation was also effective for a solid-sensing film.

Although further detailed study is required, it was found that GCMC calculation is useful for the most film-gas combinations.

4.6.4 Contributions of Electrostatic Interactions to Sorption

Understanding the contributions of the electrostatic and van der Waals forces to vapor sorption in the sensing film is important in order to understand the sorption characteristics of the sensing films, and also to elucidate the dominant interaction in the gas sorption. Although it is not possible for a single energy term to work in the actual environment, the contributions of these energy terms to the sorption can be estimated using the simulations. Since Dreiding 2.21 was used as the force field, van der Waals and electrostatic (coulomb and hydrogen bond) interactions are considered as nonbonding interactions in the simulation.

The number of sorbed gas molecules N_{vdW} without electrostatic interaction terms, are compared with N_{all} calculated using all the interaction terms. The relative contribution of electrostatic interaction is expressed by defining the percentage of the amount of sorbed gas molecules R_{elec} as

$$R_{elec} = \left(1 - \frac{N_{vdW}}{N_{all}}\right). \tag{4.12}$$

R_{elec} of alkanes, alcohols, ketones, esters and aromatics for PEG400 are shown in Fig. 4.10. The sample numbers are tabulated in Table 4.1. It can be seen that the

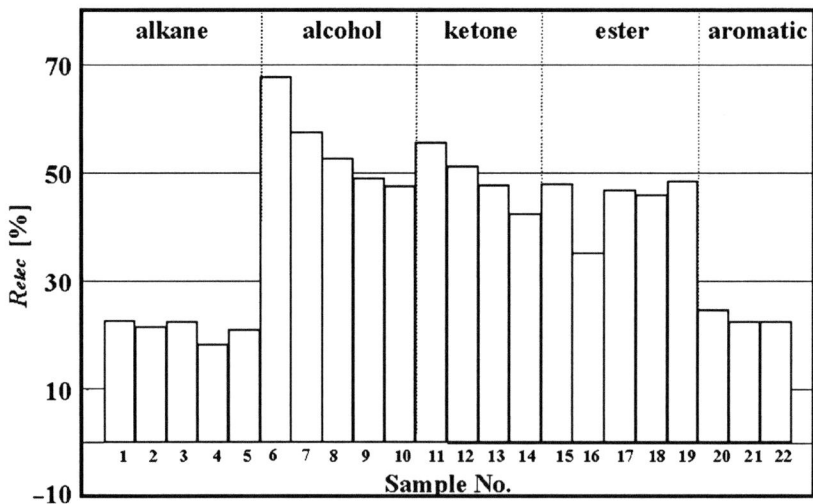

Fig. 4.10 Electrostatic contribution of various kinds of VOCs, for sensing film PEG400. Reprint with permission from ref. 31. Copyright 2000 Elsevier Science

Table 4.1 Sample numbers and names used in Fig. 4.9

Sample No	Group	Gas sample
1	Alkane	Hexane
2		Heptane
3		Octane
4		Nonane
5		Decane
6	Alcohol	Ethanol
7		1-Propanol
8		1-Butanol
9		1-Pentanol
10		1-Hexanol
11	Ketone	Acetone
12		2-Butanone
13		2-Pentanone
14		2-Hexanone
15	Ester	Ethyl formate
16		Butyl acetate
17		Propyl acetate
18		Ethyl propionate
19		Ethyl butyrate
20	Aromatic	Benzene
21		Toluene
22		Ethyl benzene

electrostatic interaction term is dominant in the case of alcohols sorbed into the polar film PEG400. R_{elec} of ketones and esters are a little smaller than those of alcohols. R_{elec} for aromatics and alkanes are the smallest among the samples. Moreover, in the case of alcohols and ketones, R_{elec} decreases in the ascending order of carbon atoms. R_{elec} for PEG400 differ among sample groups, whereas the variation of R_{elec} among those groups is small for the nonpolar squalane film.

The tendency of R_{elec} was in good agreement with our expectations. This simulation method is considered to be useful for understanding sorption mechanisms of sensing films. For examining further usefulness, R_{elec} for perfumes were calculated [31]. Even if it is hard to predict the contribution of the electrostatic force due to the complicated structure of perfume compound, the simulation enables us to understand it more easily.

4.7   Summary

A computational chemistry method, based on GCMC simulations was used for predicting QCM gas sensor responses. The predicted partition coefficients of alcohol, aromatics, ketones, esters, alkanes and perfumes for typical GC films agreed with the experimental ones. It was found that QCM sensor responses could be roughly predicted over a wide range of odor intensity, and the influence of the film polarity on the gas sorption was revealed in the simulation. Therefore, this method shows merit in predicting sensor response without experiment.

Moreover, the contributions of electrostatic interactions can be obtained using this simulation method. This would be helpful to characterize sensing film properties. However, the problem still remains when the hydrogen bond energy term was dominant. The evaluation method of conformation change due to hydrogen bonding should be further improved.

Acknowledgments The author wishes to thank his former student, Dr. K. Nakamura (Anritsu Corp) for his previous work on the odor sensing system.

References

1. Pearce, T. C.; Schiffman, S. S.; Nagle, H. T.; Gardner, J. W., Eds., Handbook of Machine Olfaction; Wiley-VCH, Weinheim, **2003**
2. Nakamoto, T.; Moriizumi, T., Artificial olfactory system using neural network, In Handbook of Sensors and Actuators; Yamazaki H., (Ed.); Elsevier, Amsterdam, **1996**, vol. 3, 263–272
3. King, W. H., Piezoelectric sorption detector, *Anal. Chem.* 36, **1964**, 1735–1739
4. Nakamoto, T.; Sasaki, S.; Fukuda, A.; Moriizumi, T., Selection method of sensing membranes in odor sensing system, *Sens. Materials* **1992**, 4, 111–119
5. Allen, M. P.; Tidesley, D. J., Computer Simulation of Liquid, Oxford Press, Blackwell, **1987**
6. Zauerbrey, G., Verwendung von Schwingquarzen zur Wagung dunner Schichten und zur Mikrowagung, *Z. Phys.* **1959**, 155, 206–222
7. Nakamoto, T.; Inadama, K.; Moriizumi, T., Study on quartz thickness-shear resonator immersed in liquid and its biosensor application, In Proceedings of 17th international Symposium on Acoustical Imaging, **1988**, 619–626
8. Kanazawa, K. K.; Gordon II, J. G., Frequency of a quartz microbalance in contact with liquid, *Anal. Chem.* **1985**, 57, 1770–1771
9. Muramatsu, H.; Tamiya, E.; Karube, I., Computation of equivalent circuit parameters of quartz crystals in contact with liquid and study of liquid properties, *Anal. Chem.* **1988**, 60, 2142–2146
10. Nakamoto, T.; Moriizumi, T., A theory of a quartz crystal microbalance based upon a mason equivalent circuit, *Jpn. J. Phys.* **1990**, 29, 963–969
11. Sugimoto, I.; Nakamura, M.; Kuwano, H., Organic gas sorption characteristics of plasma-deposited amino acid films, *Anal. Chem.* **1944**, 66, 4316–4323
12. Munos, S.; Nakamoto, T.; Moriizumi, T., A comparison between calixalene LB and cast films in odor sensing system, *Sens. Mater.* **1999**, 11, 427–435
13. Munoz, S.; Nakamoto, T.; Moriizumi, T., Study of deposition of gas sensing films on quartz crystal microbalance using an ultrasonic atomizer, *Sens. Actuators B* **2005**, 105, 144–149
14. Nakamoto, T.; Kobayashi, T., Development of circuit for measuring both Q variation and resonant frequency shift of quartz crystal microbalance, *IEEE Trans. UFFC* **1994**, 41, 806–811
15. Grate, J. W.; Patrash, S. L.; Abraham, M. H., Method for estimating polymer-coated acoutic wave vapor sensor responses, *Anal. Chem.* **1995**, 67, 2162–2169
16. Grate, J. W.; Abraham, M. H., Solubility interactions and the design of chemically selective sorbent coatings for chemical sensors and arrays, *Sens. Actuators B*, **1991**, 3, 85–111
17. Hansh, C.; Fujita, T., $\rho - \sigma - \Pi$ analysis. A method for the correlation of biological activity and chemical structures, *J. Am. Chem. Soc.* **1964**, 86, 1616–1626
18. Ohnishi, M.; Ishimoto, C.; Seto, J., The biomimetic property of gas-sensitive films for odorants constructed by the Langmuir-Blodgett technique, *Thin Solid Film* **1992**, 210, 455–457

19. Nakamoto, T.; Fukuda, A.; Moriizumi, T., Prediction method of quartz resonator gas sensor response, *Trans. IEICE* **1991**, J74-C-II 450–457 (in Japanese)
20. McReynolds, W. O., Gas chromatographic retention data, Preston Technical Abstracts Company, **1966**
21. Wison, G. M., Vapor-liquid equilibrium. XI. A new expression for the excess free energy of mixing, *J. Am. Chem. Soc.* **1964**, 86 127–130
22. Fukuda, A.; Misawa; Nakamoto, T.; Moriizumi, T.; Analysis of sorption phenomenon of vapor mixture to quartz resonator gas sensor using Wilson equation, Extended abstracts, Spring Meeting of the Japan Society of Applied Physics and related societies, **1992**, 29aQ-2 (in Japanese)
23. Okahata, Y., Molecular recognition on synthetic lipid membranes, *Membrane* **1991**, 16, 26–33 (in Japanese)
24. Small, P. A., Some factors affecting the solubility of polymers, *J. Appl. Chem.* **1953**, 3, 71–80.
25. Kurosawa, S.; Kano, N., Characteristics of sorption of various gases to plasma-polymerized copper phthalocyanine, Langmuir, **1992**, 8, 254–256.
26. Heckl, W. M.; Marassi, F. M.; Kallury, K. M. R.; Stone, D. C.; Thompson, M., Surface acoustic wave sensor response and molecular modeling: Selective binding of nitrobenzene derivatives to (aminopropyl)triethoxysilane, *Anal. Chem.* **1990**, 62, 32–37
27. Fujimoto, A.; Kanashima, T.; Okuyama, M., Molecular orbital calculation of surface reaction of SnO_2 gas sensors for aminic and carboxylic smells, Digest of Technical papers, Transducers03, **2003**, 540–543
28. Yamazaki, T.; Watanuki, I.; Ozawa, S.; Ogino, Y., An IR study on methane adsorbed on ZSM-5 type zeolites, *Nippon Kagaku kaishi* **1987**, 8, 1535–1540 (in Japanese)
29. Nakamura, K.; Nakamoto, T.; Moriizumi, T., Prediction of quartz crystal microbalance gas sensor responses using a computational chemistry method, *Sens. Actuators B* **1999**, 61, 6–11
30. Osawa, E.; Hirano, T.; Honda, K., Introduction of computational chemistry, *Koudansha* **1994**, 61–61 (in Japanese)
31. Nakamura, K.; Nakamoto, T; Moriizumi, T., Prediction of QCM gas sensor responses and calculation of electrostatic contribution to sensor responses using a computational chemistry method, *Mater. Sci. Eng. C* **2000**, 12, 3–7

Chapter 5
Computer-Aided Design of Organic Host Architectures for Selective Chemosensors

Benjamin P. Hay and Vyacheslav S. Bryantsev

Abstract Selective organic hosts provide the foundation for the development of many types of sensors. The deliberate design of host molecules with predetermined selectivity, however, remains a challenge in supramolecular chemistry. To address this issue, we have developed a de novo structure-based design approach for the unbiased construction of complementary host architectures. This chapter summarizes recent progress including improvements on a computer software program, HostDesigner, specifically tailored to discover host architectures for small guest molecules. HostDesigner is capable of generating and evaluating millions of candidate structures in minutes on a desktop personal computer, allowing a user to rapidly identify three-dimensional architectures that are structurally organized for binding a targeted guest species. The efficacy of this computational methodology is illustrated with a search for cation hosts containing aliphatic ether oxygen groups and anion hosts containing urea groups.

5.1 Introduction

The search for highly functional sensor materials is of considerable importance in various areas including food control, medical monitoring, biotechnology, environmental sciences, and nuclear industry. A general strategy for the development of molecular signaling or sensing systems is the coupling of at least two equally important functional components: the receptor site and the indicator subunit [1–8]. The latter should be able to change a physical property of the system that

B.P. Hay (✉)
Chemical Sciences Division, Oak Ridge National Laboratory, Oak Ridge, TN 37831-6119, USA
e-mail: haybp@ornl.gov

V.S. Bryantsev
Division of Chemistry and Chemical Engineering, California Institute of Technology, Pasadena, CA 91125, USA

M.A. Ryan et al. (eds.), *Computational Methods for Sensor Material Selection*, Integrated Analytical Systems, DOI 10.1007/978-0-387-73715-7_5, © Springer Science+Business Media, LLC 2009

is monitored in response to a binding event. Technological advances in polymer chemistry, material science, and analytical instrumentations combined with application of various transduction principles (optical, thermal, electrical, mechanical, and chemical) have led to development of many types of chemical sensor devices exhibiting high robustness and sensitivity [9–15]. Other characteristics, such as selectivity and reversibility, depend mainly on the molecular mechanism of binding between a receptor and a target analyte. Nonspecific physical adsorption is reversible but suffers from poor selectivity [11]. Covalent chemical bonding, in contrast, might be highly selective but will result in poor reversibility [4, 11, 17]. Therefore, molecular recognition process that involves strong and specific interaction between a host and a guest without forming or breaking covalent bonds is preferable in most cases [18–20].

There has been a large body of research in coordination and supramolecular chemistry [21, 22] aimed toward the development of organic receptors that specifically recognize various metal ions, anions, and organic molecules in competitive media (see, for example [2–6, 23–40]. A relatively new approach for the solution-based molecular recognition is to use several receptors in array formats for differential sensing of complex analytes and mixtures [41]. Nevertheless, achieving selective recognition remains a difficult challenge, and in this chapter, we systematically address this problem from the viewpoint of electronic, size, and shape complementarity using rational design and combinatorial screening tools.

A common approach for preparing host structures is to add several binding sites to an organic scaffold to yield receptors that interact with ligands by virtue of incorporated functionalities, for example, through hydrogen bonding, metal – ligand electrostatic interaction, π–π stacking interaction, van der Waal forces, etc. [21, 22]. Once a group of binding sites has been selected, a key challenge of rational host design is the structural organization of the binding sites about the guest. An optimal host will exhibit two structural properties. First, the host must be able to adopt a conformation in which all binding sites are positioned to structurally complement the guest [23]. Information on optimal arrangement of host binding sites about the guest can be obtained from electronic structure calculations and, when available, from crystallographic data. Second, the host should exhibit a limited number of stable conformations, and the binding conformation should be low in energy relative to other possible forms [24–28]. In the ideal case, the host would be preorganized such that the binding conformation is the most stable form.

The deliberate design of host structures by assembling sets of disconnected binding sites in three dimensions is not a trivial task. One approach is to generate trial structures by hand with a graphical user interface, an extremely time-consuming process. Often, it is not readily obvious which linkage structures might be best used to connect the binding sites to obtain a host cavity that compliments the guest. To attain a high degree of structural organization within the host–guest system, one needs a tool to go beyond just informed guess or chemical intuition.

Drug designers have developed computational approaches to address the inverse of the problem, in other words, how to identify molecular structures (guests) that will complement the binding site of a protein (host) [42–47]. These approaches include de novo structure-based design strategies which couple molecule building algorithms with scoring functions that are used to prioritize the candidate structures. The building algorithms assemble guest molecule structures that can physically interact with a known protein structure from pieces which are either atoms [48–53] or larger, chemically reasonable fragments [54–63]. The ability to generate large numbers of potential guest structures necessitates the use of simple scoring functions to prioritize the output. To this end, methods have been developed to estimate the binding free energy by summing free energy increments for hydrogen bond interactions, ionic interactions, lipophilic interactions, the number of rotatable bonds in the guest molecule, etc. [45–47, 64–67]. After an initial prioritization of the results, computational demand of evaluations of the host-guest complex may be made to achieve a more accurate ranking for the best candidates.

Computer programs that have been developed to perform *de novo* structure-based drug design are, in general, not applicable to the design of host molecules. These programs require input of the atomic coordinates of a protein binding site, are highly specialized to address protein–organic interactions, and do not contain scoring functions to address other types of host-guest interactions. To bring the powerful concepts embodied in de novo structure-based drug design to the field of supramolecular chemistry, we devised computer algorithms for building structures from host components and rapid methods for scoring the resulting structures with respect to their degree of organization for a guest species. The result is HostDesigner, the first structure-based design software that is specifically created for the discovery of host architectures for the complexation of small ions and molecules [68, 69]. Since the initial release of HostDesigner in 2002, there have been significant modifications both to enhance performance and extend its application. In this chapter we will review how the building and scoring algorithms work. Several examples are provided to illustrate how the de novo design approach can be used to identify host architectures organized for small ionic guests.

5.2 The LINKER Algorithm

5.2.1 Complex Fragments

A multidentate host can be dissected into two or more simpler host components. For example, the well-known 18-crown-6 macrocycle can be broken down into two triglyme components, three diglyme components, or six dimethylether components. It is possible to define the structure of a complex fragment, in other words, a piece of a host–guest complex, by combining a host component with a guest. In constructing the complex fragment, the guest is positioned relative to the host component to define a complementary geometry, that is, a geometry that would give the strongest interaction between the binding sites of the host component and the guest in an actual complex. The complementary geometry for one or more binding sites

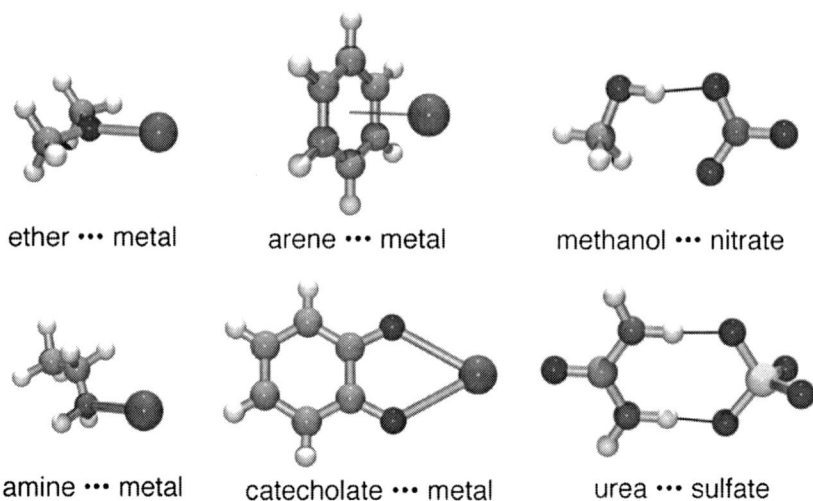

ether ··· metal   arene ··· metal   methanol ··· nitrate

amine ··· metal   catecholate ··· metal   urea ··· sulfate

Fig. 5.1 Examples of complex fragments

interacting with a guest can be obtained from examination of experimental geometries of host–guest complexes or through the careful application of electronic structure calculations. Examples of complex fragments are shown in Fig. 5.1.

5.2.2 Assembling the Pieces

LINKER builds new host structures by forming bonds between two complex fragments provided by the user and linking fragments taken from a library (*vide infra*). The user must create an input file for each complex fragment that specifies the coordinates for all the atoms, atom connectivity, and attachment vectors. Attachment vectors are indicated by listing the hydrogen atoms that can be removed from the complex fragment. Finally, the input file can contain a specification of structural degrees of freedom, in other words, distances, angles, and dihedral angles that can be varied during the building process.

The process used by LINKER to construct a new host molecule is illustrated in Fig. 5.2. In this example, two identical lithium-dimethylether complex fragments are attached to a methylene linkage. The steps are as follows:

(a) two complex fragments are defined with attachment vectors indicated
(b) linking fragment containing two attachment vectors is selected from the library
(c) bond is formed between the first complex fragment and the linking fragment by aligning the attachment vectors and setting the bond distance to an appropriate value on the basis of the identity of the bonded atoms
(d) dihedral angle about the new bond is adjusted to a specific value on the basis of hybridization and degree of substitution of the bonded atoms

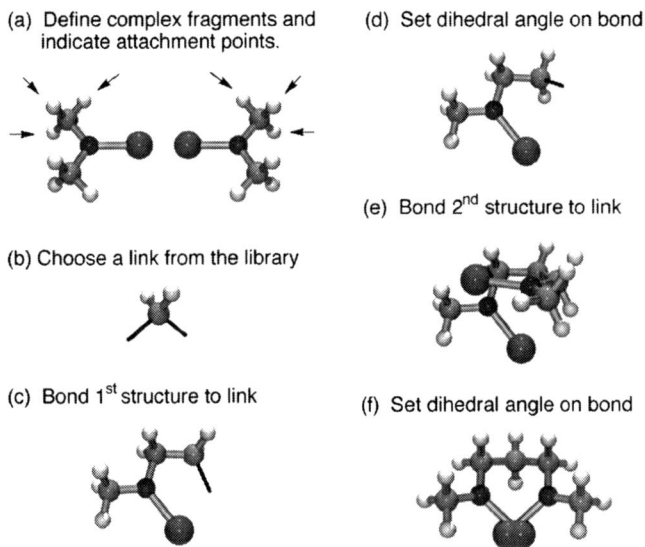

(a) Define complex fragments and
indicate attachment points.

(d) Set dihedral angle on bond

(e) Bond 2nd structure to link

(b) Choose a link from the library

(c) Bond 1st structure to link

(f) Set dihedral angle on bond

Fig. 5.2 The process of assembling fragments with the LINKER algorithm

(e) bond is formed between the second complex fragment and the remaining
    attachment vector on the linking fragment by aligning the attachment vectors
    and setting the bond distance to an appropriate value on the basis of the identity
    of the bonded atoms
(f) dihedral angle about the new bond is adjusted to a specific value on the basis of
    hybridization and degree of substitution of the bonded atoms.

Each bond formed by LINKER requires the assignment of a length and a dihedral
angle. The parameters used to make these assignments are stored in a lookup table.
Because these parameters depend on the identity of the fragments that are being
bonded to one another, the types of bonds that can be made are limited. At this time
the list includes C–C, C–N(amide), C–N(amine), C–O(ether), and C–S(thioether)
bonds. Bond distances were taken from default MM3 parameters [70]. Dihedral
angles values used for these rotations are on the basis of an examination of MM3
potential energy surfaces for rotation about the bonds that can be formed by all
chemically reasonable combinations of the 20 prototype rotors shown in Fig. 5.3.

It is generally possible to build a large number of host structures from a single
linking fragment. The ability to define multiple attachment vectors on each complex fragment gives rise to the potential for different connectivities. The presence of
chirality in either the complex fragments or in the linking fragment gives rise to the
potential for stereoisomers. Finally, the presence of multiple rotational minima for
the bonds formed between complex fragments and linking fragments gives rise to
multiple conformers for each connectivity. LINKER has been designed to build
every possible connectivity that can be made from the two complex fragments with

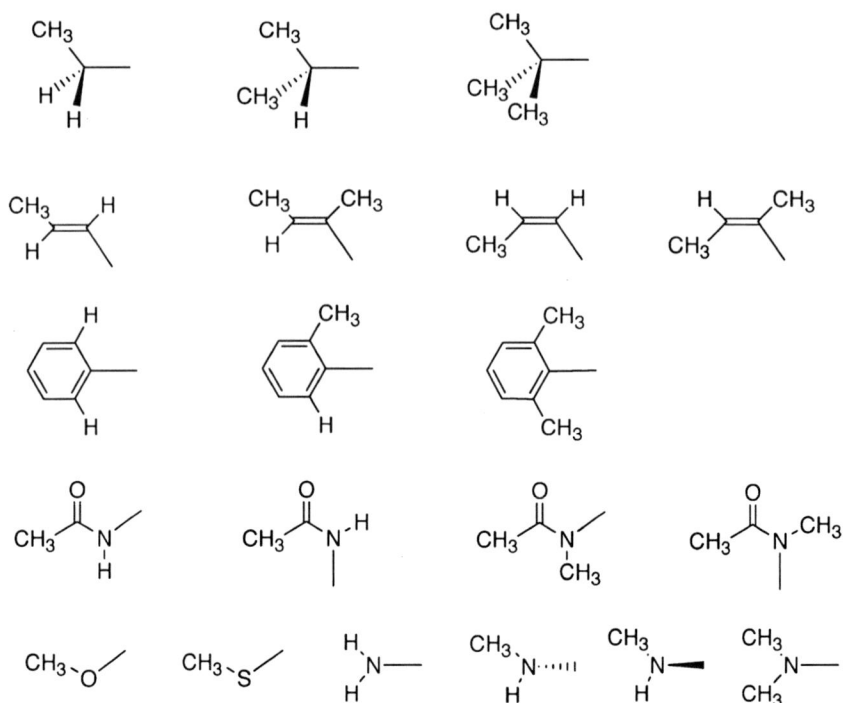

Fig. 5.3 Prototype rotors used to generate rotational potential surfaces for dihedral angle assignments

a given linking fragment including linkage isomers. In addition, when either the complex fragments or the linkage fragment is chiral, LINKER will examine all possible stereoisomers that can be made by inverting each chiral fragment. LINK-ER also examines every conformation that can be generated for each connectivity by rotation about the bonds that are formed between the linking fragment and the complex fragments.

Consider the simple example shown in Fig. 5.2. If we define the three C–H bonds on one of the methyl groups in each complex fragment as attachment vectors and we consider each of the three rotamers for the two C(sp[3])–C(sp[3]) bonds that are formed, there are potentially 81 host structures that can be made with the methylene linkage. In this case, all of these host structures are conformers of the same molecule as the connectivity remains constant. LINKER generates all 81 structures, but will retain only the first 9 structures shown in Fig. 5.4. Some of the potential structures are rejected because of close contacts between nonbonded atoms, which indicates a physically unreasonable collision or superposition of atoms (see Fig. 5.4). In addition, some of the structures are rejected because they are dupli-cates, structures that are either identical to or nonsuperimposable mirror images of previously generated structures. In other words, LINKER will retain only one member of a pair of enantiomers.

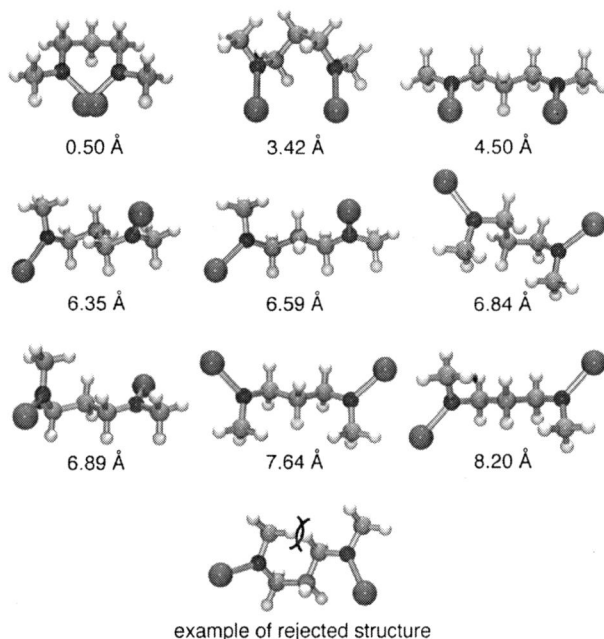

<table>
<tr><td>0.50 Å</td><td>3.42 Å</td><td>4.50 Å</td></tr>
<tr><td>6.35 Å</td><td>6.59 Å</td><td>6.84 Å</td></tr>
<tr><td>6.89 Å</td><td>7.64 Å</td><td>8.20 Å</td></tr>
</table>

example of rejected structure

Fig. 5.4 Nine unique configurations can be generated from the fragments shown in Fig. 5.2. The RMSD for guest superposition, given below each structure, is used to prioritize the structures in terms of complementarity

5.2.3   Scoring the Results

Two scoring methods are used to prioritize the structures produced from a LINKER run. The first method ranks the structures in terms of complementarity estimated using geometric parameters. The second method ranks the structures in terms of preorganization, on the basis of an estimate of the conformational energy of the host structure.

During the construction of each complex fragment, a guest is positioned relative to a host component to define a complementary geometry with the binding sites in that host component. When two complex fragments are combined, the degree of superposition of the two guests provides a simple criterion for the rapid evaluation of the degree of complementarity in the new host. This is measured by the root-mean-squared deviation, RMSD, of the distances between equivalent pairs of atoms in the two guests. Optimal complementarity would be obtained when the RMSD is zero, in other words, when the two guests representing the optimal bonding orientation with each host component are exactly superimposed. When the guest is a single atom, as in the example given in Fig. 5.4, the RMSD is simply the distance between the two guest atoms. As this example illustrates, the host structure with the smallest RMSD clearly gives the most complementary placement of the two ether binding sites. LINKER uses RMSD to score the generated host structures and outputs Cartesian coordinates for each structure in the order of increasing RMSD, in other words, in the order of decreasing complementarity for the guest.

In the example given in Fig. 5.4 each of the guests is a single lithium ion and the RMSD is simply the distance between them. These distances range from 0.50 to 8.20 Å; the host structure with the shortest Li-Li distance clearly gives the most complementary placement of the two ether binding sites. LINKER would use the Li-Li distance to score the generated host structures and output Cartesian coordinates for each structure in the order of increasing distance, in other words, in the order of decreasing complementarity for the guest. In the case of multiatom guests, the determination of RMSD requires HostDesigner to decide how the atoms from one guest should be paired with the atoms from the other guest for the superposition. This is done automatically by the code. If the guest has symmetry, there may be more than one way to pair the atoms. In such cases, LINKER will try every symmetry equivalent pairing of the atoms and report the minimal RMSD result.

After the results have been sorted by RMSD, a second prioritization is performed to rank the structures in terms of preorganization, on the basis of an estimate of the relative conformational free energy of the host structure. The conformational free energy is estimated using the following equation:

$$\Delta G_{conf} = \Delta H_{link} + \Delta H_{bondA} + \Delta H_{bondB} + N_{rot} \cdot \Delta G_{rot}.$$

The first three terms are enthalpic. A ΔH_{link} value is stored for each linking fragment. When the linking fragment has only one conformer, this value is zero. However, when the linking fragment has more than one conformer, the ΔH_{link} value is the relative enthalpy for that conformer with two methyl groups bound to the attachment vectors, obtained from MM3 calculations. The ΔH_{bondA} and ΔH_{bondB} terms are the relative rotamer energies associated with the first and second bond formed during the building process. The values assigned to each rotamer are on the basis of MM3 potential surfaces for the groups shown in Fig. 5.3. The final term, $N_{rot} \cdot \Delta G_{rot}$, is an estimate of the entropic penalty associated with restricted rotation of single bonds. The N_{rot} value is the sum of the rotatable bonds in the link, read from the LIBRARY, plus user-defined values for the attachment points in the complex fragments. The free energy per rotatable bond is set to a default value of 0.31 kcal/ mol per restricted rotation [71–74]. After LINKER sorts the output by RMSD, the list of structures is sorted again, to yield a second output file prioritized by ΔG_{conf}.

5.2.4   The Linking Fragment Library

The linking fragment library is a file from which LINKER reads the Cartesian coordinates and attributes of linking fragments that are used to connect the two complex fragments. Each linking fragment is a three-dimensional molecular structure with two specified binding vectors. In building the initial library, we decided to (1) limit the entries to molecules containing hydrogen and up to six carbon atoms, (2) limit carbon hybridization to sp[2] and sp[3], and (3) exclude three- and four-membered rings. This gave a total of 81 hydrocarbon connectivities. Subsequently, 66 additional connectivities representing all dimethylated five and six-membered rings and 59 additional bi and tri-cyclic connectivities have been added. These connectivities, Fig. 5.5, include the null case, which is used when the first complex fragment is directly bonded to the second complex fragment.

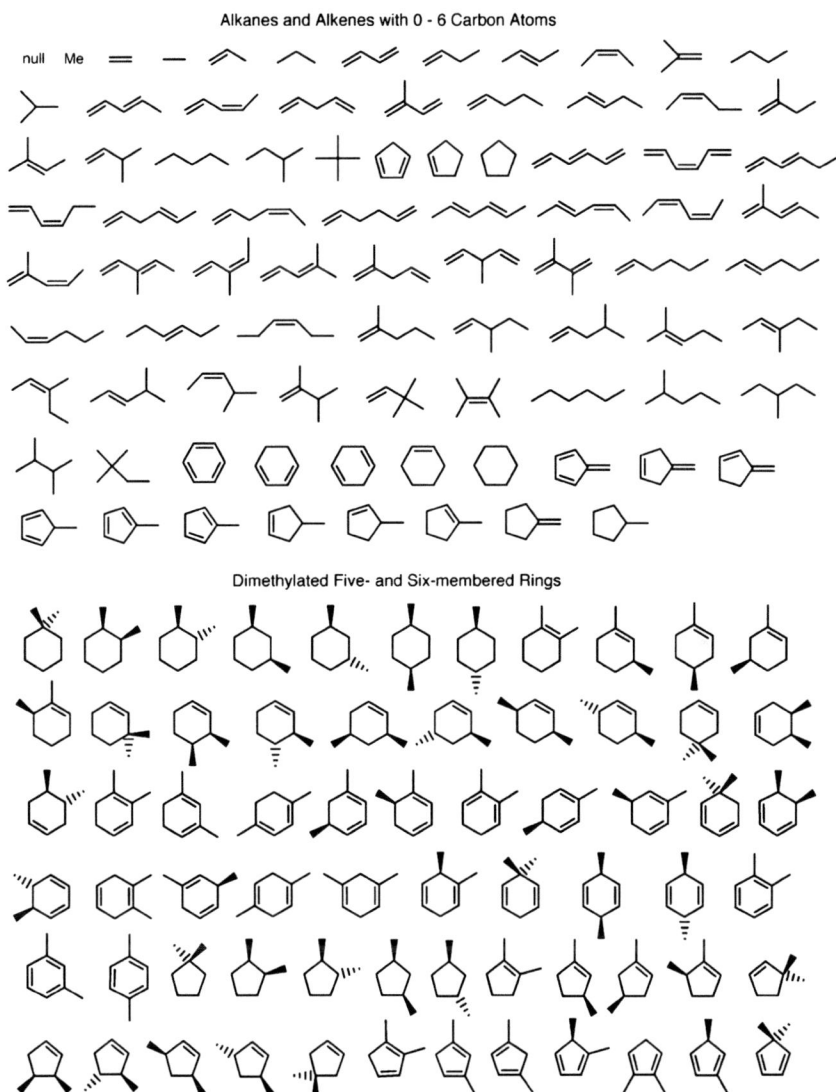

Fig. 5.5 (Continued)

Selected Polycyclic Hydrocarbons

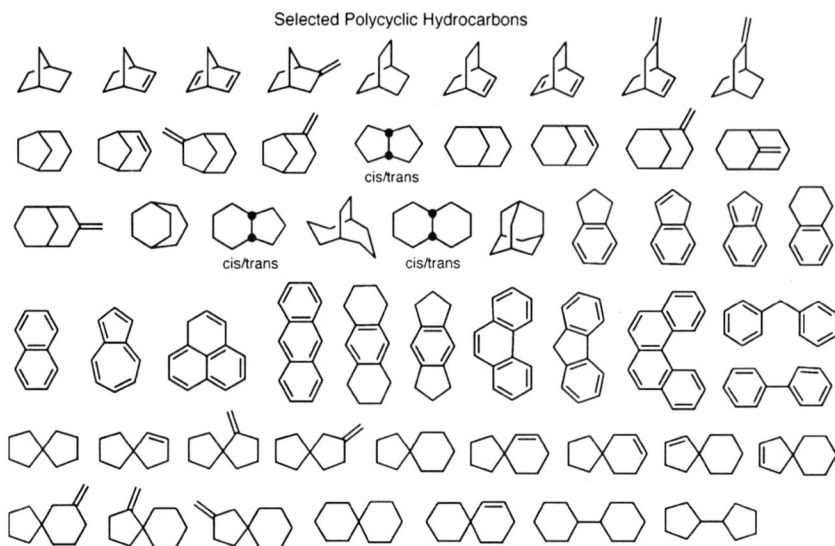

Fig. 5.5 Hydrocarbon connectivities used to generate linking fragments

A number of linking fragments were generated from each hydrocarbon connectivity by using the process illustrated in Fig. 5.6. The steps are as follows:

(a) select a connectivity
(b) perform a search to locate all stable conformers using the MM3 program
(c) choose one conformer and remove a pair of hydrogen atoms to generate a pair of bonding vectors
(d) place methyl groups at the ends of the bonding vectors
(e) optimize the structure with MM3
(f) remove the methyl groups to obtain the final linking fragment with two bonding vectors.

Steps (c)–(f) are repeated for all conformers and for all possible hydrogen atom pairs. By optimizing each structure with methyl groups attached to the bonding vectors, the linking fragments more accurately reflect the geometries that should result when attached to a substituent. Each new linking fragment is retained if it is unique or rejected if it is a duplicate of a previously generated linking fragment. In the example shown in Fig. 5.6, *n*-butane has two conformers and there are 45 hydrogen pairs per conformer. Thus, a total of 90 structures were processed for this connectivity yielding 41 unique linking fragments after removal of duplicates. A total of 11,251 linking fragments were prepared by performing this process on all 206 connectivities shown in Fig. 5.5.

HostDesigner contains a number of filters that can be invoked to limit the linking fragments that are selected during the building process. The choice of linking fragments can be limited by specifying a minimum length of the shortest

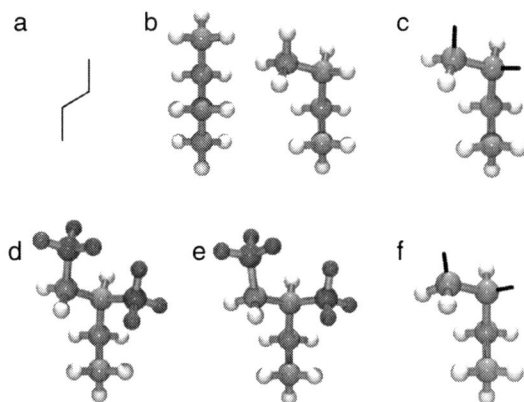

Fig. 5.6 The process used to generate a linking fragment (see text)

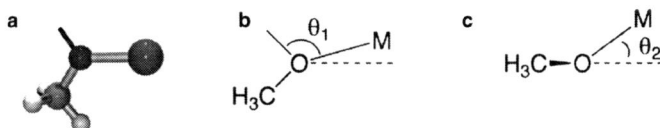

Fig. 5.7 (**a**) Complex fragment formed from dimethylether and a metal ion. Attachment vector comes off the oxygen atom. Two degrees of freedom, in-plane bending (**b**) and out-of-plane bending (**c**), were varied during the building process

connecting chain, a maximum length for the longest connecting chain, and the valence of the bonding carbon atoms. When a linking fragment derives from a connectivity with more than one conformation, the selection can be limited to only those linkages made from the lowest energy conformer. In addition, it is possible to exclude fragments on the basis of chirality, prochirality, or asymmetry.

5.2.5   *Examples of LINKER Applications*

We now present two examples to illustrate the use of LINKER. In the first example, HostDesigner was used to search for improved building blocks for macrocyclic ethers. One complex fragment used in this study, Fig. 5.7, was derived from an optimized geometry for dimethylether in which (a) one of the methyl groups was removed to define a bonding vector and (b) a metal ion was placed along the dipole moment of the ether to yield an optimal orientation [75]. The metal oxygen distance was set to 2.0 Å to complement a small metal ion such as Li^+ or Mg^{2+}. In order to account for known flexibility, two degrees of freedom were specified to vary the position of the metal ion relative to the ether. These were bending of the in-plane

M-O-C angle and the out-of-plane angle (see Fig. 5.7). The extent of the variation in each degree of freedom, based on the displacement on potential energy surfaces that would result in a 1 kcal mol^{-1} rise in energy, was $\pm 10°$ for the M–O–C angle and $\pm 20°$ for the out-of-plane angle.

In the initial run using the entire linking fragment library, HostDesigner constructed and evaluated a total of 143,985,540 geometries in less than 7 min. Subsequently, constraints were applied to exclude links that were asymmetric and to limit the size of the chelate rings formed to five or six atoms. The resulting hits were subjected to more accurate scoring methods, using molecular mechanics energies to rank them on the basis of their degree of organization for the metal ion. The best bidentate ether structures were used as building blocks for macrocycles, yielding hosts that give significantly higher binding affinities for alkali cations when compared with prototype crown ethers that contain poorly organized ethylene-bridged building blocks [75].

In a second example, HostDesigner was used to search for bis-urea podands that were organized for complexation with tetrahedral oxoanions [76]. N-methylurea was used as a starting binding unit for a perchlorate guest. The N-H group *cis* to the carbonyl oxygen was used as the attachment point to a hydrocarbon spacer. Initial structures were obtained from previous MP2 calculations on urea - anion complexes [77, 78]. Two examples of edge and vertex binding configurations and specified structural degrees of freedom are shown in Fig. 5.8. They include variation of the distance between the host and guest (± 0.2 Å), rotation about the H-axis ($\pm 20°$), rotation about the O-axis ($\pm 60°$ for edge forms or $+10$ to $-40°$ for vertex form), and, in the case of the vertex form, rotation about the Cl-O$_{vertex}$ bond ($\pm 60°$) that results in approximately a 1 kcal mol^{-1} decrease in binding energy from the equilibrium geometry. To enhance synthetic accessibility of the hits, a screening option to consider only symmetrical links was imposed.

With this input several HD runs were performed to sample all possible combinations of host–guest fragments. In a typical run, HD constructed and scored 300 million geometries within 40 min – a rate of 7.5 million geometries per minute. The

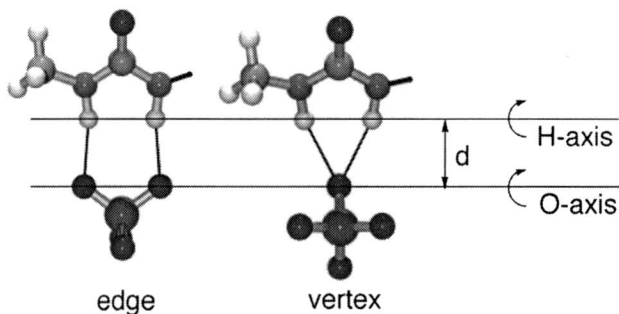

Fig. 5.8 Complex fragments formed from methylurea and perchlorate anion. Attachment vectors emanate from the nitrogen atoms. Three degrees of freedom, distance between the host and guest, rotation about the H-axis, and rotation about the O-axis are illustrated

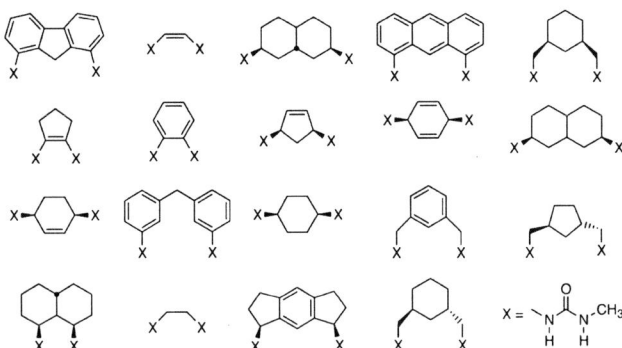

Fig. 5.9 Linking fragments that spatially arrange two urea groups, X, to complement a tetrahedral oxoanion guest

final ranking of the host structures was on the basis of molecular mechanics calculations of conformational and host-guest interaction energies. Examples of the top bis-urea podands generated by HD are shown in Fig. 5.9. While some of the spacers suggested by HD have been previously used to bridge urea and thiourea units and shown to provide organization of the binding sites for selective complexation with tetrahedral anions [79–88], there are a number of novel architectures that might possess promising complexation properties.

5.3 The OVERLAY Algorithm

5.3.1 Assembling the Pieces

OVERLAY builds new host structures by superimposing two attachment vectors on a single complex fragment with two attachment vectors on a linking fragment taken from the library (vide supra). This method is identical to that used in the program CAVEAT [89]. The user must provide an input file for a single complex fragment that specifies the coordinates for all the atoms, atom connectivity, and pairs of attachment vectors. As with LINKER, the attachment vectors are indicated by a list of hydrogen atoms that can be replaced and variation of structural degrees of freedom can be specified.

The steps used by OVERLAY to construct a new host molecule are as follows:

(a) select a linking fragment from the library
(b) adjust the lengths of the attachment vectors on the complex fragment and the linking fragment to the ideal length for the type of bond that would be formed
(c) compare geometries of the linking fragment vectors and the complex fragment vectors

(d) if the vector geometry is similar, then superimpose the attachment vectors of the linking fragment on the attachment vectors of the complex fragment to give the best overlay possible

(e) form bonds between the attaching atoms on the complex fragment and the linking fragment.

Comparison of vector geometries in step (c) involves taking the difference in three geometric parameters shown in Fig. 5.10. These are the distances, $d1$ and $d2$, and the dihedral angle, Φ. All three differences must be within user-defined tolerance limits. Smaller tolerance values give fewer results of higher quality. Larger tolerance limits give more results, but many of the structures may have distorted geometries.

The resulting structure may still be rejected even if the attachment vectors of the linking fragment are perfectly superimposed on the attachment vectors of the complex fragment. Although a perfect superposition ensures optimal distances and bond angles, the dihedral angles about each of the new bonds could have any value. Thus, after a new structure has been built, OVERLAY checks the difference between the actual dihedral angles and the dihedral angles corresponding to the nearest local minima (vide supra). If the rotational periodicity is >4, all structures are accepted. Otherwise, the structure will be rejected if the difference in dihedral angles is greater than a threshold value that depends on the periodicity of the rotational potential: twofold, 45°; threefold or fourfold 30°. Finally, the structure will be rejected if there are any close contacts between nonbonded atoms in the linking fragment and the complex fragment.

When pairs of attachment vectors in the complex fragment are related by symmetry, OVERLAY is able to build hosts by combining one complex fragment with two or more identical linking fragments. In such cases, specification of structural degrees of freedom within the complex fragment must be done in a way that the symmetry of is maintained. This is achieved by having an option to link several degrees of freedom such that they have the identical values at all times. This feature allows HostDesigner to generate symmetrical tripods and macrocycles.

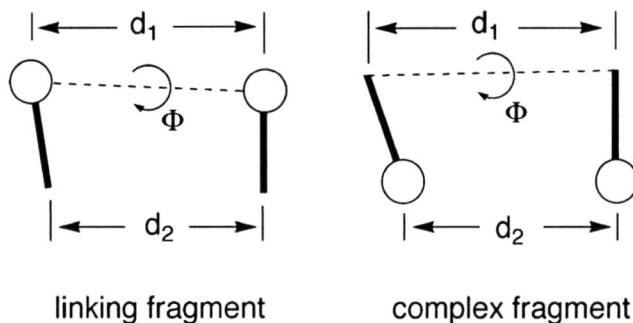

Fig. 5.10 Geometric parameters used to compare attachment vectors in the OVERLAY algorithm

5.3.2   Scoring the Results

Two scoring methods are used to prioritize the structures produced from an OVERLAY run. The first method ranks the structures in terms of how well the linking fragment fits onto the complex fragment. The second method ranks the structures in terms of preorganization, on the basis of an estimate of the conformational energy of the host structure. The degree–of–fit is measured by the RMSD of four points on the complex fragment with four points on the linking fragment, where in each case, the four points are the ends of the bonding vectors. In addition to ranking the structures by RMSD, a second method is used to rank the structures in terms of preorganization, on the basis of an estimate of the relative conformational free energy of the host structure. This estimate is obtained as described above for LINKER.

5.3.3   Example of OVERLAY Application

In the preceding examples, we used the LINKER algorithm to build simple open chain molecules from three separate pieces. However, it is not possible to build

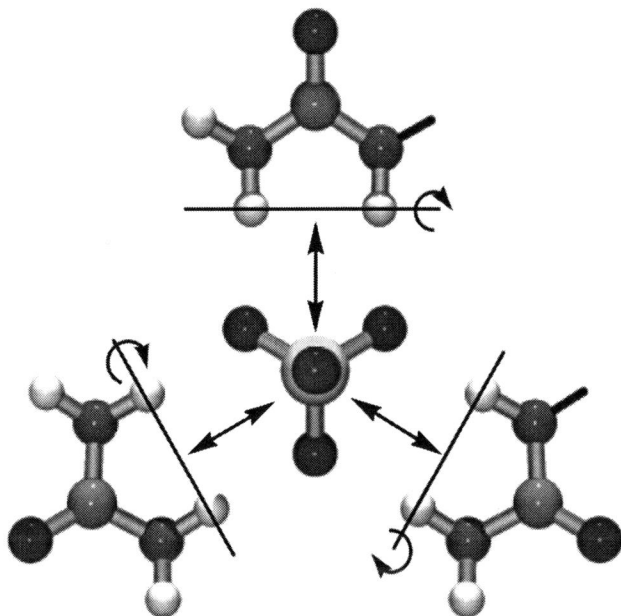

Fig. 5.11 Complex fragment formed from three urea donor groups and one sulfate anion guest has C_3 symmetry. One pair of attachment vectors, shown emanating from adjacent urea nitrogen atoms, is related by symmetry to two other pairs (not shown). Two degrees of freedom were specified for each urea group, distance between the urea and the anion and rotation about the H-axis, and linked to maintain symmetry

more elaborate structures, such as macrocycles or macrobicycles with this approach. To illustrate the use of the OVERLAY algorithm, consider the molecular design of a tris-urea macrocycle having C_3 symmetry. Cartesian coordinates for the complex fragment, Fig. 5.11, were taken from a 3:1 urea-sulfate C_3 symmetric structure optimized at the B3LYP/DZVP2 level of theory [77]. The N-H hydrogen atoms that are not involved in the interaction with sulfate specify the positions of three pairs of symmetrical attachment vectors to the structure. In order to allow for known flexibility, two degrees of freedom were specified to vary the position of each urea group relative to the anion (see Fig. 5.11). These were variation of the distance between the urea and the anion (± 0.2 Å) and rotation about the axis passing through the bound hydrogen atoms ($\pm 20°$). Variation in these degrees of freedom was linked to maintain C_3 symmetry.

With linking fragment library filters set to discard all asymmetric linkages, HostDesigner examined nearly 28,000 potential macrocyclic structures and stored 338 hits in just 6 s. Three of the architectures generated by the code, Fig. 5.12,

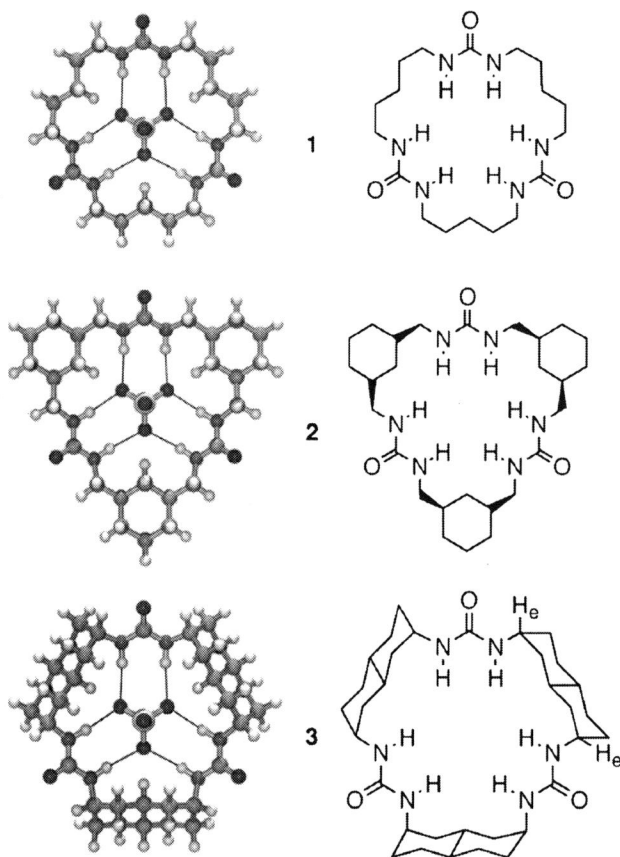

Fig. 5.12 Three tris-urea macrocycles generated with the OVERLAY algorithm

contain links with five carbon atom chains spanning the two urea nitrogen groups. It is interesting to note that the thiourea analog of **1**, with three pentane links, has been studied as a nitrate receptor [90]. The guanidinium analog of **1** also has been prepared and shown to bind phosphate in highly competitive methanol:water media [91]. Although they have the same macrocyclic conformation, structures **2** and **3**, which have not yet been studied, are conformationally constrained and may provide more organized cavities.

5.4 Summary

This chapter has presented an overview of a computer program, HostDesigner, that has been created to allow the de novo structure-based design of receptors that are structurally organized for complexation of small ionic and molecular guests. The methodology applies fundamental information about structure and bonding as a basis to search for host architectures that are highly organized for guest complexation. This software provides an efficient tool for virtual designing and screening of novel scaffolds to assist synthetic chemists in the identification of potential candidate structures before starting the experiment.

Acknowledgments BPH was supported by the Division of Chemical Sciences, Geosciences, and Biosciences, Office of Basic Energy Sciences, U.S. Department of Energy under contract number DE-AC05–00OR22725 with Oak Ridge National Laboratory (managed by UT-Battelle, LLC).

References

1. Desvergne, J. P.; Czarnik, A. W.; Eds., Chemosensors of Ion and Molecule Recognition; Kluwer, Dordrecht, Netherlands, **1997**
2. Beer, P. D.; Gale, P. A., Anion recognition and sensing: The state of the art and future perspectives, *Angew. Chem. Int. Ed.* **2001**, 40, 486–516
3. Manez-Martinez, R.; Sancenon, F., New advances in fluorogenic anion chemosensors, *J. Fluoresc.* **2005**, 15, 267–285
4. Suksai, C.; Tuntulani, T., Chromogenic anion sensors, *Chem. Soc. Rev.* **2003**, 32, 192–202
5. Gunnlaugsson, T.; Glynn, M.; Tocci, G. M. (nee Hussey); Kruger, P. E.; Pfeffer, F. M., Anion recognition and sensing in organic and aqueous media using luminescent and colorimetric sensors, *Coord. Chem. Rev.* **2006**, 250, 3094–3117
6. Nguyen, B. T.; Anslyn, E. V., Indicator-displacement assays, *Coord. Chem. Rev.* **2006**, 250, 3118–3127
7. Li, Y. Q.; Bricks, J. L.; Resch-Genger, U.; Spieles, M., Rettig, Bifunctional charge transfer operated fluorescent probes with acceptor and donor Receptors. 2. Bifunctional cation coordination behavior of biphenyl-type sensor molecules incorporating 2,2′:6′,2″-terpyridine acceptors, *J. Phys. Chem. A* **2006**, 110, 10972–10984
8. Rurack, K.; Resch-Genger, U., Rigidization, preorientation and electronic decoupling - the 'magic triangle' for the design of highly efficient fluorescent sensors and switches, *Chem. Soc. Rev.* **2002**, 31, 116–127

9. Gardner, J. W.; Bartlett, P. N., Electronic Noses: Principles and Applications; Oxford University Press, Oxford, **1999**

10. Pearce, T. C.; Schiffman, S. S.; Nagle, H. T.; Gardner, J. W., Eds., Handbook of Machine Olfaction: Electronic Nose Technology; Willey-VCH, Weinheim, **2003**

11. James, D.; Scott, S. M.; Ali, Z.; O'Hare, W. T., Chemical Sensors for electronic nose systems, *Microchim. Acta* **2005**, 149, 1–17

12. Wolfbeis, O. S., Materials for fluorescent-based optical chemical sensors, *J. Mater. Chem.* **2005**, 15, 2657–2669

13. Brogan, K. L.; Walt, D. R., Optical fiber-based sensors: Application to chemical biology, *Curr. Opin. Chem. Biol.* **2005**, 9, 494–500

14. Wolfbeis, O. S., Fiber-optic chemical sensors and biosensors, *Anal. Chem.* **2006**, 78, 3859–3874

15. Polymeric sensor materials: Toward an alliance of combinatorial and rational design tools? *Angew. Chem. Int. Ed.* **2006**, 45, 702–723

16. Frost, M.; Meyerhoff, M. E., Sensors: Tackling biocompatibility, *Anal. Chem.* **2006**, 7371–7377

17. Mohr, G. J., Covalent bond formation as an analytical tool to optically detect neutral and anionic analytes, *Sens. Actuators B* **2005**, 107, 2–13

18. Toma, H. E., Molecular materials and devices: Developing new functional systems based on the coordination chemistry approah, *J. Braz. Chem. Soc.* **2003**, 14, 845–869

19. Yang, R. H.; Wang, K. M.; Xiao, D.; Yang, X. H., A host-guest optical sensor for aliphatic amines based on lipophilic cyclodextrin, *Fresenius J. Anal. Chem.* **2000**, 367, 429–435

20. Finney, N. S., Combinatorial discovery of fluorophores and fluorescent probes, *Curr. Opin. Chem. Biol.* **2006**, 10, 238–245

21. Steed, J.; Atwood, J., Supramolecular Chemistry; Wiley, LTD, Chichester, **2000**

22. Schneider, H.-J.; Yatsimirsky, A., Principles and Methods in Supramolecular Chemistry; Wiley, LTD, Chichester, **2000**

23. Cram, D. J.; Lein, G. M. Host-guest complexation. 36. Spherand and lithium and sodium ion complexation rates and equilibria, *J. Am. Chem. Soc.* **1985**, 107, 3657–3668

24. Busch, D. H.; Farmery, K.; Goedken, V.; Katovic, V.; Melnyk, A.C.; Sperati, C. R.; Tokel, N., Chemical foundations for understanding of natural macrocyclic complexes, *Adv. Chem. Ser.* **1971**, 100, 44

25. McDougall, G. J.; Hancock, R. D.; Boeyens, J. C. A., Empirical force-field calculations of strain-energy contributions to the thermodynamics of complex formation. Part 1. The difference in stability between complexes containing five- and six-membered chelate rings, *J. Chem. Soc. Dalton Trans.* **1978**, 1438–1444

26. Anicini, A.; Fabbrizzi, L.; Paoletti, P.; Clay, R. M., A microcalorimetric study of the macrocyclic effect. Enthalpies of formation of copper(II) and zinc(II) complexes with some tetra-aza macrocyclic ligands in aqueous solution, *J. Chem. Soc. Dalton Trans.* **1978**, 577–583

27. Cram, D. J.; Kaneda, T.; Helgeson, R.C.; Brown, S. B.; Knobler, C. B.; Maverick, E.; Trueblood, K. N., Host-guest complexation. 35. Spherands, the first completely preorganized ligand systems, *J. Am. Chem. Soc.* **1985**, 107, 3645–3657

28. Stack, T. D. P.; Hou, Z.; Raymond, K. N., Rational reduction of the conformational space of a siderophore analog through nonbonded interactions: the role of entropy in enterobactin, *J. Am. Chem. Soc.* **1993**, 115, 6466–6467

29. Bianchi, A.; Bowman-James, K.; García-España, E., Eds., Supramolecular Chemistry of Anions; Wiley-VHC, New York, **1997**

30. Schmidtchen, F. P.; Berger, M., Artificial organic host molecules for anions, *Chem. Rev.* **1997**, 97, 1609–1646

31. Gale, P. A., Anion coordination and anion-directed assembly: Highlights from 1997 and 1998, *Coord. Chem. Rev.* **2000**, 199, 181–233

32. Gale, P. A., Anion receptor chemistry: highlights from 1999, *Coord. Chem. Rev.* **2001**, 213, 79–128

33. Fitzmaurice, R. J.; Kyne, G. M.; Douheret, D.; Kilburn, J. D., Synthetic receptors for carboxylic acids and carboxylates, *J. Chem. Soc.; Perkin Trans.* **2002**, 1, 841–864

34. Martínez-Máñez, R.; Sacenón, F., Fluorogenic and chromogenic chemosensors and reagents for anions, *Chem. Rev.* **2003**, 103, 4419–4476

35. Choi, K.; Hamilton, A. D., Macrocyclic anion receptors based on directed hydrogen bonding interactions, *Coord. Chem. Rev.* **2003**, 240, 101–110

36. Lambert, T. N.; Smith, B. D., Synthetic receptors for phospholipid headgroups, *Coord. Chem. Rev.* **2003**, 240, 129–141

37. Davis, A. P.; Joos, J.-B., Steroids as organising elements in anion receptors, *Coord. Chem. Rev.* **2003**, 240, 143–156

38. Gale, P. A., Anion and ion-pair receptor chemistry: Highlights from 2000 and 2001, *Coord. Chem. Rev.* **2003**, 240, 191–221

39. Moyer, B. A.; Singh, R. P., Eds., Fundamentals and Applications of Anion Separations; Kluwer, New York, **2004**

40. Severin, K., Supramolecular chemistry with organometallic half-sandwich complexes, *Chem. Commun.* **2006**, 3859–3867

41. Wright, A. T.; Anslyn, E. V., Differential receptor arrays and assays for solution-based molecular recognition, *Chem. Soc. Rev.* **2006**, 35, 14–28

42. Kuntz, I. D.; Meng, E. C.; Shoichet, B. K., Structure-based molecular design, *Acct. Chem. Res.* **1994**, 27, 117–123

43. Lybrand, T. P., Ligand-protein docking and rational drug design, *Curr. Opin. Struct. Biol.* **1995**, 5, 224–228

44. Böhm, H.-J., Computational tools for structure-based ligand design, *Prog. Biophys. Mol. Biol.* **1996**, 66, 197–210

45. Ajay, Murcko, M. A., Computational methods to predict binding free energy in ligand-receptor complexes, *J. Med. Chem.* **1995**, 38, 4953–4967

46. Eldridge, M. D.; Murray, C. W.; Auton, T. R.; Paolini, G. V.; Mee, R. P., Empirical scoring functions: I. The development of a fast empirical scoring function to estimate the binding affinity of ligands in receptor complexes, *J. Comput. Aided Mol. Des.* **1997**, 11, 425–445

47. Kitchen, D. B.; Decornez, H.; Furr, J. R.; Bajorath, J., Docking and scoring in virtual screening for drug discovery: methods and applications, *Nature Rev. Drug Discov.* **2004**, 3, 935–949

48. Nishibata, Y.; Itai, A., Automatic creation of drug candidate structures based on receptor structure. Starting point for artificial lead generation, *Tetrahedron* **1991**, 47, 8985–8990

49. Nishibata, Y.; Itai, A., Confirmation of usefulness of a structure construction program based on three-dimensional receptor structure for rational lead generation, *J. Med. Chem.* **1993**, 36, 2921–2928

50. Rotstein, S. H.; Murcko, M. A., GenStar: A method for de novo drug design, *J. Comput. Aided Mol. Des.* **1993**, 7, 23–43

51. Bohecek, R. S.; McMartin, C., Multiple highly diverse structures complementary to enzyme binding sites: Results of extensive application of a de novo design method incorporating combinatorial growth, *J. Am. Chem. Soc.* **1994**, 116, 5560–5571

52. Gehlhaar, D. K.; Moerder, K. E.; Zichi, D.; Sherman, C. J.; Ogden, R. C.; Freer, S. T., De novo design of enzyme inhibitors by Monte Carlo ligand generation, *J. Med. Chem.* **1995**, 38, 466–472

53. Luo, Z.; Wang, R.; Lai, L., RASSE: A new method for structure-based drug design *J. Chem. Inf. Comput. Sci.* **1996**, 36, 1187–1194

54. Böhm, H.-J., The computer program LUDI: A new method for the de novo design of enzyme inhibitors, *J. Comput. Aided Mol. Des.* **1992**, 6, 61–78

55. Lawrence, M. C.; Davis, P. C., CLIX: A search algorithm for finding novel ligands capable of binding proteins of known three-dimensional structure, *Proteins: Struct. Funct. Genet.* **1992**, 12, 31–41

56. Ho, C. M. W.; Marshall, G. R., SPLICE: A program to assemble partial query solutions from three-dimensional database searches into novel ligands, *J. Comput.-Aided Mol. Des.* **1993**, 7, 623–647

57. Rotstein, S. H.; Murcko, M. K., GroupBuild: A fragment-based method for de novo drug design, *J. Med. Chem.* **1993**, 36, 1700–1710

58. Tschinke, V.; Cohen, N. C., The NEWLEAD program: A new method for the design of candidate structures from pharmacophoric hypotheses, *J. Med. Chem.* **1993**, 36, 3863–3870.

59. Gillet, V. J.; Newell, W.; Mata, P.; Myatt, G. J.; Sike, S.; Zsoldos, Z.; Johnson, A. P., SPROUT: Recent developments in the de novo design of molecules, *J. Chem. Inf. Comput. Sci.* **1994**, 34, 207–217.

60. Leach, A. R.; Kilvington, S. R., Automated molecular design: A new fragment-joining algorithm, *J. Comput.-Aided Mol. Des.* **1994**, 8, 283–298.

61. Mata, P.; Gillet, V. J.; Johnson, A. P.; Lampreia, J.; Myatt, G. J.; Sike, S.; Stebbings, A. L., SPROUT: 3D structure generation using templates, *J. Chem. Inf. Comp. Sci.* **1995**, 35, 479–493.

62. Roe, D. C.; Kuntz, I. D., BUILDER v.2: Improving the chemistry of a de novo design strategy, *J. Comput.-Aided Mol. Des.* **1995**, 9, 269–282.

63. Wang, R. X.; Gao, Y.; Lai, L. H., LigBuilder: A multi-purpose program for structure-based drug design, *J. Mol. Mod.* **2000**, 6, 498–516.

64. Head, R. D.; Smythe, M. L.; Oprea, T. I.; Waller, C. L.; Green, S. M.; Marshall, G. R., VALIDATE: A new method for the receptor-based prediction of binding affinities of novel ligands, *J. Am. Chem. Soc.* **1996**, 118, 3959–3969.

65. Baxter, C. A.; Murray, C. W.; Clark, D. E.; Westhead, D. R.; Eldridge, M. D., Flexible docking using tabu search and an empirical estimate of binding affinity, *Proteins: Struct. Funct. Genet.* **1998**, 33, 367–382.

66. Wang, R.; Liu, L.; Lai, L.; Tang, Y., SCORE: A new empirical method for estimating the binding affinity of a protein-ligand complex, *J. Mol. Model.* **1998**, 4, 379–394.

67. Böhm, H.-J.; Schneider, G., Virtual screening and fast automated docking methods, *Drug Discov. Today* **2002**, 7, 64–70.

68. Hay, B. P.; Firman, T. K., HostDesigner: A program for the de novo structure-based design of molecular receptors with binding sites that complement metal ion guests, *Inorg. Chem.* **2002**, 41, 5502–5512.

69. Hay, B. P.; Firman, T. K.; Bryantsev, V. S., HostDesigner User's Manual, PNNL-13850, Pacific Northwest National Laboratory, Richland, WA, **2006**. *HostDesigner software and User's Manual can be obtained free of charge by contacting BPH (haybp@ornl.gov)*

70. Allinger, N. L.; Lii, J.-H., Molecular mechanics. The MM3 force field for hydrocarbons. 1, *J. Am. Chem. Soc.* **1989**, 111, 8551–8566.

71. Eblinger, F.; Schneider, H.-J., Stabilities of hydrogen-bonded supramolecular complexes with various numbers of single bonds: Attempts to quantify a dogma in host-guest chemistry, *Angew. Chem. Int. Ed.* **1998**, 37, 826–829.

72. Mammen, M.; Shakhnovich, E. I.; Whitesides, G. M., Using a convenient, quantitative model for torsional entropy to establish qualitative trends for molecular processes that restrict conformational freedom, *J. Org. Chem.* **1998**, 63, 3168–3175.

73. Houk, K. N.; Leach, A. G.; Kim, S. P.; Zhang, X., Binding affinities of host-guest, protein-ligand, and protein-transition-state complexes, *Angew. Chem. Int. Ed.* **2003**, 42, 4872–4897.

74. Deanda, F.; Smith, K. M.; Liu, J.; Pearlman, R. S., GSSI, a general model for solute–solvent interactions. 1. Description of the model, *Mol. Pharm.* **2004**, 1, 23–39.

75. Hay, B. P.; Oliferenko, A. A.; Uddin, J.; Zhang, C.; Firman, T. K., Search for improved host architectures: Application of the de novo structure-based design and high-throughput screening methods to identify optimal binding blocks for multidentate ethers, *J. Am. Chem. Soc.* **2005**, 127, 17043–17053.

76. Bryantsev, V. S.; Hay, B. P., De novo structure-based design of bisurea hosts for tetrahedral oxoanion guests, *J. Am. Chem. Soc.* **2006**, 128, 2035–2042.

77. Hay, B. P.; Firman, T. K.; Moyer, B. A., Structural design criteria for anion hosts: Strategies for achieving anion shape recognition through the complementary placement of urea donor groups, *J. Am. Chem. Soc.* **2005**, 127, 1810–1819.

78. Bryantsev V. S.; Hay, B. P., Using the MMFF94 model to predict structures and energies for hydrogen–bonded urea–anion complexes, *J. Mol. Struct. (THEOCHEM)* **2005**, 725, 177–182.
79. Albert, J. S.; Hamilton, A. D., Synthetic analogs of the ristocetin binding site: Neutral, multidentate receptors for carboxylate recognition, *Tetrahedron Lett.* **1993**, 34, 7363–7366.
80. Nishizawa, S.; Bühlmann, P.; Iwao, M.; Umezawa, Y., Anion recognition by urea and thiourea groups: Remarkably simple neutral receptors for dihydrogenphosphate, *Tetrahedron Lett.* **1995**, 36, 6483–6486.
81. Kwon, J. Y.; Jang, Y. J.; Kim, S. K.; Lee, K.-H.; Kim, J. S.; Yoon, J., Unique hydrogen bonds between 9-anthracenyl hydrogen and anions, *J. Org. Chem.* **2004**, 69, 5155–5157.
82. Brooks, S. J.; Gale, P. A.; Light, M. E., Carboxylate complexation by 1,1′-(1,2-phenylene)bis (3-phenylurea) in solution and the solid state, *Chem. Commun.* **2005**, 4696–4698.
83. Amendola, V.; Boicchi, M.; Esteban-Gomez, D.; Fabbrizzi, L.; Monzani, E., Chiral receptors for phosphate ions, *Org. Biomol. Chem.* **2005**, 3, 2632–2639.
84. Hamann, B. C.; Branda, N. R.; Rebek, J., Jr., Multipoint recognition of carboxylates by neutral hosts in non-polar solvents, *Tetrahedron Lett.* **1993**, 34, 6837–6840.
85. Bühlmann, P.; Nishizawa, S.; Xiao, K. P.; Umezawa, Y., Strong hydrogen bond-mediated complexation of $H_2PO_4^-$ by neutral bis-thiourea hosts, *Tetrahedron* **1997**, 53, 1647–1654.
86. Tobe, Y.; Sasaki, S.; Mizuno, M.; Naemura, K., Synthesis and anion binding ability of metacyclophane-based cyclic thioureas, *Chem. Lett.* **1998**, 8, 835–836.
87. Nishizawa, S.; Kamaishi, T.; Yokobori, T.; Kato, R.; Cui, Y.-Y.; Shioya, T.; Teramae, N., Facilitated sulfate transfer across the nitrobenzene-water interface as mediated by hydrogen-bonding ionophores, *Anal. Sci.* **2004**, 20, 1559–1566.
88. Nishizawa, S.; Rokobori, T.; Kato, R.; Yoshimoto, K.; Kamaishi, T.; Teramae, N., Hydrogen-bond forming ionophore for highly efficient transport of phosphate anions across the nitrobenzene-water interface, *Analyst* **2003**, 128, 663–669.
89. Lauri, G.; Bartlett, P. A., CAVEAT: A program to facilitate the design of organic molecules, *J. Comp. Aided Mol. Design* **1994**, 8, 51–66.
90. Herges, R.; Dikmans, A.; Jana, U.; Köhler, F.; Jones, P. G.; Dix, I.; Fricke, T.; König, B., Design of a neutral macrocyclic ionophore: Synthesis and binding properties for nitrate and bromide anions, *Eur. J. Org. Chem.* **2002**, 3004–3014.
91. Dietrich, B; Fyles, T. M.; Lehn, J. M.; Pease L. G.; Fyles, D. L., Anion receptor molecules - synthesis and some anion binding properties of macrocyclic guanidinium salts, *J. Chem. Soc. Chem. Comm.* **1978**, 934–936.

Chapter 6
First Principles Molecular Modeling of Sensing Material Selection for Hybrid Biomimetic Nanosensors

Mario Blanco, Michael C. McAlpine, and James R. Heath

Abstract Hybrid biomimetic nanosensors use selective polymeric and biological materials that integrate flexible recognition moieties with nanometer size transducers. These sensors have the potential to offer the building blocks for a universal sensing platform. Their vast range of chemistries and high conformational flexibility present both a problem and an opportunity. Nonetheless, it has been shown that oligopeptide aptamers from sequenced genes can be robust substrates for the selective recognition of specific chemical species. Here we present first principles molecular modeling approaches tailored to peptide sequences suitable for the selective discrimination of small molecules on nanowire arrays. The modeling strategy is fully atomistic. The excellent performance of these sensors, their potential biocompatibility combined with advanced mechanistic modeling studies, could potentially lead to applications such as: unobtrusive implantable medical sensors for disease diagnostics, light weight multi-purpose sensing devices for aerospace applications, ubiquitous environmental monitoring devices in urban and rural areas, and inexpensive smart packaging materials for active in-situ food safety labeling.

M. Blanco (✉) and J.R. Heath
Division of Chemistry and Chemical Engineering, California Institute of Technology, Pasadena, CA 91125, USA
e-mail: mario@wag.caltech.edu

M.C. McAlpine
Department of Mechanical and Aerospace Engineering, Princeton University, Princeton, NJ 08544, USA

M.A. Ryan et al. (eds.), *Computational Methods for Sensor Material Selection*,
Integrated Analytical Systems,
DOI 10.1007/978-0-387-73715-7_6, © Springer Science+Business Media, LLC 2009

6.1 Hybrid Biomimetic Nanosensors: A Universal Sensing Platform

The development of a universal sensing platform for selective and sensitive discrimination of chemical and biochemical compounds could stimulate exciting opportunities in fundamental research and a revolution in technological applications. We envisioned such a platform as a series of technologies whereby a given target compound is physically provided at the start of such sequence of steps, followed by some automated (combinatorial) process for sensing material optimization (synthetic or biological) to produce a highly sensitive, highly selective device that can detect sub-parts per billion concentrations of the input compound. The whole process should take place without major human intervention, within costs constraints, and in less than 24 h.

Semiconducting nanowire arrays, such as doped silicon nanowires (SiNW) offer some promise as an element of this universal sensing platform. Doped SiNW can provide very high sensitivity due to their large surface-to-volume ratios and the unique electronic properties of these nanodevices. The doped nanowires act as field-effect transistors (FET), their resistivity is sensitive to changes on their surface charge distribution as a result of molecules or ions present in the environment. SiNW arrays are fabricated through established methods [1]. All fabrication is done within a class 1000 or class 100 clean room environment. A typical array of nanowires fabricated by this technique is shown in Fig. 6.1.

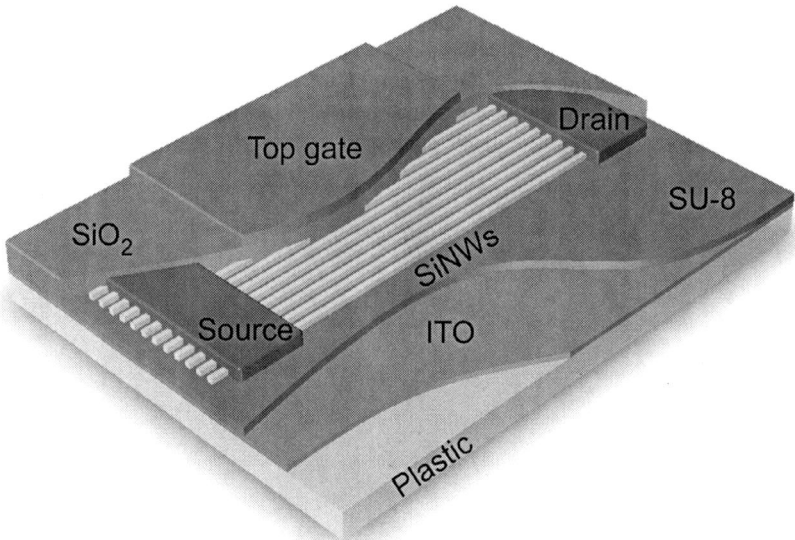

Fig. 6.1 Optical image of microfluidic functionalization channels (vertical conduits) intersecting nanowire sensor devices. The nanowire islands (*horizontal bars*) are electrically contacted by metal leads (*white lines*). (*Inset*) Scanning electron micrograph of the nanowire film

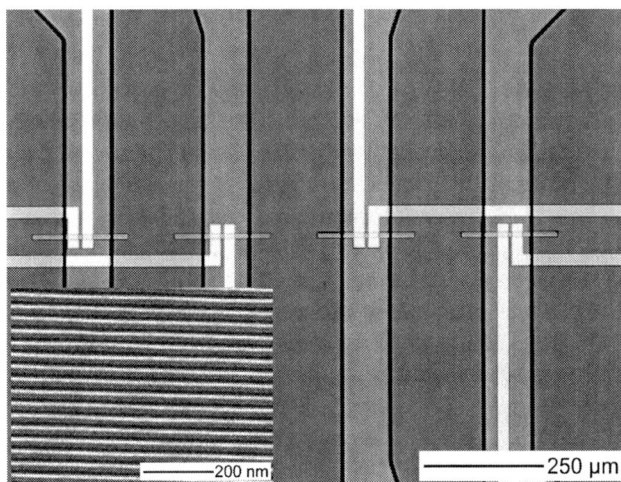

Fig. 6.2 Schematic representation of silicon nanowire (SiNW) array. The resulting device acts as a p-type field-effect transistor

Dopants are diffused into the silicon film using rapid thermal processing (RTP) at 800°C for 3 min. Four-point resistivity measurements, correlated with tabulated values, yield a doping level of $\sim 10^{18}/cm^3$. The finished nanosensor is schematically shown in Fig. 6.2.

Without further modification these nanowire arrays are not chemically selective. A specific amount of surface charge density will result in the same electrical resistivity change regardless of the type of molecule binding to the surface. To add selectivity to these devices, one needs to incorporate synthetic polymers or biopolymer monolayer to trap the target compound on the surface of the nanowire array and nothing else if possible. This process has already provided nanosensor arrays for small vapor bound molecules [2]. The surface of the SiNW array is modified with peptides synthesized on Fmoc-Rink Amide MBHA resin (0.67 mmol/g, Anaspec) using conventional solid-phase synthesis strategy with Fmoc protection chemistry [3].

A general scheme for achieving a high degree of binding specificity, to a target molecule of interest, has been an elusive goal. The problem is due in part to the high degree of molecular flexibility that these sensing materials possess. Peptide sequences were selected from encoding regions of the P30953 human olfactory receptor [6], without any guarantees that these sequences were selective towards the compounds of interest. However, (MacAlpine 2008) found that, depending on the peptide sequence used, these hybrid materials exhibit orthogonal sensing to significantly different compounds, such as acetic acid and ammonia vapors, and can even detect traces of these gases in complex, "chemically camouflaged" mixtures. A peptide modeling protocol was developed to help explain the measurements and level of selectivity of these hybrid nanosensors [14].

6.2 Modeling Peptide/Analyte Molecular Interactions

The observed selective changes in resistivity, upon exposure to various analytes, require a theoretical explanation. A first principles model to explain the sensitivity of synthetic amorphous polymers to vapor bound compounds has been previously developed [4]. Molecular Dynamics protocols for modeling of small molecule interactions with polymer sensing materials rely on statistical mechanical distributions of molecular structural features [5]. The 3D structure of peptide sensing material and the presence of tertiary structure, offers new challenges. We first focus on the molecular interactions between the chosen peptide sequence and a specific analyte. We use fully atomistic simulations with a few parameters (particularly atomic charges) calculated using quantum mechanics. The objective is to model the system from first principles, in a general way such that we can apply the method, after validation, to any arbitrary peptide sequence and analyte without the need for experimental information. This is in contrast with QSAR approaches that by necessity require a significant number of experimental input data to make predictions. Typically QSAR predictions apply only within a short excursion outside the initial training set. The process begins by modeling the aminoacid sequences of the peptides used in the experiment. These and the composition of the analyte compound are the only two pieces of experimental information that enter the modeling.

6.2.1 Modeling Protocol for 3D Structure Peptide Sequence Determination

The two peptide sequences, using single letter aminoacid identification, are DLESFLD and RVNEWVID. As it is customary the sequences are given from the N to the C terminus. With the exception of the D residue in the C terminus (Aspartic acid was added as coupling agent to covalently bind the residue to the SiNW array), both the sequences belong to the trans-membrane region of the P30953 G-protein coupled human olfactory receptor. These are believed to be in the binding region of the GPCR receptor 6.

To obtain a predicted 3D structure for these two peptides we designed the following protocol:

1. Force Field: We use a general force field, Dreiding [7] in all searches of the global minimum, the stable 3D peptide conformations, as well as the binding geometries of analyte and peptide.
2. Electrostatics: All electrostatic interaction pairs were included in the calculation without the use of cutoffs or spline functions. As customary, nearest neighbor and next nearest neighbor coulomb terms, so called 1-2 and 1-2-3 interactions, were excluded.
3. Atomic charges: from Mulliken populations were obtained using full geometry minimization of the primary peptide structure using the Becke-Lee-Yang-Par

[8, 9] hybrid density functional method (B3LYP). We employed the 6-31g** basis set, 1160 and 1450 basis functions for DLESFLD and RVNEWVID respectively.

4. Net charges: With the exception of ARG1 and GLU4 in RVNEWVID all other aminoacids were modeled with a net molecular charge of zero because there is no solvent present in the gas phase to stabilize the acid/base side-chains of some aminoacids as it occurs in solution. This choice was further validated when we observed proton transfer occurring freely during quantum mechanical geometry minimization of DLEFSLD from the positively charged amino-terminus and the deprotonated aspartic acid.

5. Salt-Bridges: One exception to the neutral charge rule in step 4 is the ARG1(+)-GLU4(−) salt-bridge, which was calculated by B3LYP DFT to be 11.5 Kcal/mol more stable in vacuum than compared to the neutral forms of these two aminoacids in RVNEWVID.

6. Boltzmann-Jump Conformational Search: quantum mechanically optimized geometries and charges were used as input for a Boltzmann-Jump search [10]. The search was done over all ϕ and ψ angles on the peptide's backbone as well as all side chain rotatable bonds ω. The protocol for finding the lowest energy peptide conformation consisted of a hybrid procedure involving two steps:

7. Annealed molecular dynamics: 10 anneal cycles, each starting from 200 to 500 K in steps of 100 K, followed by quenching minimization, with an anneal period of 1,000 time steps, each 1fs, followed by 2,000 Boltzmann Jump sequences, with an average of 30 perturbations per sequence, an adjustable dihedral window of 10° for all rotatable bonds, and an acceptance maximum temperature of 5,000 K relative to the current minimum. The lowest 100 energy conformations were examined and a cluster analysis showed that these were all virtually identical, within the expected window room mean square difference of 10°.

Further details of the Boltzmann Jump sequence procedure are described below:

7.1. Minimize the energy of the current conformation. This minimized conformation is referred to as the Reference Conformation.

7.2. The Reference conformation is made the Working Conformation.

7.3. The Working Conformation is perturbed by randomly altering each VARIABLE torsion angle randomly within a specified window. The energy of the perturbed conformation is computed. If the change in Energy, ΔE, is negative, the perturbed conformation is selected and retained as the new Working Conformation. If, on the other hand, the energy increases, the Boltzmann factor $F = \exp(-\Delta E/RT)$ is computed. A random number N (0,1) between 0 and 1 is then generated. If $N < F$, the perturbed conformation is selected making it the new Working conformation; otherwise the perturbed conformation is rejected retaining the Working conformation.

7.4. Step 7.5 is repeated a specified number of times (10), keeping track of the number of selections and the number rejections.

7.5. The working conformation is then energy minimized and stored into the specified output file. The RMS difference between this conformation and the Reference conformation is computed for plotting purposes. Then this minimized conformation is made the new Reference conformation.

7.6. Optionally at this step, the angular window employed is adjusted to give on an average a roughly 50% rate of acceptance.

7.7. Steps 7.4–7.6 are repeated a specified number of times (2,000).

Figure 6.3 shows the results of the Boltzmann Jump search protocol. The lowest 20 conformations for RVNEWVID are superimposed. Figure 6.4 shows the energy progression during the conformational search using simulated annealing and the Boltzmann Jump method. The Boltzmann Jump method is about 9 times faster in searching conformational space than simulated annealing (80 vs. 9 conformations found per cpu hour), as well as more effective in lowering the energy of the original reference conformation (-174 kcal/mol/cpu-hour versus -47 kcal/mol/cpu-hour). However, simulated annealing is much more efficient in the first few iterations, as shown in Fig. 6.4. After this tests case we employed a combination of the two methods with 2 steps of simulated annealing prior to a full Boltzmann Jump search with a maximum of 2000 conformations. This was the final protocol used with the remaining peptide sequence DLESFLD.

The lowest, most stable, conformation of RVNEWVID is shown on the right side of Fig. 6.4. It includes a very stable salt-bridge. Surprisingly the salt-bridge is stable at temperatures well above room temperature in vacuum (see Fig. 6.5).

Figure 6.6 shows the most stable conformation for DLESFLD. In this case the sequence shows a strong preference for hydrophobic aminoacid residues (LFL) lining up on the opposite site of hydrophilic residues (DESD).

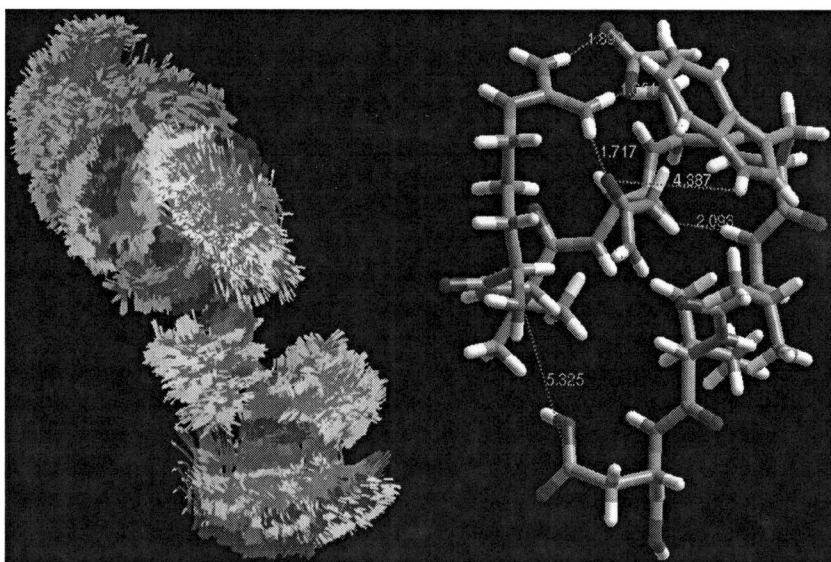

Fig. 6.3 The twenty lowest energy conformations of RVNEWVID peptide are shown superimposed. The most stable conformation is shown to the *right*

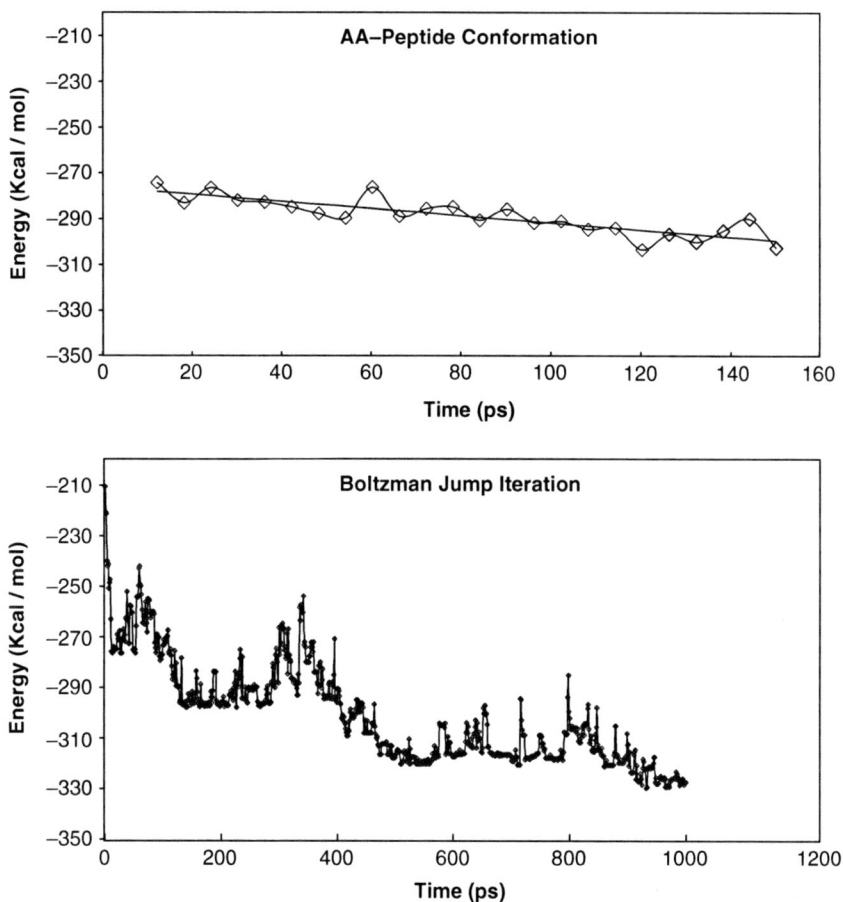

Fig. 6.4 Energy progression in the simulated annealing method (*top*) and the Boltzmann Jump method (*bottom*) for low energy conformations of RVNEWVID

6.2.2 Modeling Protocol for Analyte/Peptide Non-Reactive Interactions

The two peptide sequences were used to fabricate SiNW sensors and tested for their specificity to acetic acid and ammonia. At this stage we use atomistic simulations to predict the degree of association, assuming no reactions take place, of these two gas phase analytes with the low energy conformations of the peptide sequences chemically bound to the SiNW arrays.

The 3D modeled peptide structures as determined with the previous protocol are simulated in the presence of ammonia and acetic acid. These two small molecules could potentially bind into a large number of exposed sites on the surface of these peptides in a variety of molecular orientations. We need a method to sample these

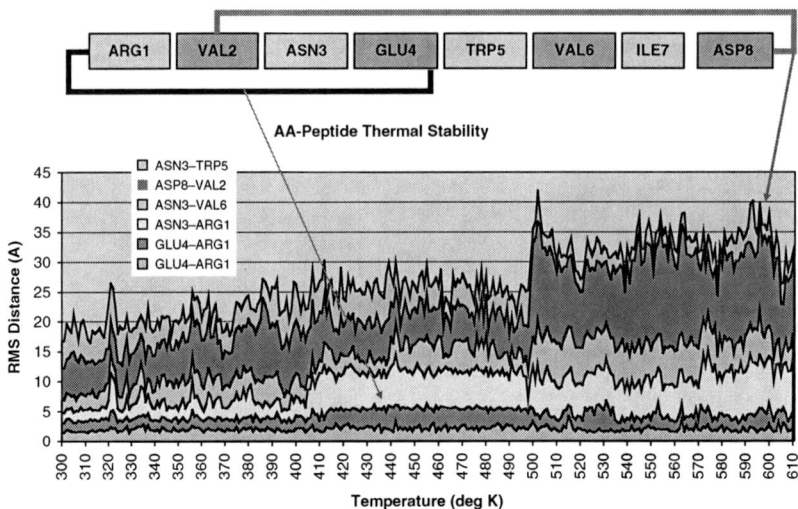

Fig. 6.5 Various inter-residue distances are monitored as a function of temperature in molecular dynamics simulations of peptide sequence RVNEWVID. The GLU4-ARG1 distances are stable, even at high temperatures, indicating the strong stability of this salt-bridge vacuum

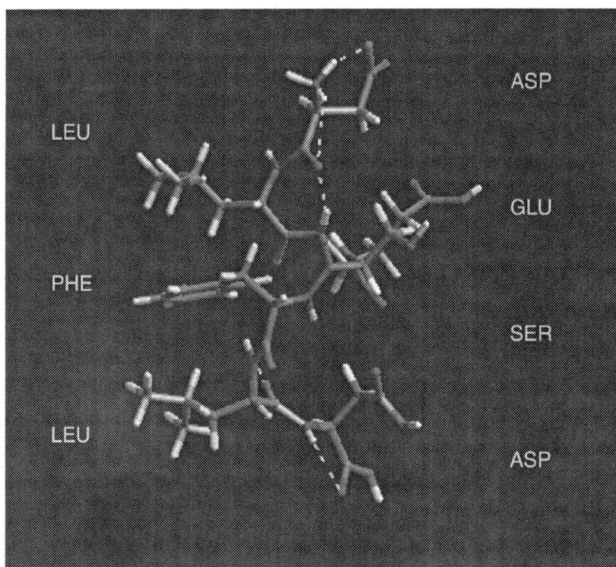

Fig. 6.6 Lowest energy conformation for DLESFLD. Hydrophobic and hydrophilic residues line up on opposite sides of the peptide

efficiently, more efficiently than molecular dynamics methods, which might get trapped into local minima or the wrong analyte orientation for long periods of time.

We employ a Monte Carlo method, the Molecular Silverware algorithm [11], to sample up to 180 points on the surfaces of the peptides. For each of these points we sample the binding energy of these molecules with 120 distinct Euler orientations of the analyte compound. Because of the concave topology of the peptides, multiple solutions exist for most binding orientations and the total number of sampled peptide/analyte pairs exceeded 47,000.

Figure 6.7 shows an example of the binding energetics obtained with the Molecular Silverware method in the form of a histogram.

The average binding energy was calculated using Boltzmann factors

$$E_{binding} = \frac{\sum_{i=1}^{N} E_i e^{-\left(E_i/RT\right)}}{\sum_{i=1}^{N} e^{-\left(E_i/RT\right)}}. \tag{6.1}$$

The non-reactive binding energy values for ammonia and acetic acid with the modeled peptides calculated in this manner are reported in the results section.

Molecular dynamics, Molecular Silverware Monte Carlo and Boltzmann Jump searches were conducted with the Software Developer's Kit of the Cerius^2 package [12].

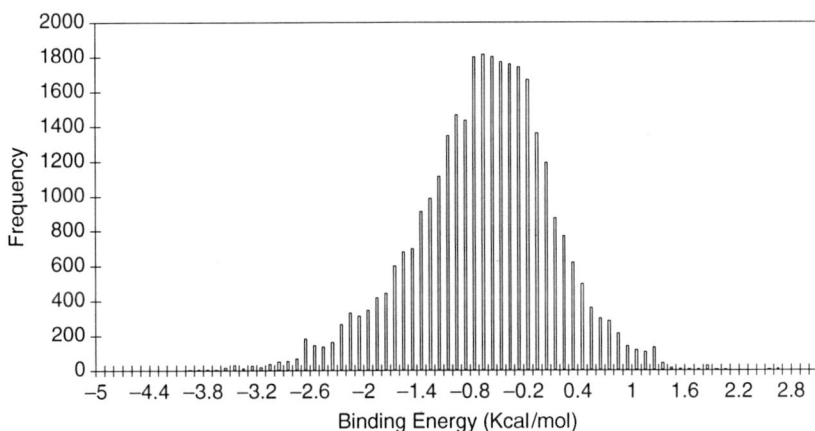

Fig. 6.7 Frequency histogram of the non-reactive binding energies of over 47,000 configurations of ammonia (NH_3) on the surface of DLESFLD peptide

6.2.3 Analyte/Peptide Reactive Interactions

Because the experiments employed rather reactive analytes, and given the presence of acid and basic groups in the peptide sequences, we investigated the possibility of acid/base reactions as part of the detection mechanism. We sought to calculate the energy of the following reactions:

$$DLESFLD + NH_3 \rightarrow [DLESFLD]^- + NH_4^+$$

$$DLESFLD + CH_3COOH \rightarrow [DLESFLD]^+ + CH_3COO^-$$

and

$$RVNEWVID + NH_3 \rightarrow [RVNEWVID]^- + NH_4^+$$

$$RVNEWVID + CH_3COOH \rightarrow [RVNEWVID]^+ + CH_3COO^-$$

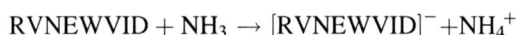

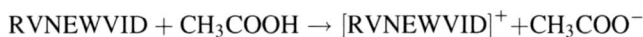

The size of the peptides makes a full quantum mechanical calculation exceed most computer resources and is not necessary. Instead we looked for models that could be employed to represent the full peptide. For example, the non-reactive binding energetics of each peptide/analyte pair were sorted and the strongest binding configurations were chosen for quantum mechanical B3LYP DFT calculations using a local motif, a shorter version of the peptide. For RVNEWVID the N-terminus ARG1-GLU4 motif was found to be most strongly bound to ammonia and acetic acid and for DLESFLD the ASP1-GLU3 motif was selected. These motifs are shown in Fig. 6.8.

Whenever needed, internal constraint coordinates were employed to keep reactants apart while the minimization proceeded to find a local energy minimum geometry, but once the minimized geometry was found, these constraints were set free. The difference in energy between the neutral and the charge/transfer products of the acid/base reactions are reported in the following section for the fully minimized geometries using the same level of theory as that used to compute atomic charges (B3LYP/6-31g**). Figure 6.9 shows the non-reactive (classical, full peptide) and the reactive (quantum motif) analyte/peptide most stable geometries. All quantum calculations were performed using Jaguar quantum program suite [13].

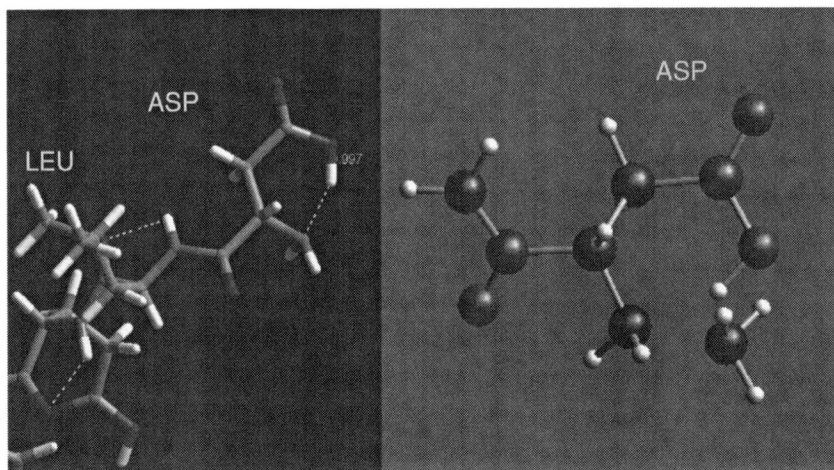

Fig. 6.8 Quantum mechanical reaction energy calculations of DLESFLD (*left figure*) proceed with the use of a shorter N-terminus motif (*right figure*), shown here with the initial un- reacted most stable geometry with ammonia

6.3   Results and Discussion

Table 6.1 presents the non-reactive binding energies and the reaction energies for the various combinations of peptides and analytes. Negative values indicate exothermic favorable interactions.

Figure 6.10 shows the experimental resistivities for the same combinations in the same order. Negative changes in resistivity mean that the charge density on the surface of the SiNW array is such that it increases the conductivity of the field effect transistor. This is dependent on the dipole orientation of the generated charge distribution upon binding. To know precisely this orientation one would need to model an entire monolayer of the peptide on the SiNW surface and predict the molecular axis alignment at the experimental surface coverage values. We have not attempted to look at dipole orientations in the present work. Instead we focus solely on the magnitude of the resistivity changes and not their sign.

We note that the non-reactive binding energies for RVNEWVID are not correlated with the experimental findings. In particular, the favorable interaction with ammonia (-3.14 Kcal/mol) is comparable to the favorable interaction with acetic acid (-3.29 Kcal/mol) indicating that the magnitude of the response should be somewhat the same. However, the RVNEWVID is unresponsive to ammonia.

On the other hand the reaction energetics are well correlated with the resistivity measurements. For example, for RVNEWVID the highly endothermic reaction (15.1 Kcal/mol) with ammonia will indicate that this peptide is unresponsive to the presence of ammonia, while the favorable (-5.54 Kcal/mol) exothermic

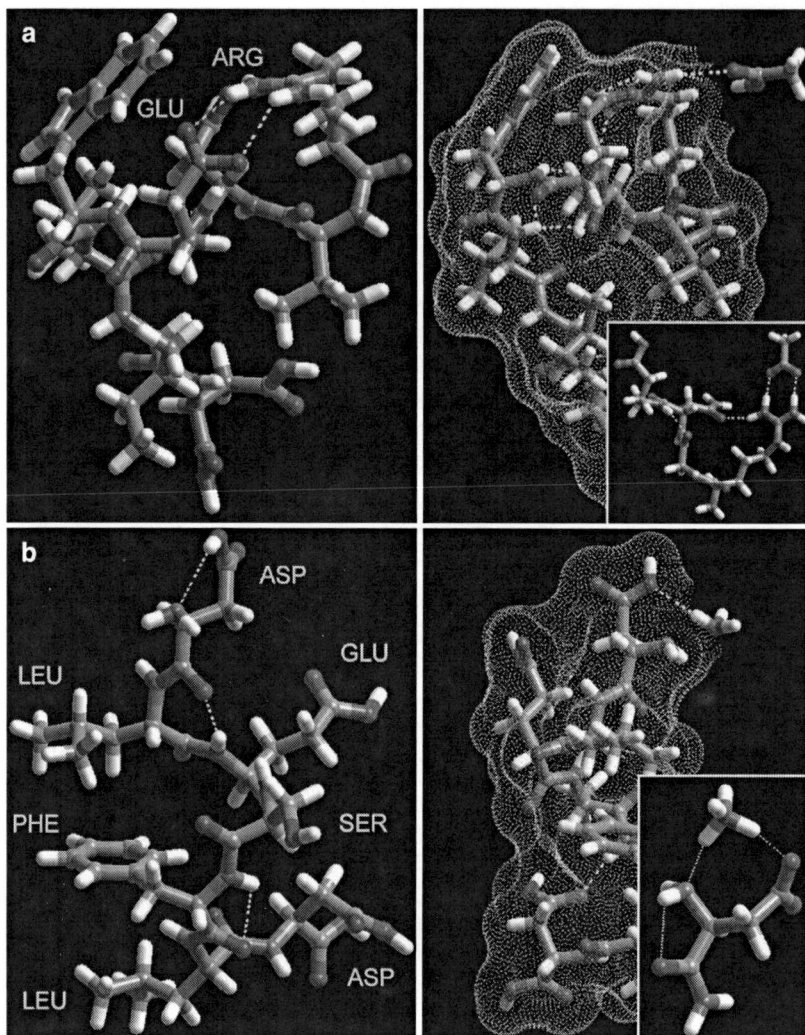

Fig. 6.9 (a) This panel shows the lowest conformation of RVNEWVID peptide, which contains a GLU4-ARG1 salt bridge (*dashed lines*). The right panel shows preferential binding of acetic acid to the ARG1 N-terminus. (*Inset*) The reaction products: neutral GLU4 and protonated ARG1, stabilized by acetate. (**b**) The *left panel* shows the lowest conformation of the DLESFLD peptide. The polar and non-polar amino acids align on opposing sides. The *right panel* shows ammonia binding at the neutral N-terminus ASP1. (*Inset*) Ammonium stabilized by hydrogen bonds to the deprotonated aspartic acid and the N-terminus

reaction with acetic acid will predict good selectivity towards this analyte. These predictions match well with the <1% and >6% changes in SiNW array resistivity for ammonia and acetic acid with RVNEWVID respectively. For DLESFLD a

Table 6.1 Calculated values for non-reactive (average binding energies) and reactive (reaction energies) of analytes with the putative sensing peptides

Analyte peptide	Ammonia DLESFLD	Ammonia RVNEWVID	Acetic Acid DLESFLD	Acetic Acid RVNEWVID
Non-reactive	−8.17	−3.14	−6.79	−3.29
Reactive	−7.4	−15.1	0.42	−5.54

All energies are in Kcal/mol

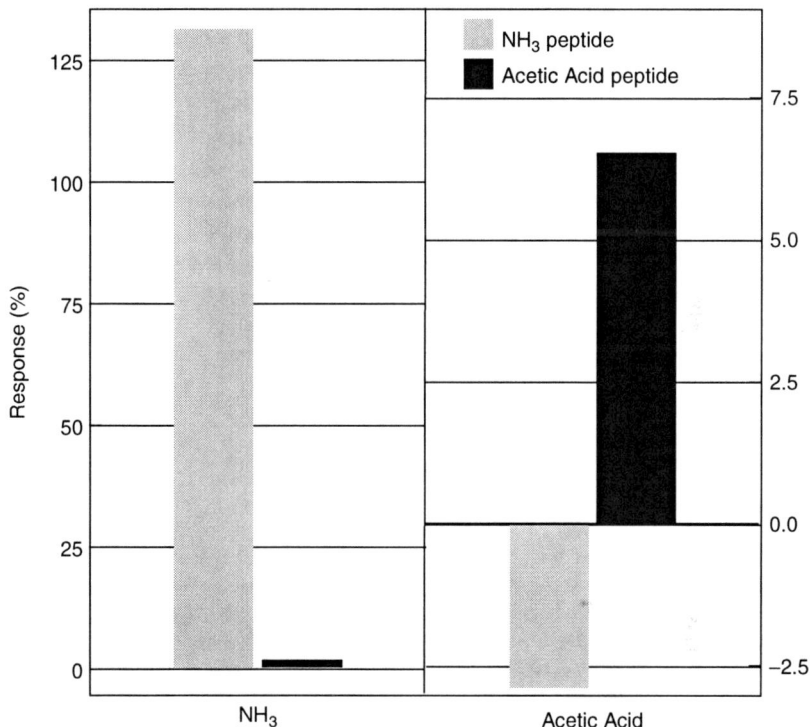

Fig. 6.10 Conductance responses of the peptide-nanowire hybrid sensors, averaged over a 5-min time window of target vapor exposure (starting 10 min after the analyte gas exposure), and normalized to an amine-terminated sensor. The abscissa is labeled with the analyte vapors [2]

favorable reaction energy (−7.4 kcal/mol) with ammonia correlates well with the observed + 125% in resistivity. The slightly endothermic (0.45 Kcal/mol) reaction of DLESFLD with acetic acid leads to 1/3 reacted and 2/3 unreacted peptide at room temperature, predicting a moderate response to acetic acid. The change in resistivity is indeed moderate (−2.5%) as predicted by the acid/base equilibrium.

6.4   Conclusions

Doped semiconducting nanowire arrays offer high sensitivity (ppb) to small compounds in the gas phase due to their field-effect transistor characteristics. Without surface chemical modification these devices are not selective. Surface modifications by peptides offer the opportunity of a large space of possibilities. Given the extraordinary choice of potential peptides, with say, less than twenty aminoacids, as well as the existing chemistry to prepare them with state-of-the-art, fully automatic, solid state peptide synthesis, these hybrid materials could form the basis for a universal sensing platform, capable of adapting to a variety of analytes in a highly automated way. Of course, the search space is too large to be explored by brute force alone, although combinatorial chemistry and high throughput screening could provide an empirical answer. Here we investigate the use of molecular modeling, both using classical mechanics as well as quantum mechanics, to provide an effective sensing material selection procedure. The protocol presented here was capable of yielding theoretical rationalizations of observed chemical selectivity in these nanosensors. It was found that acid/base reactions correlate closely with the observed changes in resistivity. Strategies to identify non-reactive materials capable of high selectivity towards specific analytes are underway.

Acknowledgments This work was partly supported by the Materials and Process Simulation Center, Beckman Institute at the California Institute of Technology.

References

1. McAlpine, M. C.; Ahmad, H.; Wang, D.; Heath, J. R., Highly ordered nanowire arrays on plastic substrates for ultrasensitive flexible chemical sensors, *Nat. Mat.* **2007**, 6, 379–384
2. McAlpine, M. C.; Agnew, H. D.; Rohde, R. D.; Blanco, M.; Ahmad, H.; Stuparu, A. D.; Goddard, W. A.; Heath, J. R., Peptide-nanowire hybrid materials for selective sensing of small molecules, *J. Am. Chem. Soc.* **2008**, 130, 9583–9589
3. Chan, W. C.; White, P. D., Fmoc Solid Phase Peptide Synthesis: A Practical Approach; Oxford University Press, Oxford, **2000**
4. Belmares, M.; Blanco, M.; Goddard, W. A.; Ross, R. B.; Caldwell, G.; Chou, S. H.; Pham, J.; Olofson, P. M.; Thomas, C., Hildebrand and Hansen solubility parameters from molecular dynamics with applications to electronic nose polymer sensors, *J. Comput. Chem.* **2004**, 25, 1814–1826
5. Cozmuta, I.; Blanco, M.; Goddard, W. A., Gas sorption and barrier properties of polymeric membranes from molecular dynamics and Monte Carlo simulations, *J. Phys. Chem. B* **2007**, 111, 3151–3166
6. Wu, T.-Z.; Lo, Y.-R.; Chan, E.-C., Exploring the recognized bio-mimicry materials for gas sensing *Biosens. Bioelectron.* **2001**, 16, 945–953
7. Mayo, S. L.; Olafson, B. D.; Goddard, W. A., Dreiding – A generic force-field for molecular simulations. *J. Phys. Chem.* **1990**, 94, 8897–8909
8. Lee, C. T.; Yang, W. T.; Parr, R. G., Development of the colle-salvetti correlation-energy formula into a functional of the electron-density, *Phys. Rev. B* **1988**, 37, 785–789

Part II
Statistical And Multivariate Methods For Materials Evaluation

Chapter 7
Development of New Sensing Materials Using Combinatorial and High-Throughput Experimentation

Radislav A. Potyrailo and Vladimir M. Mirsky

Abstract New sensors with improved performance characteristics are needed for applications as diverse as bedside continuous monitoring, tracking of environmental pollutants, monitoring of food and water quality, monitoring of chemical processes, and safety in industrial, consumer, and automotive settings. Typical requirements in sensor improvement are selectivity, long-term stability, sensitivity, response time, reversibility, and reproducibility. Design of new sensing materials is the important cornerstone in the effort to develop new sensors. Often, sensing materials are too complex to predict their performance quantitatively in the design stage. Thus, combinatorial and high-throughput experimentation methodologies provide an opportunity to generate new required data to discover new sensing materials and/or to optimize existing material compositions. The goal of this chapter is to provide an overview of the key concepts of experimental development of sensing materials using combinatorial and high-throughput experimentation tools, and to promote additional fruitful interactions between computational scientists and experimentalists.

7.1 Introduction

While basic concepts of design of sensing materials are understood, the quantitative details of materials performance on both, short- and long-term scales remain difficult to predict using existing knowledge. Here are several examples of performance of sensing materials that are, at present, difficult to quantitatively predict:

R.A. Potyrailo (✉)
Chemical and Biological Sensing Laboratory, Materials Analysis and Chemical Sciences, General Electric Company, Global Research Center New York 12309 Niskayuna, USA
e-mail: potyrailo@crd.ge.com

V.M. Mirsky
Department of Nanobiotechnology, Lausitz University of Applied Sciences, Senftenberg, Germany
e-mail: vmirsky@fh-lausitz.de

M.A. Ryan et al. (eds.), *Computational Methods for Sensor Material Selection*, 151
Integrated Analytical Systems,
DOI 10.1007/978-0-387-73715-7_7, © Springer Science+Business Media, LLC 2009

- How will the gas-selectivity pattern change upon doping of mixed metal oxides?
- How will the analyte sensitivity and selectivity of nanowires and nanofibers vary upon aging?
- What is a common solvent for all components for a formulated sensing coating?
- What is the shelf- and operation-lifetime of a polymeric sensing film on a micro- or nano-transducer in the presence or absence of dewetting?
- How will the ion response of optical sensors or ion-selective electrodes be affected by different additives?
- What is the optimal ratio of different polymerizable monomers, cross-linker, and template for synthesis of molecularly imprinted polymers?

Despite essential progress in understanding the physics and chemistry of sensing processes and development of corresponding theoretical approaches, these and many other questions are still being answered by careful experiments. For example, an "extensive systematic study" of more that 500 compositions to optimize vapor sensing polymeric materials was performed by Cammann and co-workers [1]. Walt and co-workers [2] reported screening over 100 polymer candidates in a search for "their ability to serve as sensing matrices" for solvatochromic reagents. Seitz and co-workers [3] have investigated the influence of multicomponent compositions on the properties of pH-swellable polymers by designing $3 \times 3 \times 3 \times 2$ factorial experiments.

Combinatorial and high-throughput materials screening technologies have been introduced to make the search for new materials more intellectually rewarding [4]. Numerous academic groups that were involved in the development of new sensing materials turned to combinatorial methodologies to speed up knowledge discovery [5–11]. From results achieved using combinatorial and high-throughput methods, the most successful have been in the areas of molecular imprinting, polymeric compositions, catalytic metals for field-effect devices, and metal oxides for con-ductometric sensors. In those materials, the desired selectivity and sensitivity have been achieved by the exploration of multidimensional chemical composition and process parameters space at a previously unavailable level of detail, at a fraction of time required for conventional one-at-a-time experiments. These new tools provided the opportunity for the more challenging, yet more rewarding explorations that previously were too time-consuming to pursue.

As the amount and depth of knowledge related to the development of sensing materials are expanding, so are the capabilities to compute certain properties of materials, based on developed empirical structure-function relationships or on advances in molecular modeling. The goal of this chapter is to provide an overview of the key concepts of experimental development of sensing materials using combinatorial and high-throughput experimentation tools and to promote additional fruitful interactions between computational scientists and experimen-talists. A much more detailed treatise of the development of new sensing materials using combinatorial and high-throughput experimentation is provided in the book "Combinatorial methods for chemical and biological sensors", included in this book series [12].

7.2   Basics of Combinatorial and High-Throughput Screening of Sensing Materials

Sensing materials can be categorized into three general groups that include inorganic, organic, and biological materials. We define inorganic sensing materials as materials that have inorganic receptor- or signal-generation components (e.g., metals, metal oxides, semiconductor nanocrystals) that may or may not be further incorporated into a matrix. Organic sensing materials include molecular artificial receptors, chemo-sensitive indicator dyes, polymer/reagent compositions, conjugated polymers and molecularly imprinted polymers. Biological sensing materials include such receptors as aptamers, peptides, antibodies, enzymes, etc. Rational design of sensing materials based on prior knowledge is a very attractive approach because it could avoid time-consuming synthesis and testing of numerous materials candidates [13–15].

However, to provide quantitative prediction, rational design [16–21] requires detailed knowledge regarding not only the relation of intrinsic properties of sensing materials to their performance properties (e.g., affinity to an analyte and interferences, kinetic constants of analyte binding and dissociation, long term stability, shelf life, resistance to poisoning, best operation temperature, decrease of reagent performance after immobilization, etc.) but also regarding changes of these properties due to interaction of different components of sensing materials with one another. This knowledge is often obtained from extensive experiments which are difficult to perform using traditional approaches or from theoretical models which should also be validated in extensive experiments. Conventionally, detailed experimentation with sensing materials candidates for their screening and optimization consumes tremendous amount of time without adding to "intellectual satisfaction". Fortunately, new synthetic and measurement principles and instrumentation significantly accelerate the development of new materials. The practical challenges in rational design of sensor materials also provide tremendous prospects for combinatorial materials research.

In materials development including sensing materials, the materials' properties depend not only on composition, but also on morphology, microstructure and other parameters related to the material-preparation conditions, and on the end-use environment. As a result of this complexity, true combinatorial experimentation is rarely performed with a complete set of materials and process variables rarely explored. Instead, carefully selected subsets of the parameters are often explored in an automated parallel or rapid sequential fashion using high-throughput experimentation (HTE). Nevertheless, the terms "combinatorial chemistry" and "combinatorial materials science" are often applied for all types of automated parallel and rapid sequential materials and process parameters evaluation processes. Thus, an adequate definition of combinatorial and high-throughput materials science is a process that couples the capability for parallel production of large arrays of diverse materials together with different high-throughput measurement techniques for various intrinsic and performance properties followed by navigation in the collected data for identifying "lead" materials [22–30].

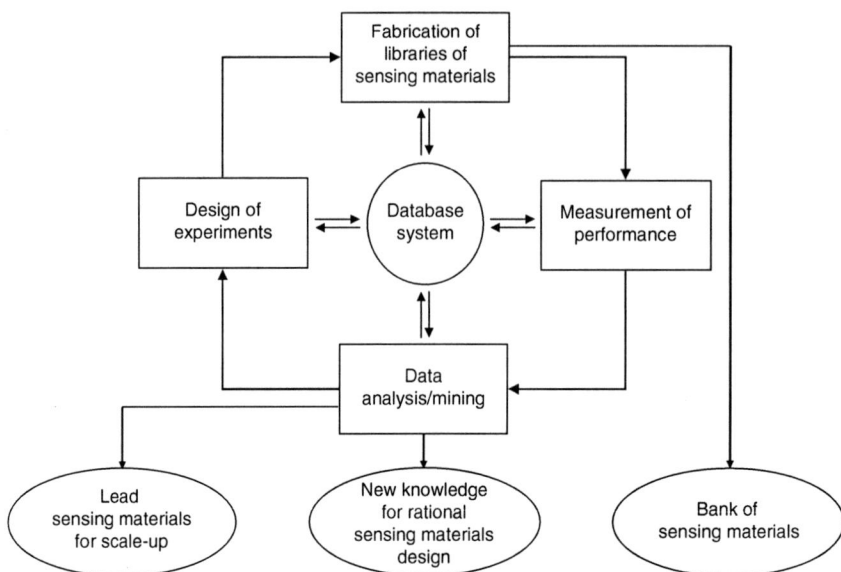

Fig. 7.1 Typical cycle in combinatorial and high-throughput approach for sensing materials development

A typical combinatorial materials development cycle is outlined in Fig. 7.1. This workflow has several important aspects such as planning of experiments, data mining, and scale up. In combinatorial screening of materials, concepts originally thought to be highly automated have been recently refined to have more human input, with only an appropriate level of automation. For the throughput of 50–100 materials formulations per day, it is acceptable to perform certain aspects of the process manually [31, 32]. To address numerous materials-specific properties, a variety of high-throughput characterization tools are required. Characterization tools are used for rapid and automated assessment of single or multiple properties of the large number of samples fabricated together as a combinatorial array or "library" [26, 33, 34].

In addition to the parallel synthesis and high-throughput characterization instrumentation that significantly differ from conventional equipment, the data management approaches also differ from conventional data evaluation [30]. In an ideal combinatorial workflow, one should "analyze in a day what is made in a day" [35]. This requires significant computational assistance. In the combinatorial workflow, design and syntheses protocols for materials libraries are computer assisted, materials synthesis and library preparation are carried out with computer-controlled manipulators, and property screening and materials characterization are also software controlled. Further, materials synthesis data as well as property and characterization data are collected into a materials database. This database contains information on starting components, their descriptors, process conditions, materials testing algorithms, and performance properties of libraries of sensing materials.

Table 7.1 Parameters relevant to sensor applications, which are difficult or impossible to quantitatively predict and calculate using existing knowledge and rational approaches

Group of sensing materials	Parameter	Ref.
Inorganic	Fundamental effects of volume dopants on base metal oxide materials	[44]
Inorganic	Hot spots in surface-enhanced Raman scattering	[45]
Inorganic	Room-temperature gas response of nanomaterials	[46]
Organic	Selectivity of prospective ionophores	[38]
Organic	Design of polymers for selective complexation with desired metal ions	[37]
Organic	Collective effects of preparation method of conducting polymers, polymer morphology, properties of the substrate/film interface	[41]
Organic	Surface ratio of different types of molecules in mixed self-assembled monolayers formed in quasi-equilibrium conditions	[47]
Biological	Oligonucleotide sequence in aptamers for specific and sensitive target binding	[48]
Biological	Activity of immobilized bioreceptors	[7]
Biological	Selection of chemical coupling reagents for immobilization of antibodies	[49]
Biological	Long-term stability of enzyme-containing polymer films	[42]
Biological	Affinity properties of antibodies (kinetic association and dissociation constants, binding constant)	[50]

Data in such a database is not just stored, but is also processed with the proper statistical analysis, visualization, modeling, and data-mining tools. Combinatorial synthesis of materials provides a good opportunity for formation of banks of combinatorial materials. Such banks can be used, for example, for the further investigation of materials of interest for some new applications or as reference materials. However, this approach has not yet found wide application in combinatorial materials science.

It is difficult to predict quantitatively, numerous sensor material parameters using rational approaches [7, 8, 36–43]. Table 7.1 provides examples of parameters relevant to sensor applications, which are difficult or impossible to quantitatively predict and calculate using existing knowledge and rational approaches. Relevant descriptors must be determined to understand the details of sensor materials design and to establish quantitative structure-function relationships. While the application of combinatorial and high-throughput screening tools can accelerate this process, the number of purely serendipitous combinations of sensor materials components is simply too large to handle even with the "ultra-high-throughput screening" in a time- and cost-effective manner. It is not feasible to synthesize all possible molecules and apply all possible process conditions to characterize materials function. In this case, a methodology that allows the most promising candidates to be shortlisted should be applied. In materials science, focused libraries are often designed, produced, and tested in a high-throughput mode for a subset of a truly "combinatorial" space where the initial subset selection is performed using rational and intuitive approaches.

In drug discovery, a binding or inhibition test is performed as a secondary screening with a much while a functional investigation of biological effects is performed as a secondary screening with a much smaller library of selected

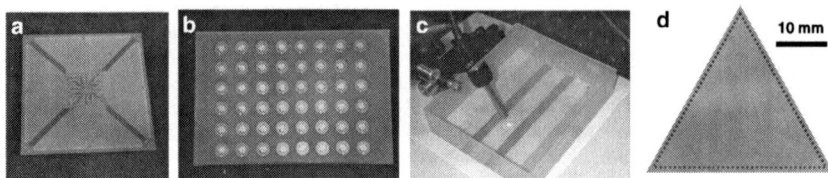

Fig. 7.2 Examples of discrete (**a, b**) and gradient (**c, d**) sensing materials arrays: (**a**) Electro-polymerized conductive polymers on a 96-element interdigital electrode array; (**b**) Formulated solid polymer electrolyte compositions on a 48-element array of radio-frequency identification sensors; (**c**) Gradient thickness sensing films with three concentrations of a colorimetric indicator; (**d**) Ternary combination of gradient concentrations of fluorescent reagents in a formulated optical sensor film

compounds; this replacement of biological experiments by biochemical ones allows minimization of process costs. In the case of materials screening for development of sensor materials, a primary binding assay is typical only in the development of biological sensing materials. This is probably the only way because biological sensing materials possess only receptor properties and require a separate transducing system. Many organic and inorganic materials (except artificial monomolecular receptors and molecularly imprinted polymers) have intrinsic transducing properties; therefore, a functional screening including a characterization of both receptor and transducer properties is performed as the primary screening.

Variations of individual parameters of sensing materials result in a modification of a number of response parameters. Thus, the goal of combinatorial and high-throughput development of sensing materials is to determine the structure-function relationships in sensing materials. In combinatorial screening, discrete [51, 52] and gradient [53–57] arrays (libraries) of sensing materials are employed. Selected examples of discrete and gradient sensing materials arrays are presented in Fig. 7.2. A specific type of library layout will depend on the required density of space to be explored, available library-fabrication capabilities, and capabilities of high-throughput characterization techniques.

Discrete or gradient sensor regions are attractive for a variety of different specific applications and can be produced with a variety of tools. Discrete sensor regions can be produced using a variety of approaches, for example chemical vapor deposition, pulsed-laser deposition, dip coating, spin coating, electropolymerization, slurry dispensing, liquid dispensing, screen printing, and others. Sensor material optimization can be performed using gradient sensor materials. Spatial gradients in sensing materials can be generated by varying the nature and concentration of starting components, processing conditions, thickness, and so on. Gradient sensor regions can be produced using a variety of approaches, for example, in situ photopolymerization, microextrusion, solvent casting, self-assembly, temperature-gradient chemical vapor deposition, thickness-gradient chemical vapor deposition, 2-D thickness gradient evaporation of metals, gradient surface coverage and gradient particle size, and some others. Once the gradient or discrete sensing materials array is fabricated, it is exposed to an environment of interest, and steady-state or dynamic measurements are acquired. Serial scanning mode of analysis (e.g., optical or impedance spectroscopies), rather than parallel analysis (e.g., imaging) is often performed to provide more detailed information about the materials' property.

7.3 Diversity in Needs for Combinatorial Development of Sensing Materials

A typical sensing material is based on a successful combination of two somewhat contradictory material requirements. The first requirement is to have the desired material response to the changes in the concentration of species of interest in a sample. It is desirable for most applications that the response is fast and selective; this entails that the kinetic association constant and the binding constant (the ratio of the kinetic association and kinetic dissociation constants) should be high. The second requirement is to have a reversible sensor response; therefore, the kinetic dissociation constant should also be high. Thus, development of sensing materials needs knowledge on how to fulfill these controversial requirements or to find a suitable compromise.

For diverse applications, specifications on sensing materials are often weighted differently according to the application. High reliability, adequate long-term stability, and resolution top the priority list for industrial sensor users, while often the size and maturity of the technology are the least important factors [58–60]. The low false positive rate is very critical for the first responders [61]. In contrast, medical users focus on cost for disposable sensors. Specific requirements for medical in vivo sensors include blood compatibility and minute size [62]. Resistance to gamma radiation during sterilization, drift-free performance, and cost are the most critical specific requirements for sensors in disposable bioprocess components [63]. The importance of continuous monitoring also differs from application to application. For instance, glucose sensing should be performed 2–4 times a day using home blood glucose biosensors, while blood-gas sensors for use in intensive care should be capable of continuous monitoring with sub-second time resolution [64, 65].

Table 7.2 illustrates parameters that should be optimized and controlled for successful development of inorganic, organic, and biological sensing materials. Inorganic sensing materials include catalytic metals for field-effect devices [5, 66, 67], metal oxides for conductometric [9, 10, 44, 68–82] and cataluminescent [83, 84] sensors, plasmonic [85–93], and semiconductor nanocrystal [94–96] materials. Organic sensing materials include indicator dyes (free, polymer immobilized, and surface-confined) [97–111], polymeric compositions [6, 8, 36, 112–114], homo- and copolymers [115–121], conjugated polymers [11, 122–125], and molecularly-imprinted polymers [47, 126–129]. Biological materials include surface- and matrix-immobilized bioreceptors [7, 122, 130].

Table 7.2 Parameters that should be optimized and controlled for successful development of inorganic, organic, and biological sensing materials

Group of sensing materials	Type of sensing material	Optimized and controlled material parameters
Inorganic	Catalytic metals	Surface additives Porosity Layered structure Grain size Alloying Deposition method
Inorganic	Metal oxide materials	Base single or mixed metal oxides Deposition method and conditions of base metal oxide(s) Annealing method and conditions Dopant(s) Doping method and conditions Purity of materials
Inorganic	Plasmonic nanostructures and nanoparticles	Substrate type Nanoparticle material Nanoparticle shape, size, morphology Nanoparticles arrangement Surface functionality
Inorganic	Plasmonic nanoparticles in polymers	Size of nanoparticle Strength of polymer/particle interaction Polymer grafting density Polymer chain length
Organic	Indicators	Binding constant pH-influence Redox state Selectivity Toxicity Poisoning agents
Organic	Polymeric compositions	Analyte- responsive reagent Polymer matrix Analyte specific ligand Plasticizer Other agents (stabilizing, phase transfer, etc.) Common solvent
Organic	Conjugated polymers	Polymerization conditions Types of heterocycles Additive(s) Side groups Dopant Oxidation state Electrode material Thickness Morphology

(*continued*)

Table 7.2 (continued)

Group of sensing materials	Type of sensing material	Optimized and controlled material parameters
Organic	Molecularly imprinted polymers	Functional monomer(s) Template concentration Cross-linker Porogen Monomer(s)/template ratio Physical conditions during polymerization
Biological	Surface-immobilized bioreceptors	Immobilization technique Receptor-surface spacer Receptor surface density Material between bioreceptors
Biological	Matrix-immobilized bioreceptors	Immobilization technique Receptor density Matrix hydrophilicity Matrix charge Matrix chemical content Matrix thickness

7.4 Outlook

The future advances in combinatorial development of sensing materials will be related to several key remaining unmet needs that prevent researchers from having a complete combinatorial workflow and to "analyze in a day what is made in a day" [35].

First, new fabrication methods of combinatorial libraries of sensing materials will be implemented including those adapted from other materials synthesis and fabrication approaches [131, 132] and those developed specifically for sensing applications [133].

Second, new screening tools will be developed for high-throughput characterization of intrinsic materials' properties in order to keep up with the rates of performance screening of sensing materials candidates.

Third, applications of data mining in sensing materials [118, 134–136] will expand. Designing the high-throughput experiments to discover relevant descriptors will become more attractive [137].

Fourth, predictive models of behavior of sensing materials under realistic conditions over long periods of time will be proposed. These modeling efforts will require inputs not only from screening of the performance and intrinsic properties of sensing materials but also from screening of the effects of interfaces between sensing materials and transducers.

References

1. Buhlmann, K.; Schlatt, B.; Cammann, K.; Shulga, A., Plasticised polymeric electrolytes: New extremely versatile receptor materials for gas sensors (VOC monitoring) and electronic noses (odour identification:discrimination), *Sens. Actuators B*. **1998**, 49, 156–165

2. Walt, D. R.; Dickinson, T.; White, J.; Kauer, J.; Johnson, S.; Engelhardt, H.; Sutter, J.; Jurs, P., Optical sensor arrays for odor recognition, *Biosens. Bioelectron*. **1998**, 13, 697–699

3. Conway, V. L.; Hassen, K. P.; Zhang, L.; Seitz, W. R.; Gross, T. S., The influence of composition on the properties of pH-swellable polymers for chemical sensors, *Sens. Actuators B*. **1997**, 45, 1–9

4. Potyrailo, R. A.; Mirsky, V. M., Combinatorial and high-throughput development of sensing materials: The first ten years, *Chem. Rev*. **2008**, 108, 770–813

5. Lundström, I.; Sundgren, H.; Winquist, F.; Eriksson, M.; Krantz-Rülcker, C.; Lloyd-Spetz, A., Twenty-five years of field effect gas sensor research in Linköping, *Sens. Actuators B*. **2007**, 121, 247–262

6. Dickinson, T. A.; Walt, D. R.; White, J.; Kauer, J. S., Generating sensor diversity through combinatorial polymer synthesis, *Anal. Chem*. **1997**, 69, 3413–3418

7. Cho, E. J.; Tao, Z.; Tang, Y.; Tehan, E. C.; Bright, F. V.; Hicks, W. L., Jr.; Gardella, J. A., Jr.; Hard, R., Tools to rapidly produce and screen biodegradable polymer and sol-gel-derived xerogel formulations, *Appl. Spectrosc*. **2002**, 56, 1385–1389

8. Apostolidis, A.; Klimant, I.; Andrzejewski, D.; Wolfbeis, O. S., A combinatorial approach for development of materials for optical sensing of gases, *J. Comb. Chem*. **2004**, 6, 325–331

9. Simon, U.; Sanders, D.; Jockel, J.; Heppel, C.; Brinz, T., Design strategies for multielectrode arrays applicable for high-throughput impedance spectroscopy on novel gas sensor materials, *J. Comb. Chem*. **2002**, 4, 511–515

10. Frantzen, A.; Scheidtmann, J.; Frenzer, G.; Maier, W. F.; Jockel, J.; Brinz, T.; Sanders, D.; Simon, U., High-throughput method for the impedance spectroscopic characterization of resistive gas sensors, *Angew. Chem. Int. Ed*. **2004**, 43, 752–754

11. Mirsky, V. M.; Kulikov, V.; Hao, Q.; Wolfbeis, O. S., Multiparameter high-throughput characterization of combinatorial chemical microarrays of chemosensitive polymers, *Macromol. Rapid Commun*. **2004**, 25, 253–258

12. Potyrailo, R. A.; Mirsky, V. M., Eds., Combinatorial Methods for Chemical and Biological Sensors; Springer, New York, NY, **2009**

13. Njagi, J.; Warner, J.; Andreescu, S., A bioanalytical chemistry experiment for undergraduate students: Biosensors based on metal nanoparticles, *J. Chem. Educ*. **2007**, 84, 1180–1182

14. Shtoyko, T.; Zudans, I.; Seliskar, C. J.; Heineman, W. R.; Richardson, J. N., An attenuated total reflectance sensor for copper: An experiment for analytical or physical chemistry, *J. Chem. Educ*. **2004**, 81, 1617–1619

15. Honeybourne, C. L., Organic vapor sensors for food quality assessment, *J. Chem. Educ*. **2000**, 77, 338–344

16. Newnham, R. E., Structure-property relationships in sensors, *Cryst. Rev*. **1988**, 1, 253–280

17. Akporiaye, D. E., Towards a rational synthesis of large-pore zeolite-type materials? *Angew. Chem. Int. Ed*. **1998**, 37, 2456–2457

18. Ulmer II, C. W.; Smith, D. A.; Sumpter, B. G.; Noid, D. I., Computational neural networks and the rational design of polymeric materials: The next generation polycarbonates, *Comput. Theor. Polym. Sci*. **1998**, 8, 311–321

19. Suman, M.; Freddi, M.; Massera, C.; Ugozzoli, F.; Dalcanale, E., Rational design of cavitand receptors for mass sensors, *J. Am. Chem. Soc*. **2003**, 125, 12068–12069

20. Lavigne, J. J.; Anslyn, E. V., Sensing a paradigm shift in the field of molecular recognition: From selective to differential receptors, *Angew. Chem. Int. Ed*. **2001**, 40, 3119–3130

21. Hatchett, D. W.; Josowicz, M., Composites of intrinsically conducting polymers as sensing nanomaterials, *Chem. Rev*. **2008**, 108, 746–769

22. Jandeleit, B.; Schaefer, D. J.; Powers, T. S.; Turner, H. W.; Weinberg, W. H., Combinatorial materials science and catalysis, *Angew. Chem. Int. Ed.* **1999**, 38, 2494–2532
23. Maier, W.; Kirsten, G.; Orschel, M.; Weiß, P.-A.; Holzwarth, A.; Klein, J., Combinatorial chemistry of materials, polymers, and catalysts, In Combinatorial Approaches to Materials Development; Malhotra, R., Ed.; American Chemical Society, Washington, DC, **2002**, Vol. 814, 1–21
24. Takeuchi, I.; Newsam, J. M.; Wille, L. T.; Koinuma, H.; Amis, E. J., Eds., Combinatorial and Artificial Intelligence Methods in Materials Science; Materials Research Society, Warrendale, PA, **2002**, Vol. 700
25. Xiang, X.-D.; Takeuchi, I., Eds., Combinatorial Materials Synthesis; Marcel Dekker, New York, NY, **2003**
26. Potyrailo, R. A.; Amis, E. J., Eds., High-throughput Analysis: A Tool for Combinatorial Materials Science; Kluwer Academic/Plenum Publishers, New York, NY, **2003**
27. Koinuma, H.; Takeuchi, I., Combinatorial solid state chemistry of inorganic materials, *Nat. Mater.* **2004**, 3, 429–438
28. Potyrailo, R. A.; Karim, A.; Wang, Q.; Chikyow, T., Eds., Combinatorial and Artificial Intelligence Methods in Materials Science II; Materials Research Society, Warrendale, PA, **2004**, Vol. 804
29. Potyrailo, R. A.; Takeuchi, I., Eds., Special Feature on Combinatorial and High-Throughput Materials Research; Measurement Science Technology. **2005**, Vol. 16, 316
30. Potyrailo, R. A.; Maier, W. F., Eds., Combinatorial and High-Throughput Discovery and Optimization of Catalysts and Materials; CRC Press, Boca Raton, FL, **2006**
31. Potyrailo, R. A.; Chisholm, B. J.; Olson, D. R.; Brennan, M. J.; Molaison, C. A., Development of combinatorial chemistry methods for coatings: High-throughput screening of abrasion resistance of coatings libraries, *Anal. Chem.* **2002**, 74, 5105–5111
32. Potyrailo, R. A.; Chisholm, B. J.; Morris, W. G.; Cawse, J. N.; Flanagan, W. P.; Hassib, L.; Molaison, C. A.; Ezbiansky, K.; Medford, G.; Reitz, H., Development of combinatorial chemistry methods for coatings: High-throughput adhesion evaluation and scale-up of combinatorial leads, *J. Comb. Chem.* **2003**, 5, 472–478
33. MacLean, D.; Baldwin, J. J.; Ivanov, V. T.; Kato, Y.; Shaw, A.; Schneider, P.; Gordon, E. M., Glossary of terms used in combinatorial chemistry, *J. Comb. Chem.* **2000**, 2, 562–578
34. Potyrailo, R. A.; Takeuchi, I., Role of high-throughput characterization tools in combinatorial materials science, *Meas. Sci. Technol.* **2005**, 16, 1–4
35. Cohan, P. E., Combinatorial materials science applied – mini case studies, lessons and strategies, In 2002 Combi – The 4th Annual International Symposium on Combinatorial Approaches For New Materials Discovery, Knowledge Foundation, Arlington, VA, **2002**
36. Matzger, A. J.; Lawrence, C. E.; Grubbs, R. H.; Lewis, N. S., Combinatorial approaches to the synthesis of vapor detector arrays for use in an electronic nose, *J. Comb. Chem.* **2000**, 2, 301–304
37. Lu, Y.; Liu, J.; Li, J.; Bruesehoff, P. J.; Pavot, C. M.-B.; Brown, A. K., New highly sensitive and selective catalytic DNA biosensors for metal ions, *Biosens. Bioelectron.* **2003**, 18, 529–540
38. Bakker, E.; Bühlmann, P.; Pretsch, E., Carrier-based ion-selective electrodes and bulk optodes. 1. General characteristics, *Chem. Rev.* **1997**, 97, 3083–3132
39. Deans, R.; Kim, J.; Machacek, M. R.; Swager, T. M., A poly(p-phenyleneethynylene) with a highly emissive aggregated phase, *J. Am. Chem. Soc.* **2000**, 122, 8565–8566
40. Lavastre, O.; Illitchev, I.; Jegou, G.; Dixneuf, P. H., Discovery of new fluorescent materials from fast synthesis and screening of conjugated polymers, *J. Am. Chem. Soc.* **2002**, 124, 5278–5279
41. Janata, J.; Josowicz, M., Conducting polymers in electronic chemical sensors, *Nat. Mater.* **2002**, 2, 19–24
42. Rege, K.; Raravikar, N. R.; Kim, D.-Y.; Schadler, L. S.; Ajayan, P. M.; Dordick, J. S., Enzyme-polymer-single walled carbon nanotube composites as biocatalytic films, *Nano Lett.* **2003**, 3, 829–832

43. Zhou, Y.; Freitag, M.; Hone, J.; Staii, C.; Johnson, A. T.; Pinto, N. J.; MacDiarmid, A. G., Fabrication and electrical characterization of polyaniline-based nanofibers with diameter below 30 nm, *Appl. Phys. Lett.* **2003**, 83, 3800–3802

44. Siemons, M.; Koplin, T. J.; Simon, U., Advances in high-throughput screening of gas sensing materials, *Appl. Surf. Sci.* **2007**, 254, 669–676

45. Qin, L.; Zou, S.; Xue, C.; Atkinson, A.; Schatz, G. C.; Mirkin, C. A., Designing, fabricating, and imaging Raman hot spots, Proc. Natl. Acad. Sci. USA. **2006**, 103, 13300–13303

46. Paulose, M.; Varghese, O. K.; Mor, G. K.; Grimes, C. A.; Ong, K. G., Unprecedented ultra-high hydrogen gas sensitivity in undoped titania nanotubes, *Nanotechnology.* **2006**, 17, 398–402

47. Hirsch, T.; Kettenberger, H.; Wolfbeis, O. S.; Mirsky, V. M., A simple strategy for preparation of sensor arrays: Molecularly structured monolayers as recognition elements, *Chem. Commun.* **2003**, 3, 432–433

48. Hermann, T.; Patel, D. J., Adaptive recognition by nucleic acid aptamers, *Science.* **2000**, 287, 820–825

49. Mirsky, V. M.; Riepl, M.; Wolfbeis, O. S., Capacitive monitoring of protein immobilization and antigen-antibody reaction on the mono-molecular films of alkylthiols adsorbed on gold electrodes, *Biosens. Bioelectron.* **1997**, 12, 977–989

50. Kramer, K.; Hock, B., Antibodies for biosensors, In Ultrathin Electrochemical Chemo- and Biosensors; Mirsky, V. M., Ed.; Springer: Berlin, Germany, **2004**

51. Birina, G. A.; Boitsov, K. A., Experimental use of combinational and factorial plans for optimizing the compositions of electronic materials, *Zavodskaya Laboratoriya (in Russian).* **1974**, 40, 855–857

52. Xiang, X.-D.; Sun, X.; Briceño, G.; Lou, Y.; Wang, K.-A.; Chang, H.; Wallace-Freedman, W. G.; Chen, S.-W.; Schultz, P. G., A combinatorial approach to materials discovery, *Science.* **1995**, 268, 1738–1740

53. Kennedy, K.; Stefansky, T.; Davy, G.; Zackay, V. F.; Parker, E. R., Rapid method for determining ternary-alloy phase diagrams, *J. Appl. Phys.* **1965**, 36, 3808–3810

54. Hanak, J. J., The "multiple-sample concept" in materials research: Synthesis, compositional analysis and testing of entire multicomponent systems, *J. Mater. Sci.* **1970**, 5, 964–971

55. Bever, M. B.; Duwez, P. E., Gradients in composite materials, *Mater. Sci. Eng.* **1972**, 10, 1–8

56. Shen, M.; Bever, M. B., Gradients in polymeric materials, *J. Mater. Sci.* **1972**, 7, 741–746

57. Pompe, W.; Worch, H.; Epple, M.; Friess, W.; Gelinsky, M.; Greil, P.; Hempel, U.; Scharnweber, D.; Schulte, K., Functionally graded materials for biomedical applications, *Mater. Sci. Eng. A.* **2003**, 362, 40–60

58. Hirschfeld, T.; Callis, J. B.; Kowalski, B. R., Chemical sensing in process analysis, *Science.* **1984**, 226, 312–318

59. Wolfbeis, O. S., Ed., Fiber Optic Chemical Sensors and Biosensors; CRC Press, Boca Raton, FL, **1991**

60. Taylor, R. F.; Schultz, J. S., Ed., Handbook of Chemical and Biological Sensors; IOP Publishing, Bristol, UK, **1996**

61. Carrano, J. C.; Jeys, T.; Cousins, D.; Eversole, J.; Gillespie, J.; Healy, D.; Licata, N.; Loerop, W.; O 'Keefe, M.; Samuels, A.; Schultz, J.; Walter, M.; Wong, N.; Billotte, B.; Munley, M.; Reich, E.; Roos, J., Chemical and biological sensor standards study (CBS3), In Optically Based Biological and Chemical Sensing for Defence; Carrano, J. C.; Zukauskas, A. Eds.; SPIE – The International Society for Optical Engineering: Bellingham, WA, **2004**, Vol. 5617, xi–xiii

62. Meyerhoff, M. E., In vivo blood-gas and electrolyte sensors: Progress and challenges, *Trends Anal. Chem.* **1993**, 12, 257–266

63. Clark, K. J. R.; Furey, J., Suitability of selected single-use process monitoring and control technology, *BioProcess Int.* **2006**, 4, S16–S20

64. Newman, J. D.; Turner, A. P. F., Home blood glucose biosensors: A commercial perspective, *Biosens. Bioelectron.* **2005**, 20, 2435–2453

65. Pickup, J. C.; Alcock, S., Clinicians' requirements for chemical sensors for in vivo monitoring: A multinational survey, *Biosens. Bioelectron.* **1991**, 6, 639–646

66. Klingvall, R.; Lundström, I.; Löfdahl, M.; Eriksson, M., A combinatorial approach for field-effect gas sensor research and development, *IEEE Sens. J.* **2005**, 5, 995–1003

67. Eriksson, M.; Klingvall, R.; Lundström, I., A combinatorial method for optimization of materials for gas sensitive field-effect devices, In Combinatorial and High-Throughput Discovery and Optimization of Catalysts and Materials; Potyrailo, R. A.; Maier, W. F., Eds.; CRC Press: Boca Raton, FL, **2006**, 85–95

68. Sysoev, V. V.; Kiselev, I.; Frietsch, M.; Goschnick, J., Temperature gradient effect on gas discrimination power of a metal-oxide thin-film sensor microarray, *Sensors.* **2004**, 4, 37–46

69. Goschnick, J.; Koronczi, I.; Frietsch, M.; Kiselev, I., Water pollution recognition with the electronic nose KAMINA, *Sens. Actuators B.* **2005**, 106, 182–186

70. Mazza, T.; Barborini, E.; Kholmanov, I. N.; Piseri, P.; Bongiorno, G.; Vinati, S.; Milania, P.; Ducati, C.; Cattaneo, D.; Li Bassi, A.; Bottani, C. E.; Taurino, A. M.; Siciliano, P., Libraries of cluster-assembled titania films for chemical sensing, *Appl. Phys. Lett.* **2005**, 87, 103108

71. Korotcenkov, G., Gas response control through structural and chemical modification of metal oxide films: State of the art and approaches, *Sens. Actuators B.* **2005**, 107, 209–232

72. Franke, M. E.; Koplin, T. J.; Simon, U., Metal and metal oxide nanoparticles in chemiresistors: Does the nanoscale matter? *Small* **2006**, 2, 36–50

73. Barsan, N.; Koziej, D.; Weimar, U., Metal oxide-based gas sensor research: How to? *Sens. Actuators B* **2007**, 121, 18–35

74. Taylor, C. J.; Semancik, S., Use of microhotplate arrays as microdeposition substrates for materials exploration, *Chem. Mater.* **2002**, 14, 1671–1677

75. Semancik, S. Correlation of chemisorption and electronic effects for metal oxide interfaces: Transducing principles for temperature programmed gas microsensors. Final technical report project number: EMSP 65421, grant number: 07-98ER62709; US Department of Energy Information Bridge: 2002, http://www.osti.gov/bridge/

76. Semancik, S., Temperature-dependent materials research with micromachined array platforms, In Combinatorial Materials Synthesis; Xiang, X.-D.; Takeuchi, I., Eds.; Marcel Dekker, New York, NY, **2003**, 263–295

77. Aronova, M. A.; Chang, K. S.; Takeuchi, I.; Jabs, H.; Westerheim, D.; Gonzalez-Martin, A.; Kim, J.; Lewis, B., Combinatorial libraries of semiconductor gas sensors as inorganic electronic noses, *Appl. Phys. Lett.* **2003**, 83, 1255–1257

78. Simon, U.; Sanders, D.; Jockel, J.; Brinz, T., Setup for high-throughput impedance screening of gas-sensing materials, *J. Comb. Chem.* **2005**, 7, 682–687

79. Sanders, D.; Simon, U., High-throughput gas sensing screening of surface-doped In_2O_3, *J. Comb. Chem.* **2007**, 9, 53–61

80. Siemons, M.; Simon, U., Preparation and gas sensing properties of nanocrystalline la-doped $CoTiO_3$, *Sens. Actuators B.* **2006**, 120, 110–118

81. Siemons, M.; Simon, U., Gas sensing properties of volume-doped $CoTiO_3$ synthesized via polyol method, *Sens. Actuators B.* **2007**, 126, 595–603

82. Siemons, M.; Simon, U., High-throughput screening of the propylene and ethanol sensing properties of rare-earth orthoferrites and orthochromites, *Sens. Actuators B.* **2007**, 126, 181–186

83. Nakagawa, M.; Okabayashi, T.; Fujimoto, T.; Utsunomiya, K.; Yamamoto, I.; Wada, T.; Yamashita, Y.; Yamashita, N., A new method for recognizing organic vapor by spectroscopic image on cataluminescence-based gas sensor, *Sens. Actuators B.* **1998**, 51, 159–162

84. Nakagawa, M.; Yamashita, N., Cataluminescence-based gas sensors, *Springer Ser. Chem. Sens. Biosens.* **2005**, 3, 93–132

85. Baker, B. E.; Kline, N. J.; Treado, P. J.; Natan, M. J., Solution-based assembly of metal surfaces by combinatorial methods, *J. Am. Chem. Soc.* **1996**, 118, 8721–8722

86. Kahl, M.; Voges, E.; Kostrewa, S.; Viets, C.; Hill, W., Periodically structured metallic substrates for SERS, *Sens. Actuators B.* **1998**, 51, 285–291

87. Han, M. S.; Lytton-Jean, A. K. R.; Oh, B.-K.; Heo, J.; Mirkin, C. A., Colorimetric screening of DNA-binding molecules with gold nanoparticle probes, *Angew. Chem. Int. Ed.* **2006**, 45, 1807–1810

88. Dovidenko, K.; Potyrailo, R. A.; Grande, J., Focused ion beam microscope as an analytical tool for nanoscale characterization of gradient-formulated polymeric sensor materials, In Combinatorial Methods and Informatics in Materials Science. Materials Research Society Symposium Proceedings; Fasolka, M.; Wang, Q.; Potyrailo, R. A.; Chikyow, T.; Schubert, U. S.; Korkin, A., Eds.; Materials Research Society, Warrendale, PA, **2006**, Vol. 894, 231–236

89. Bhat, R. R.; Genzer, J., Combinatorial study of nanoparticle dispersion in surface-grafted macromolecular gradients, *Appl. Surf. Sci.* **2005**, 252, 2549–2554

90. Bhat, R. R.; Tomlinson, M. R.; Wu, T.; Genzer, J., Surface-grafted polymer gradients: Formation, characterization and applications, *Adv. Polym. Sci.* **2006**, 198, 51–124

91. Bhat, R. R.; Genzer, J., Tuning the number density of nanoparticles by multivariant tailoring of attachment points on flat substrates, *Nanotechnology* **2007**, 18, 025301

92. Demers, L. M.; Mirkin, C. A., Combinatorial templates generated by dip-pen nanolithography for the formation of two-dimensional particle arrays, *Angew. Chem. Int. Ed.* **2001**, 40, 3069–3071

93. Ivanisevic, A.; McCumber, K. V.; Mirkin, C. A., Site-directed exchange studies with combinatorial libraries of nanostructures, *J. Am. Chem. Soc.* **2002**, 124, 11997–12001

94. Potyrailo, R. A.; Leach, A. M., Gas sensor materials based on semiconductor nanocrystal/ polymer composite films, In Proceedings of Transducers '05, The 13th International Conference On Solid-State Sensors, Actuators and Microsystems, Seoul, Korea, June 5–9 **2005**, 1292–1295

95. Potyrailo, R. A.; Leach, A. M., Selective gas nanosensors with multisize cdse nanocrystal/ polymer composite films and dynamic pattern recognition, *Appl. Phys. Lett.* **2006**, 88, 134110

96. Leach, A. M.; Potyrailo, R. A., Gas sensor materials based on semiconductor nanocrystal/ polymer composite films, In Combinatorial Methods and Informatics in Materials Science. Materials Research Society Symposium Proceedings; Wang, Q.; Potyrailo, R. A.; Fasolka, M.; Chikyow, T.; Schubert, U. S.; Korkin, A., Eds.; Materials Research Society, Warrendale, PA, **2006**, Vol. 894, 237–243

97. Singh, A.; Yao, Q.; Tong, L.; Still, W. C.; Sames, D., Combinatorial approach to the development of fluorescent sensors for nanomolar aqueous copper, *Tetrahedron Lett.* **2000**, 41, 9601–9605

98. Szurdoki, F.; Ren, D.; Walt, D. R., A combinatorial approach to discover new chelators for optical metal ion sensing, *Anal. Chem.* **2000**, 72, 5250–5257

99. Castillo, M.; Rivero, I. A., Combinatorial synthesis of fluorescent trialkylphosphine sulfides as sensor materials for metal ions of environmental concern, *ARKIVOC.* **2003**, 11, 193–202

100. Mello, J. V.; Finney, N. S., Reversing the discovery paradigm: A new approach to the combinatorial discovery of fluorescent chemosensors, *J. Am. Chem. Soc.* **2005**, 127, 10124–10125

101. Hagihara, M.; Fukuda, M.; Hasegawa, T.; Morii, T., A modular strategy for tailoring fluorescent biosensors from ribonucleopeptide complexes, *J. Am. Chem. Soc.* **2006**, 128, 12932–12940

102. Wang, S.; Chang, Y.-T., Combinatorial synthesis of benzimidazolium dyes and its diversity directed application toward GTP-selective fluorescent chemosensors, *J. Am. Chem. Soc.* **2006**, 128, 10380–10381

103. Buryak, A.; Severin, K., Dynamic combinatorial libraries of dye complexes as sensors, *Angew. Chem. Int. Ed.* **2005**, 44, 7935–7938

104. Buryak, A.; Severin, K., Easy to optimize: Dynamic combinatorial libraries of metal-dye complexes as flexible sensors for tripeptides, *J. Comb. Chem.* **2006**, 8, 540–543

105. Li, Q.; Lee, J.-S.; Ha, C.; Park, C. B.; Yang, G.; Gan, W. B.; Chang, Y.-T., Solid-phase synthesis of styryl dyes and their application as amyloid sensors, *Angew. Chem. Int. Ed.* **2004**, 46, 6331–6335

106. Rosania, G. R.; Lee, J. W.; Ding, L.; Yoon, H.-S.; Chang, Y.-T., Combinatorial approach to organelle-targeted fluorescent library based on the styryl scaffold, *J. Am. Chem. Soc.* **2003**, 125, 1130–1131

107. Shedden, K.; Brumer, J.; Chang, Y. T.; Rosania, G. R., Chemoinformatic analysis of a supertargeted combinatorial library of styryl molecules, *J. Chem. Inf. Comput. Sci.* **2003**, 43, 2068–2080

108. Basabe-Desmonts, L.; Beld, J.; Zimmerman, R. S.; Hernando, J.; Mela, P.; Garcia Parajo, M. F.; van Hulst, N. F.; van den Berg, A.; Reinhoudt, D. N.; Crego-Calama, M., A simple approach to sensor discovery and fabrication on self-assembled monolayers on glass, *J. Am. Chem. Soc.* **2004**, 126, 7293–7299

109. Basabe-Desmonts, L.; Zimmerman, R. S.; Reinhoudt, D. N.; Crego-Calama, M., Combinatorial method for surface-confined sensor design and fabrication, *Springer Ser. Chem. Sens. Biosens.* **2005**, 3, 169–188

110. Basabe-Desmonts, L.; Reinhoudt, D. N.; Crego-Calama, M., Combinatorial fabrication of fluorescent patterns with metal ions using soft lithography, *Adv. Mater.* **2006**, 18, 1028–1032

111. Basabe-Desmonts, L.; Reinhoudt, D. N.; Crego-Calama, M., Design of fluorescent materials for chemical sensing, *Chem. Soc. Rev.* **2007**, 36, 993–1017

112. Chojnacki, P.; Werner, T.; Wolfbeis, O. S., Combinatorial approach towards materials for optical ion sensors, *Microchim. Acta.* **2004**, 147, 87–92

113. Potyrailo, R. A., Expanding combinatorial methods from automotive to sensor coatings, *Polymeric Mater. Sci. Eng. Polymer Preprints* **2004**, 90, 797–798

114. Hassib, L.; Potyrailo, R. A., Combinatorial development of polymer coating formulations for chemical sensor applications, *Polymer Preprints* **2004**, 45, 211–212

115. Potyrailo, R. A.; Morris, W. G.; Wroczynski, R. J., Acoustic-wave sensors for high-throughput screening of materials, In High-Throughput Analysis: A Tool for Combinatorial Materials Science; Potyrailo, R. A.; Amis, E. J., Eds.; Kluwer Academic/Plenum Publishers: New York, NY, **2003**; Chap. 11

116. Potyrailo, R. A.; Morris, W. G.; Wroczynski, R. J., Multifunctional sensor system for high-throughput primary, secondary, and tertiary screening of combinatorially developed materials, *Rev. Sci. Instrum.* **2004**, 75, 2177–2186

117. Potyrailo, R. A.; McCloskey, P. J.; Ramesh, N.; Surman, C. M., Sensor devices containing co-polymer substrates for analysis of chemical and biological species in water and air, US Patent Application 2005133697: **2005**

118. Potyrailo, R. A.; McCloskey, P. J.; Wroczynski, R. J.; Morris, W. G., High-throughput determination of quantitative structure-property relationships using resonant multisensor system: Solvent-resistance of bisphenol a polycarbonate copolymers, *Anal. Chem.* **2006**, 78, 3090–3096

119. Potyrailo, R. A.; Morris, W. G., Wireless resonant sensor array for high-throughput screening of materials, *Rev. Sci. Instrum.* **2007**, 78, 072214

120. Wu, X.; Kim, J.; Dordick, J. S., Enzymatically and combinatorially generated array-based polyphenol metal ion sensor, *Biotechnol. Prog.* **2000**, 16, 513–516

121. Kim, D.-Y.; Wu, X.; Dordick, J. S., Generation of environmentally compatible polymer libraries via combinatorial biocatalysis, In Biocatalysis in Polymer Science. American Chemical Society, Washington, DC, **2003**, Vol. 840, 34–49

122. Mirsky, V. M.; Kulikov, V., Combinatorial electropolymerization: Concept, equipment and applications, In High-Throughput Analysis: A Tool for Combinatorial Materials Science; Potyrailo, R. A.; Amis, E. J., Eds.; Kluwer Academic/Plenum Publishers, New York, NY, **2003**, Chap. 20

123. Kulikov, V.; Mirsky, V. M., Equipment for combinatorial electrochemical polymerization and high-throughput investigation of electrical properties of the synthesized polymers, *Meas. Sci. Technol.* **2004**, 15, 49–54

124. Kulikov, V.; Mirsky, V. M.; Delaney, T. L.; Donoval, D.; Koch, A. W.; Wolfbeis, O. S., High-throughput analysis of bulk and contact conductance of polymer layers on electrodes, *Meas. Sci. Technol.* **2005**, 16, 95–99

125. Xiang, Y.; LaVan, D., Parallel microfluidic synthesis of conductive biopolymers, In *Proceedings of the 2nd IEEE/ASME International Conference on Mechatronic and Embedded Systems and Applications,* **2006**; 1–5

126. Mirsky, V. M.; Hirsch, T.; Piletsky, S. A.; Wolfbeis, O. S., A spreader-bar approach to molecular architecture: Formation of stable artificial chemoreceptors, *Angew. Chem. Int. Ed.* **1999**, 38, 1108–1110

127. Lahav, M.; Katz, E.; Willner, I., Photochemical imprint of molecular recognition sites in two-dimensional monolayers assembled on au electrodes: Effects of the monolayer structures on the binding affinities and association kinetics to the imprinted interfaces, *Langmuir.* **2001**, 17, 7387–7395

128. Prodromidis, M. I.; Hirsch, T.; Mirsky, V. M.; Wolfbeis, O. S., Enantioselective artificial receptors formed by the spreader-bar technique, *Electroanalysis.* **2003**, 15, 1795–1798

129. Tappura, K.; IVikholm-Lundin, I.; Albers, W. M., Lipoate-based imprinted self-assembled molecular thin films for biosensor applications, *Biosens. Bioelectron.* **2007**, 22, 912–919

130. Cho, E. J.; Tao, Z.; Tehan, E. C.; Bright, F. V., Multianalyte pin-printed biosensor arrays based on protein-doped xerogels, *Anal. Chem.* **2002**, 74, 6177–6184

131. de Gans, B.-J.; Wijnans, S.; Woutes, D.; Schubert, U. S., Sector spin coating for fast preparation of polymer libraries, *J. Comb. Chem.* **2005**, 7, 952–957

132. Egger, S.; Higuchi, S.; Nakayama, T., A method for combinatorial fabrication and characterization of organic/inorganic thin film devices in UHV, *J. Comb. Chem.* **2006**, 8, 275–279

133. Potyrailo, R. A.; Morris, W. G.; Leach, A. M.; Hassib, L.; Krishnan, K.; Surman, C.; Wroczynski, R.; Boyette, S.; Xiao, C.; Shrikhande, P.; Agree, A.; Cecconie, T., Theory and practice of ubiquitous quantitative chemical analysis using conventional computer optical disk drives, *Appl. Opt.* **2007**, 46, 7007–7017

134. Frenzer, G.; Frantzen, A.; Sanders, D.; Simon, U.; Maier, W. F., Wet chemical synthesis and screening of thick porous oxide films for resistive gas sensing applications, *Sensors.* **2006**, 6, 1568–1586

135. Villoslada, F. N.; Takeuchi, T., Multivariate analysis and experimental design in the screening of combinatorial libraries of molecular imprinted polymers, *Bull. Chem. Soc. Japan.* **2005**, 78, 1354–1361

136. Mijangos, I.; Navarro-Villoslada, F.; Guerreiro, A.; Piletska, E.; Chianella, I.; Karim, K.; Turner, A.; Piletsky, S., Influence of initiator and different polymerisation conditions on performance of molecularly imprinted polymers, *Biosens. Bioelectron.* **2006**, 22, 381–387

137. Potyrailo, R. A. High-throughput experimentation in early 21st century: Searching for materials descriptors, not for a needle in the haystack, *6th DPI Workshop on Combinatorial and High-Throughput Approaches in Polymer Science, September 10–11*, Darmstadt, Germany, **2007**

Chapter 8
Chemical Sensor Array Response Modeling Using Quantitative Structure-Activity Relationships Technique

Abhijit V. Shevade, Margaret A. Ryan, Margie L. Homer, Hanying Zhou, Allison M. Manfreda, Liana M. Lara, Shiao -Pin S. Yen, April D. Jewell, Kenneth S. Manatt, and Adam K. Kisor

Abstract We have developed a Quantitative Structure-Activity Relationships (QSAR) based approach to correlate the response of chemical sensors in an array with molecular descriptors. A novel molecular descriptor set has been developed; this set combines descriptors of sensing film-analyte interactions, representing sensor response, with a basic analyte descriptor set commonly used in QSAR studies. The descriptors are obtained using a combination of molecular modeling tools and empirical and semi-empirical Quantitative Structure-Property Relationships (QSPR) methods. The sensors under investigation are polymer-carbon sensing films which have been exposed to analyte vapors at parts-per-million (ppm) concentrations; response is measured as change in film resistance. Statistically validated QSAR models have been developed using Genetic Function Approximations (GFA) for a sensor array for a given training data set. The applicability of the sensor response models has been tested by using it to predict the sensor activities for test analytes not considered in the training set for the model development. The validated QSAR sensor response models show good predictive ability. The QSAR approach is a promising computational tool for sensing materials evaluation and selection. It can also be used to predict response of an existing sensing film to new target analytes.

8.1 Introduction

The ability to predict sensor responses accurately is of great help in characterizing and selecting sensing materials. Typically, sensor arrays use response libraries to develop signal processing algorithms in order to identify and quantify chemical

Abhijit V. Shevade (✉)
Jet Propulsion Laboratory, California Institute of Technology, 4800 Oak Grove Drive, Pasadena, CA 91109, USA
e-mail: Abhijit.Shevade@jpl.nasa.gov

M.A. Ryan et al. (eds.), *Computational Methods for Sensor Material Selection*, Integrated Analytical Systems, DOI 10.1007/978-0-387-73715-7_8, © Springer Science+Business Media, LLC 2009

species; these response libraries consist of laboratory data cataloguing individual sensor responses to analytes of interest. Developing a response library for a sensor array with a set of analytes and under a set of environmental conditions (temperature, pressure, and humidity) is time consuming; in addition, developing response libraries and calibration information may impinge on the useful lifetime of the sensors. Sensor response models that provide such predictive ability would be of great help in sensing materials evaluation and selection prior to experiments.

Sensor response modeling for polymer-carbon composite systems is of interest to us. Sensor response modeling in polymer films exposed to chemical vapors has been investigated using solubility parameters [1], Linear Solvation Energy Relationships (LSER) [2, 3] and sorption described by Monte Carlo molecular modeling approach [4]. Good correlations between calculated and measured responses have been reported with these approaches. In the case of polymer composite sensing films, such as polymer-carbon films previous molecular modeling approaches [5, 6] have taken into account only the effect of polymer-analyte interactions and assumed that neither the carbon or the carbon-analyte interactions in the film plays a role in sorbing analyte molecules or contributes to the sensing film response. This sensor response model may not represent a complete picture of response in polymer-carbon composite sensors, especially at concentrations of single to tens of parts-per-million (ppm) or sub-ppm analyte in air. Some of these approaches for sensor response modeling are discussed in detail in other chapters in this book volume such as the LSER (Chap. 9), solubility parameter based (Chap. 3) and sorption described by Monte Carlo molecular modeling approach (Chap. 4).

Recently we have developed a multivariate statistical approach [7] based on Quantitative Structure-Activity Relationships (QSAR) using Genetic Function Approximation (GFA) to correlate sensor activity with molecular descriptors that describe the physical and chemical properties of sensing film components, analytes and interactions of the two. The sensors considered were polymer-carbon sensing films in the JPL Electronic Nose (ENose) sensing array [8–19]; these sensors were exposed to several target vapors at ppm to sub-ppm concentrations and response measured as change in film resistance. The methodology for the QSAR sensor activity model will be discussed in detail in subsequent sections. In short, the approach for the development of the QSAR sensor response model for a given polymer-carbon sensor begins with the creation of molecular structures for the polymer, carbon black, water and analytes. This is followed by creating a QSAR study table that consists of experimental inputs to be correlated to molecular descriptors that consist of analyte properties as well as various interaction energies of the target molecules with polymer-carbon sensing film components. These QSAR study table descriptors are calculated using empirical methods, semi-empirical predictive methods and molecular simulation methods. Once the QSAR table is ready, GFA is used to generate sensor activity models by correlating experimental inputs with molecular descriptors, with different number of terms and functionality (linear, quadratic, spline or combination etc). Among the several equations developed, the statistically most significant equation that contains the polymer-analyte interaction energy term is chosen to represent the sensor response.

The QSAR technique that we have used for developing our sensor activity models has been used extensively in biochemical, medical and environmental remediation fields for drug-receptor screening and in evaluating phenomenological models [20–27]. In the development of QSAR models, a set of experimental data that is used to develop the model is commonly referred as the training set, while a subset of the data which is used to compare model predictions are called the test set. The training sets for QSAR studies in pharmaceutical and biological systems typically include more than 100 compounds. These larger training sets produce, in general, more robust models. As a result of the application of the JPL ENose, which has been developed as an event monitor for spacecraft air, the target analyte set of the QSAR model developed for the JPL ENose polymer-carbon composite sensors is small, 20–25 analytes. Developing a statistically significant and validated QSAR model that has predictive capability is a challenge for these scenarios of limited training data set applications.

QSAR methodology as described in this chapter is used not only to characterize materials for chemical sensing but also to select sensing materials prior to experimentation. Our goal is to develop one representative equation for each polymer-carbon sensing film in the array. The development of "n" equations to describe the ENose sensing array will facilitate the generation of virtual response libraries for any given sensor array for analytes that may not be easily tested, such as highly toxic or explosive compounds. The predictions can also be used to generate parameters for the identification and quantification software. Subsequently, fewer experimental tests will need to be run on any given sensor array. QSAR is a powerful tool that can help us to achieve our goal.

Statistical methods available with QSAR include data and regression analysis methods [28]. Regression methods include simple and Multiple Linear Regression methods (MLR), stepwise multiple linear regression using Genetic Function Approximation (GFA) methods, and Partial Least Square methods (PLS). Methods such as cluster analysis methods (e.g., hierarchical cluster analysis) and Principal Component Analysis (PCA) are included in data analysis. Cluster analysis methods are aimed towards partitioning a data set into classes or categories consisting of elements of comparable similarity. PCA aims at representing large amounts of multi-dimensional data as a more intuitive, low-dimensional representation [28].

GFA is widely used to develop QSAR models. Like genetic evolution, during GFA thousands of candidate models are created and tested, and only the superior models survive [24]. These models are then used as "parents" for the creation of the next generation candidate models. GFA can thus select the optimum number of descriptors in linear regression analysis automatically; it also constructs multiple linear regression models with any possible combinations of terms (linear, higher order polynomials, splines and gaussians). The advantage of using GFA is that multiple models are produced and subsequently analyzed by GFA, in contrast to other common statistical methods, such as MLR and PLS, that focus on a single "best statistical" model.

The basic descriptors commonly used in QSAR studies describe the intrinsic analyte properties. These default descriptors are predicted using empirical and semi-empirical predictive methods such as Quantitative Structure-Property Relationships (QSPR) [28, 29]. In applying QSAR to sensor response modeling, the default descriptors alone may not be sufficient to describe sensor response; therefore, understanding the sensing mechanism is important to determine descriptors that would describe it. In polymer-carbon sensing films, sensor responses are measured as a change in resistance, and the response can be attributed to the swelling of the polymer film [30]; other mechanisms may also contribute to the sensor response. So the polymer-carbon sensing film response to a given analyte molecule is based on the understanding of how the sensing film components (polymer and carbon black) in the polymer-carbon composite film interacts with the analyte molecules. The QSAR equation that we have chosen to represent a given sensor is selected from a set of cross-validated equations generated by the GFA algorithm. The selected equation is the statistically most significant one (largest $\mathbf{r}^2$ value) of the equation set which also contains the polymer-analyte term. The selection of the QSAR equation containing the polymer-analyte term is not only based on the above discussion of the sensing mechanism in polymer-carbon sensing films, but also on the fact that the sensing array used in our study consists of different polymer-carbon composite films; the selection of QSAR equation containing polymer-analyte term, would provide us with a equation for each polymer-carbon sensor type. We performed initial QSAR runs to determine whether we could use only basic descriptors, consisting of analyte properties, to describe the sensor activity. Although the intrinsic analyte descriptors provide a statistically significant fit, the QSAR sensor response model developed using the default descriptors would not be representative of the different polymer types used in the JPL ENose polymer-carbon sensing films.

We have previously investigated LSER methods to model responses of three polymer-composite sensors to six different analytes [10]. Comparison to experimental data showed LSER to be a poor predictor for sensor response. This is not surprising as the equation considers only polymer-analyte interactions and does not account for the adsorption of analyte on the carbon conductive medium dispersed in the film, for the analyte molecule diffusion within the film, the film thickness, or the hydration of the film. As described in the previous discussion, the sensor response for polymer based sensors to target analytes is a function of its ability to interact with the sensors. By using only descriptors of analyte properties, a good statistical fit is possible, but not in terms of representing a sensor response. The analyte descriptors are constant in value and will not change with the sensor type, but the sensor-analyte interactions change. So, interaction energy terms are needed in the descriptor set to distinguish between the sensor responses of different sensor materials types. Accordingly, the descriptor sets developed in our QSAR study combine descriptors for analyte properties as well as those that describe interactions between the sensing film and the analytes.

In the following sections, we use polyethylene oxide-carbon sensing film as an example to describe the approaches for molecular descriptor calculations and demonstrate QSAR methodology for sensor activity model development for the polymer-carbon composite sensing film. The terms used in the QSAR model for polyethylene oxide-carbon sensing film will be discussed in detail. The approach will be extended to other polymer-carbon sensing films used in the JPL ENose array and discussed in the results and discussion section. The promise of using QSAR sensor response models for sensing materials evaluation and selection will be demonstrated and discussed in the sections following this.

8.2 Experimental Data

8.2.1 Sensor Response Measurements and QSAR Conditions

Experimental data for QSAR studies to develop sensor response models for the JPL ENose polymer-carbon composite sensors was obtained by delivering measured analyte concentrations with controlled constant humidity to the sensors. The experiments were repeated several times, over a range of analyte concentrations [8–13, 18]. The investigated analytes and concentration ranges considered for this QSAR study are shown in Table 8.1.

Sensor response is expressed as a normalized change in sensor resistance and is plotted against delivered concentration. Response data for each analyte and each sensor are fit to an equation of the form $y = A_1 x + A_2 x^2$, where x is the analyte

Table 8.1 Analyte list and concentration range tested in parts-per-million (ppm) for the JPL Electronic Nose operation (760 Torr, 23°C)

Analyte	Concentration tested Low - High (ppm)
1. Acetone	64–600
2. Ammonia	6–60
3. Chlorobenzene	3–30
4. Dichloromethane	10–150
5. Ethanol	200–6000
6. Isopropanol	30–400
7. Xylenes (mixed)	33–300
8. Tetrahydrofuran	13–120
9. Trichloroethane	7–200
10. Acetonitrile	1–25
11. Ethylbenzene	20–180
12. Freon113	15–500
13. Hexane	15–150
14. Methyl ethyl ketone	15–150
15. Methane	1,600–50,000
16. Methanol	6–100
17. Toluene	5–50
18. Benzene	10–100
19. Indole	25–450
20. Dichloroethane	10–100

The experimental data was taken at a constant humidity of 5,000 ppm water (~18% relative humidity). Reproduced by permission of ECS – The Electrochemical Society. Ref. [7]

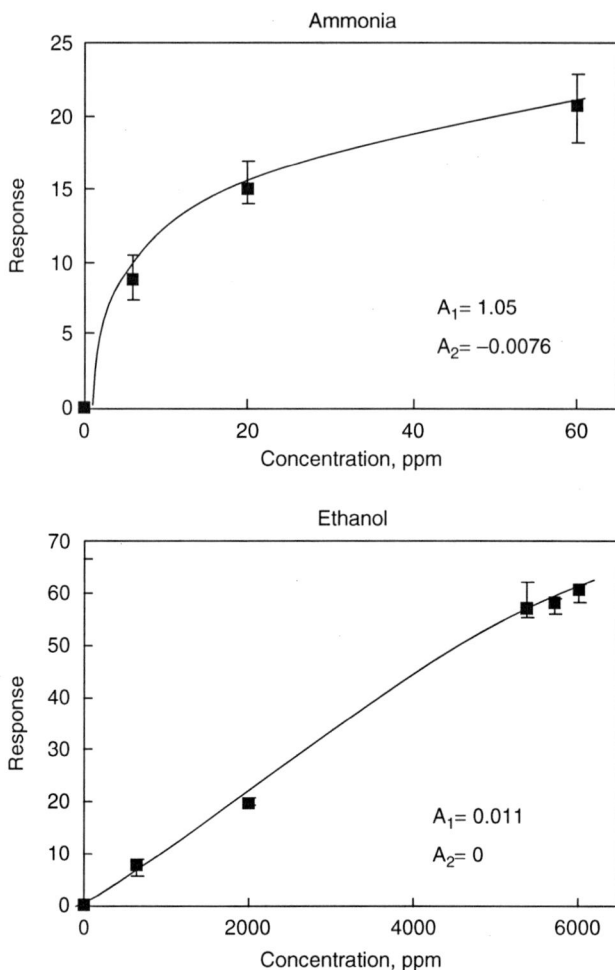

Fig. 8.1 Shows the JPL ENose experimental data and fitted concentration-response equation ($y = A_1x + A_2x^2$) for the polyethylene oxide carbon sensing film. The lines are drawn to guide the eye

concentration and y is the normalized change in resistance [8, 9, 19]. Figure 8.1 shows experimental data and the fitted concentration-response equations for polyethylene oxide-carbon sensing films and two representative analytes. As can be seen in the figure, the coefficient for the quadratic term, A_2, is generally three orders of magnitude smaller than the coefficient of the linear term, A_1. For this investigation, we have focused mainly on the linear term coefficient. Using only the linear term will necessarily mean that there is some error introduced into the measure of sensor response. The purposes of this study is to determine whether an adequate model can be developed using QSAR, for which the linear term alone is sufficient.

It is important to develop a single sensor activity correlation that is independent of analyte concentrations for a given sensor type (e.g., polymer-carbon composite), to avoid multiple correlations for a single sensor. By selecting coefficient A_1 from the sensor response equation as the activity to be used in QSAR studies, we can develop an approach to calculate a concentration independent sensor response. Analytes# 1–17 (Table 8.1) are used as the training set to develop the QSAR equations, and the remaining analytes (#18–20) are used as the test set.

In addition, because the sensor response is expressed as a quadratic equation ($y = A_1x + A_2x^2$), and this study does not consider the coefficient A_2 there will necessarily be some degree of error in the calculated A_1. As the coefficient A_2 is generally three orders of magnitude smaller than A_1, the quadratic term does not contribute strongly to sensor response at low concentration. We are trying to provide a methodology to develop sensor response models that are concentration independent and which can provide us an ability to screen sensors by predicting sensor responses to new untested analytes prior to experiments.

8.3 QSAR

8.3.1 QSAR Descriptors: Analyte Properties and Sensing Film-Analyte Interactions

The default analyte property descriptors in QSAR studies fall into electronic, thermodynamic, spatial, structural, and topological categories, as shown in Table 8.2. Descriptors for the polymer-carbon sensing film response include interaction energies of the sensing film components, polymer and carbon-black, with analyte molecules and water. The physical descriptors that are considered for our QSAR study include those that describe intrinsic analyte properties as well as sensor-analyte interactions. The goal is to correlate coefficient A_1 of a polymer-carbon composite sensor response with these combined physio-chemical molecular descriptors, consisting of analyte properties and sensing film-analyte interactions.

8.3.1.1 Approaches to Calculating Sensing Film-Analyte Interactions

Sensing film-analyte interactions for organic (*e.g.* polymer based) or inorganic (*e.g.* metal, metal oxide) sensing elements can be calculated using two approaches that require non-periodic (cluster) or periodic calculations. Descriptors calculated using cluster or a non-periodic approach, Approach I, are pair interaction or binding energy. The descriptor calculated using a periodic approach, Approach II, is heat of sorption. The contributions to the interactions between the sensing film components and analyte come from van der Waals, electrostatic, and hydrogen bonding forces. We will discuss these approaches for a polymer-carbon sensing film and its interactions with an analyte.

Table 8.2 Default analyte descriptor set used in QSAR. Reproduced by permission of ECS - The Electrochemical Society. Ref. [7]

Default analyte descriptors	Description
A_{pol}	Sum of atomic polarizabilities
Dipole-mag, Dipole-X, Y, Z	Dipole moment magnitude and X, Y, and Z components
R_G	Radius of Gyration
Area	Molecular surface area
MW	Molecular weight
V_m	Molecular volume
Density	Density
PMI-mag, PMI-X, Y, Z	Principal moment of inertia magnitude and X, Y, and Z components
Rotlbonds	Number of rotatable bonds
HB_A	Number of hydrogen bond acceptors
HB_D	Number of hydrogen bond donors
AlogP	Log of the octanol/water partition coefficient
MR	Molar refractivity

In cluster model calculations (Approach I) for polymer-carbon sensing films, the individual components of the polymer-carbon composite system, polymer and carbon black, are used to calculate the binding energies between two analyte molecules and between an analyte molecule and the polymer-carbon composite sensing film components. In Approach I, the pair interaction energies [31] which represent the polymer-carbon composite sensor response descriptors include: polymer-analyte, carbon black-analyte, polymer-water, carbon black-water, analyte-analyte, and analyte-water. Van der Waals, coulombic and hydrogen bonding forces contribute to these energies. The general notation used here to represent these descriptors is of the form E_{xy}. The suffixes x and y could be polymer (p), carbon black (cb), analyte (a), or water (w). For example, the interaction energy between the polymer (p) and the target analyte (a) is denoted E_{pa}. The sensor response descriptor set using Approach I for polyethylene oxide-carbon sensing film is shown in Table 8.3. It includes the sensing film-analyte descriptors E_{pa}, E_{pw}, $E_{p\text{-}cb}$, $E_{cb\text{-}a}$, $E_{cb\text{-}cb}$, E_{aa}, and E_{aw}. Cluster based calculations for polymer-carbon sensor response descriptors is a rapid approach that takes into account the chemical nature of the individual components, *i.e.* electronic and thermodynamic characteristics of the monomer, carbon black and the analyte.

In periodic model calculations, Approach II, the interaction of the analyte with the polymer-carbon composite film is described by calculating the isosteric heat of sorption of analyte molecules in the polymer-carbon composite. Heat of sorption (H_{sorpt}) calculations need a 3-D periodic model of the polymer-carbon composite film. We developed such a model based on the sensing film formulation process [32]. Heat of sorption calculations give a more complete representation of sensing film-analyte interactions, taking into account the structural aspect of the polymer composite film as well as the sorption process that occur. This approach mimics the

Table 8.3 QSAR study table showing polyethylene oxide-carbon sensing film-analyte interaction energy descriptors, calculated used various approaches: cluster and periodic. The sensing film-analyte interactions energies are in kcal/mol units and the standard deviations are shown in parentheses for the calculated E_{xy} and H_{sorpt} descriptors. The suffixes x and y in E_{xy} are polymer (p), carbon black (cb), analyte (a), or water (w)

Analyte	Approach I							Approach II
	E_{pa}	E_{pw}	E_{aw}	E_{aa}	E_{cba}	E_{p-cb}	E_{cb-cb}	H_{sorpt}
Acetone	−0.50 (0.56)	−0.28 (0.74)	−0.33 (0.75)	−0.65 (0.54)	−0.95 (0.36)	−0.79 (0.30)	−1.68 (0.74)	9.96 (0.10)
Ammonia	−0.16 (0.34)	−0.28	−0.11 (0.51)	−0.05 (0.23)	−0.28 (0.10)	−0.79	−1.68	3.65 (0.65)
Chlorobenzene	−0.69 (0.75)	−0.28	−0.43 (0.82)	−1.14 (1.08)	−1.28 (0.58)	−0.79	−1.68	14.72 (2.00)
Dichloromethane	−0.57 (0.63)	−0.28	−0.36 (0.83)	−0.77 (0.76)	−1.09 (0.41)	−0.79	−1.68	11.54 (1.72)
Ethanol	−0.42 (0.51)	−0.28	−0.25 (0.67)	−0.41 (0.59)	−0.78 (0.30)	−0.79	−1.68	9.88 (1.27)
Isopropanol	−0.48 (0.53)	−0.28	−0.28 (0.66)	−0.49 (0.70)	−0.90 (0.33)	−0.79	−1.68	10.58 (1.37)
o-Xylene	−0.62 (0.42)	−0.28	−0.30 (0.52)	−0.94 (0.63)	−1.20 (0.55)	−0.79	−1.68	13.79 (0.44)
Tetrahydrafuran	−0.55 (0.52)	−0.28	−0.32 (0.63)	−0.53 (0.65)	−1.19 (0.53)	−0.79	−1.68	11.72 (0.68)
Trichloroethane	−0.65 (0.60)	−0.28	−0.43 (0.79)	−1.27 (0.77)	−1.26 (0.49)	−0.79	−1.68	13.97 (2.55)
Acetonitrile	−0.44 (0.34)	−0.28	−0.29 (0.47)	−0.46 (0.25)	−0.83 (0.32)	−0.79	−1.68	7.97 (1.13)
Ethylbenzene	−0.63 (0.42)	−0.28	−0.37 (0.46)	−0.91 (0.58)	−1.19 (0.53)	−0.79	−1.68	13.03 (1.40)
Freon113	−0.65 (0.67)	−0.28	−0.45 (0.55)	−0.17 (1.70)	−1.32 (0.50)	−0.79	−1.68	12.67 (0.86)
Hexane	−0.55 (0.40)	−0.28	−0.32 (0.42)	−0.69 (0.40)	−1.06 (0.43)	−0.79	−1.68	12.05 (0.90)
Methyl ethyl ketone	−0.52 (0.57)	−0.28	−0.33 (0.73)	−0.64 (0.73)	−1.01 (0.39)	−0.79	−1.68	11.43 (1.12)
Methane	−0.32 (0.10)	−0.28	−0.17 (0.10)	−0.23 (0.06)	−0.61 (0.22)	−0.79	−1.68	3.72 (0.41)
Methanol	−0.34 (0.57)	−0.28	−0.23 (0.82)	−0.29 (0.60)	−0.65 (0.25)	−0.79	−1.68	8.57 (0.87)
Toluene	−0.62 (0.39)	−0.28	−0.36 (0.45)	−0−0.90 (0.57)	−1.17 (0.53)	−0.79	−1.68	11.46 (2.56)

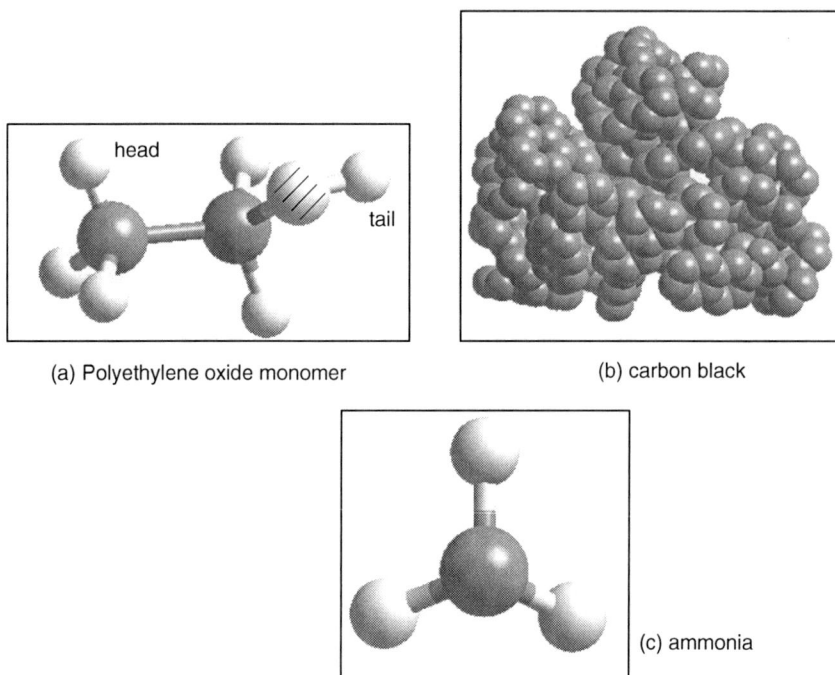

(a) Polyethylene oxide monomer   (b) carbon black

(c) ammonia

Fig. 8.2 Molecular models to calculate sensing film–analyte interactions using the cluster (nonperiodic) approach for polyethylene oxide-carbon sensing film. Shown (**a**) polyethylene oxide monomer unit (oxygen atom shaded), (**b**) carbon black and (**c**) ammonia molecule (representative model). These models are used to calculate sensing film–analyte interaction energy E_{xy}, polymer (p), carbon black (cb), analyte (a), or water (w)

process that happens when an analyte is sorbed by a sensor. The heat of sorption represents the total interaction of the analyte with the polymer and carbon black components.

The molecular models used in Approaches I and II for polyethylene oxide-carbon are shown in Figs. 8.2 and 8.3. A combined descriptor set that includes the default analyte descriptors along with sensor response descriptors calculated was used for QSAR studies to correlate the sensor coefficients A_1 with the molecular descriptors.

8.3.2   Methodology

8.3.2.1   QSAR Descriptors: Calculations

Default QSAR analyte descriptors were predicted by empirical and semi-empirical Quantitative Structure Property Relationships (QSPR) using the commercial software [33] Cerius^2 on a Silicon Graphics O2 workstation.

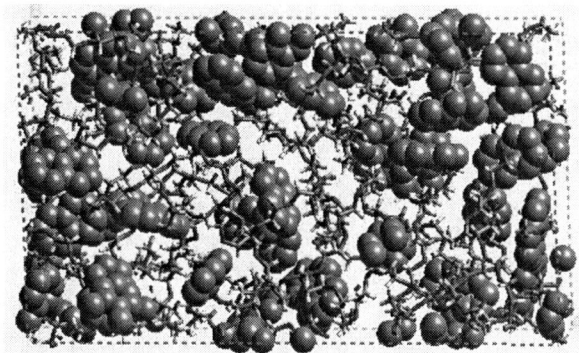

Fig. 8.3 Molecular model to calculate sensing film-analyte interactions using the periodic approach for polyethylene oxide-carbon sensing film. Seen is the three dimensional molecular model of polyethylene oxide-carbon composite sensing film (density $= 0.90$ g/cm^3). The cylindrical chains represent the polymer chains and the clusters represents the carbon black. This model is used to calculate sensing film-analyte interactions, by calculating heat of sorption (H_{sorpt}) of the analyte

Polymer-analyte interactions have been calculated using Quantum Mechanical (QM) and atomistic level calculations [5]. Quantum mechanical calculations are the more accurate; however, the time needed for QM simulation generally scales as N^3–N^5 for characteristic methods, where N is the system size. Information derived from QM is used to develop a force field, an empirical functional form, for atomistic level simulations for use in Molecular Dynamics (MD) techniques. Atomistic level simulations are faster than QM calculations and are the choice for our descriptor calculations. Chapter 3 in this volume describes the use of QM for rapid screening of chemical sensing materials.

For Approach I, cluster calculations, the polymer was modeled using its basic unit, the monomer, as shown in Fig. 8.2. Carbon black was modeled as naphthalene rings with no hydrogen (small graphite sheets) and zero charge on the carbon atom [32]. A cluster of 32 naphthalenes was used to represent the carbon black. Analyte models, except for water, were constructed using the drawing tools provided in the software, and all atoms were assigned charges and equilibrated according to the methodology discussed below. The Single Point Charge (SPC) model was used for water [34]. Charges on the monomer and analyte atoms were assigned by the charge equilibration method (Qeq) [35]. The Dreiding force field [36] was used for polymer and analyte molecules, and graphite parameters were assigned to carbon black atoms [37]. Equilibration was achieved by molecular mechanics and then by molecular dynamics simulations at 300 K. Molecular models used were developed using the Cerius^2 software.

To calculate sensor response descriptors in Approach I, sensing film-analyte interaction energies were calculated using Monte Carlo simulation techniques. Interaction energy between the polymer and analyte molecule, E_{pa}, was calculated by fixing the polymer structure in space and sampling the analyte molecule around the polymer, then averaging the energy calculated over all the random analyte configurations generated around the polymer. In this study we used 10^5 configurations. Similarly E_{pw}, E_{p-cb}, E_{cb-a}, E_{cb-cb}, E_{aa}, and E_{aw} descriptors were calculated. The BLENDS module in the Cerius^2 software performs the calculations based on the methodology described above. The sensor response descriptor set using Approach I for the QSAR study includes, E_{pw}, E_{p-cb}, E_{cb-a}, E_{cb-cb}, E_{aa}, and E_{aw} descriptors. The E_{xy} values calculated using this approach for polyethylene oxide-carbon sensing film with various analytes is shown in Table 8.3.

For Approach II, periodic calculations, the isosteric heat of sorption (H_{sorpt}) [38] of the analyte in the polymer-carbon composite is the descriptor that represents the combined interactions of the analyte with the polymer and carbon components. This term incorporates the combined effect of the separate terms used in Approach I, E_{pa} and E_{cb-a}. H_{sorpt} values of analyte molecules in the polyethylene oxide-carbon sensing film are shown in Table 8.3. As discussed in our previous work [7], the sorption simulations for this work were performed at constant analyte loading of one molecule. The SORPTION module in the Cerius^2 software was used to accomplish the task. The program generates random points in the polymer and tries to insert an analyte molecule. Insertion attempts that involve the overlapping of the analyte molecule with the polymer structure are discarded. After the insertion step, each subsequent configuration is generated by either a random translation or rotation of the analyte molecule in the polymer matrix, taken in the usual Metropolis Monte Carlo manner. The isosteric heat of sorption value is calculated at the end of the run. The sensor response descriptor set using Approach II for the QSAR study includes, H_{sorpt}, E_{aa}, and E_{aw}.

Approach I is computationally less expensive than Approach II; thus, Approach I was used to calculate the sensing film-analyte interaction for the polymers under consideration. The performance of the QSAR sensor response models obtained by the combined descriptors (sensor response and default analyte properties) developed in this work and default analyte properties were found to be similar.

8.3.2.2 QSAR Equation: Term Selection and Functional Form

QSAR model development begins with investigating the number of terms (N_{term}) and the functional forms. The functional form used to form the model can be linear or linear-quadratic. Spline terms are not considered for small training sets owing to the possibility of over-fitting, as in our case where the training set is fewer than 20 data points.

The reliability and significance of QSAR models can be determined using statistical parameters such as F [28] and the correlation coefficient (r^2). The F value is a ratio of explained to unexplained variance. The significance of the QSAR equation becomes greater as the F value increases. The r^2 value describes the goodness of fit of the data to the QSAR model. The prediction ability of the QSAR equations can be estimated using the leave-one out cross-validation technique. In this procedure, new regression coefficients are generated for a given

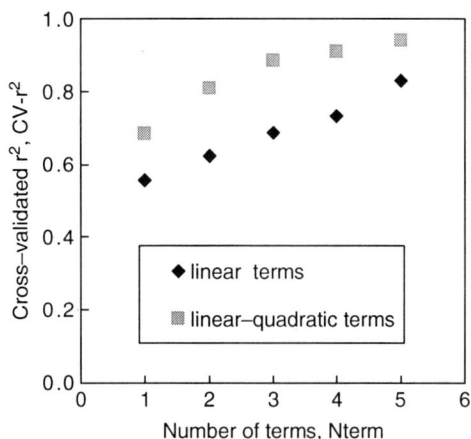

Fig. 8.4 Number of QSAR terms and functionality to be used in the model, by plotting cross-validated r^2 ($CV-r^2$) vs. number of descriptor (N_{term}) for the training analyte set for the polyethylene oxide-carbon sensing film models. The QSAR runs were performed using complete descriptor set (analyte and sensor response)

model after systematically removing one analyte sample at a time from the training data set. This new regression model is then used to predict the activity of the removed sample.

As an example, we describe a QSAR sensor activity model for polyethylene oxide-carbon composite films. The QSAR term and functional form selection for polyethylene oxide-carbon sensing films was made using a training set made up of analytes# 1–17. The QSAR studies were performed with no constant by varying the number of terms from 1 to 5, with linear and quadratic functionality using 5,000 crossovers. The leave-one out cross-validation was performed for each of the 17 analytes. The series of predictions was then used to calculate a new value for the cross-validated r^2 ($CV-r^2$). To determine the number of terms and equation type for the model, $CV-r^2$ is plotted against the number of terms for both linear and linear-quadratic equation types as shown in Fig. 8.4. As seen in the figure, the linear-quadratic terms have a greater $CV-r^2$ compared to the linear terms. It can also be seen from the figure that, for the linear-quadratic equation, no substantial increase in $CV-r^2$ is achieved by increasing the number of terms from 3 to 4. Since the training data set is limited to 17 analytes, any increase in the number of terms in the model would be an over-fit to the data.

8.3.2.3   QSAR Equation: Representation of Sensor Activity

The polymer carbon sensing film response to a given analyte molecule is based on how the sensing film components (polymer and carbon black) in the polymer-carbon composite film interacts with the analyte molecules. The QSAR equation that we have chosen to represent a given sensor is selected from a set of cross-validated equations generated by the GFA algorithm. The selected equation is the statistically most significant one (largest $\mathbf{r}^2$ value) of the equation set which also contains the polymer-analyte (E_{pa}) term, as discussed in the previous section. The objective of this modeling effort is to develop an equation for each sensor in the ENose array, each equation describing the response of that particular sensor to analytes. For a given polymer-carbon composite sensing film, it should be noted that the interaction descriptors $E_{p\text{-}cb}$, E_{pw}, and $E_{cb\text{-}cb}$ do not vary with analyte types and hence will not appear in the QSAR equations.

8.4 Results and Discussion

The general methodology for QSAR sensor response model development using GFA is shown in Fig. 8.5. As discussed in the previous sections, the QSAR sensor response models with the basic analyte descriptors only would not take into account the interactions between analyte and polymer and carbon black; these interactions play an important role in the sensing process in the polymer-carbon sensing film. Thus, a combined descriptor set that consists of analyte properties and sensing film-analyte interactions was used to formulate the QSAR sensor activity relationships.

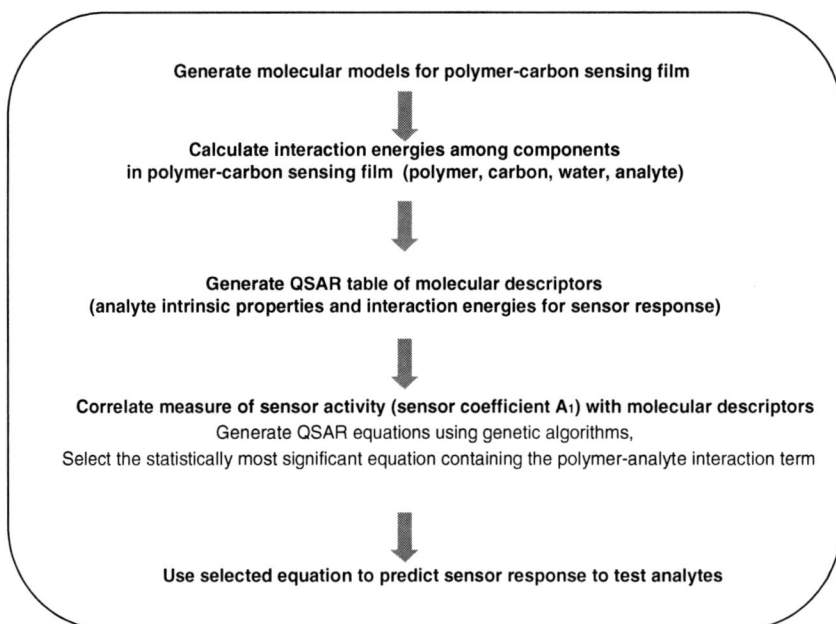

Fig. 8.5 Flow chart showing the QSAR methodology used for the polymer-carbon composite sensor response modeling

In the following section, we describe the QSAR model results for polyethylene oxide-carbon sensing film using the combined descriptor set. Following this, we will discuss the results for this approach when extended to other polymer sensors in the JPL ENose array. We also provide an example of the prediction capabilities of the QSAR model; such predictions could be used in sensor evaluation leading to selection prior to experimentation.

8.4.1   QSAR Sensor Activity Model: Polyethylene Oxide-Carbon Sensing Film

As discussed in the QSAR methodology section, for the polyethylene-oxide polymer carbon sensing film case, we have used a three term linear-quadratic equation form. A QSAR study table showing the sensing film-analyte interaction energy descriptors calculated using various approaches is shown in Table 8.3. A comparison of QSAR generated partial equation sets for the two approaches used for calculating the sensor response descriptors for polyethylene oxide-carbon sensing film is shown in Table 8.4. As described previously, Approaches I and II, cluster or non-periodic and periodic, represent different ways of calculating polymer-analyte interaction energy descriptors. The equation selected to represent the sensor activity in both approaches is the statistically significant equations with the term E_{pa} for Approach I and the term H_{sorpt} for Approach II.

8.4.1.1   Approach I

The calculated activity for the polyethylene-oxide polymer carbon sensing film, coefficient A_1; using Approach I is plotted versus experimental values as seen in Fig. 8.6. The statistically most significant equation containing the descriptor E_{pa} is:

Table 8.4 QSAR generated partial equation set for the two approaches used for polymer-analyte descriptor calculations: Approach I (cluster or non-periodic) and Approach II (periodic), containing E_{pa} and H_{sorp} terms, respectively. The statistically significant equation is the one chosen to represent the sensor activity

Calculated Activity (A_1) =	r^2	F
Approach-I (non-periodic or cluster calculations)		
$0.15207\ E_{pa} + 0.116727\ HB_D^2 + 0.000241\ MR^2$	0.86	40.6
$-0.283563\ E_{pa}^2 + 0.114330\ HB_D^2 + 0.000256\ MR^2$	0.86	38.7
$0.286937\ E_{pa}^2 + 0.182699\ HB_D^2 - 0.200424\ HB_D$	0.84	34.6
Approach-II (periodic calculations)		
$-0.008856\ H_{sorp} + 0.118308\ HB_D^2 + 0.000264\ MR^2$	0.87	44.4
$-0.000801\ H_{sorp}^2 + 0.115485\ HB_D^2 + 0.000284\ MR^2$	0.87	43.1
$-0.022602\ H_{sorp} + 0.120321 HB_D^2 + 0.013894\ MR$	0.86	40.3

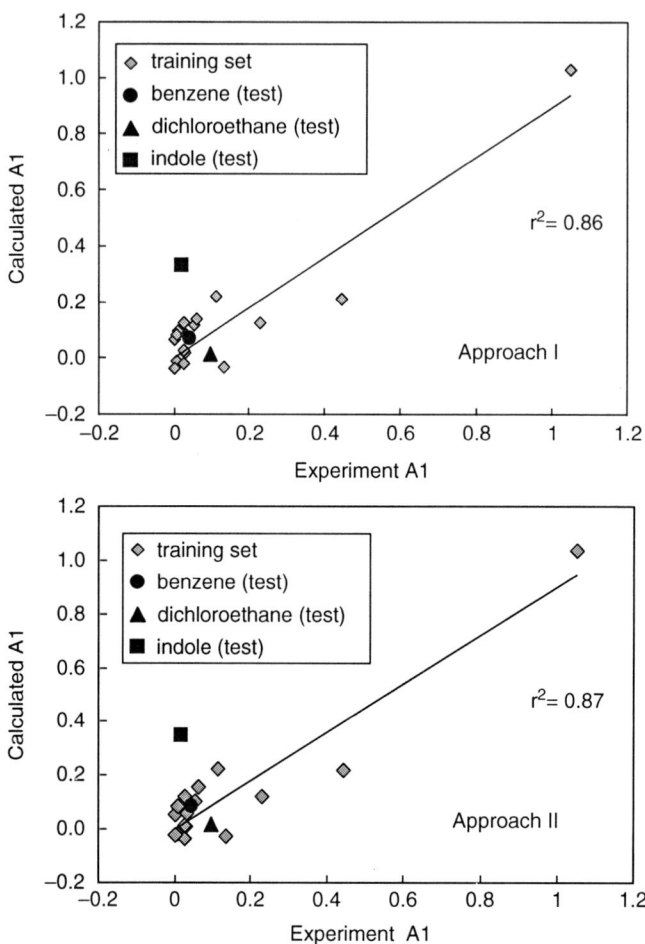

Fig. 8.6 Comparison of QSAR models for two approaches used to calculate sensor response descriptors for polyethylene oxide-carbon sensing film. QSAR studies were performed using the combined descriptor set (analyte and sensor response). Plots shows calculated vs. experimental sensor activity for training and test analyte set. The approaches are different ways to calculate polymer-carbon sensing film-analyte interactions: cluster (Approach I) and periodic (Approach II)

$$Calculated\ activity(A_1) = 0.15207\,E_{pa} + 0.116727\,HB_D^2 + 0.000241\,MR^2 \quad (8.1)$$

$(r^2 = 0.86, F = 40.6)$.

The QSAR sensor activity equation (8.1) was developed using the training set. It was further used to predict the sensor coefficients for test analytes (analytes# 18–20): benzene, dichloroethane and indole, that were not a part of the training set (analytes# 1–17) of Table 8.1. The aim of this exercise is to determine whether the equation can be used to predict sensor activity without performing exhaustive and

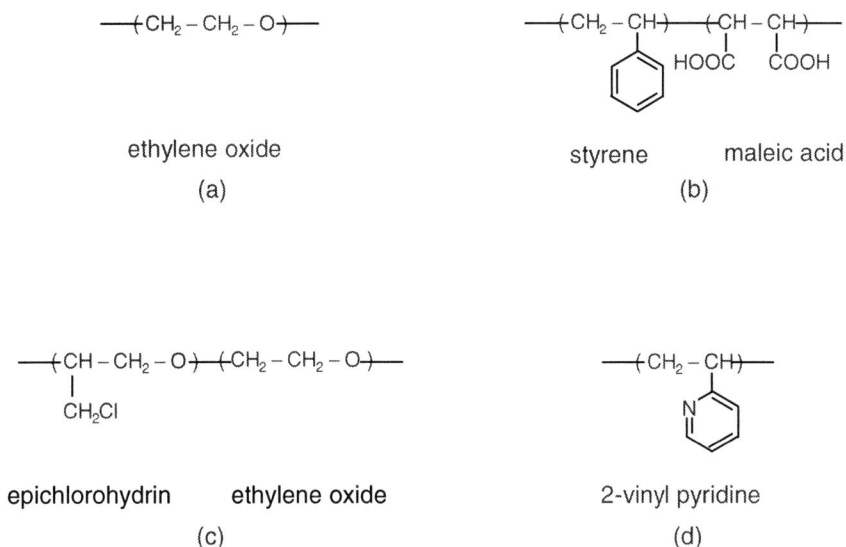

ethylene oxide

(a)

styrene   maleic acid

(b)

epichlorohydrin   ethylene oxide

(c)

2-vinyl pyridine

(d)

Fig. 8.7 Chemical structures of polymer monomer units for (**a**) Polyethylene oxide, (**b**) Poly (styrene-co-maleic acid), (**c**) Poly (epichlorohydrin-co-ethylene oxide), and (**d**) Poly (2-vinyl pyridine)

time-consuming experiments if, for example, new analytes are added to the target list. It can be seen that the model works satisfactorily for benzene and dichloroethane but over predicts the activity for indole. Indole is the least volatile analyte on our list. It is also the only aromatic that has a hydrogen bond donor site among our analytes.

In QSAR sensor activity (8.1), we observe that the analyte descriptors that appear along with the polymer-analyte interaction term (E_{pa}) are hydrogen bond donor site (HB_D) and molar refractivity (MR). The polyethylene oxide monomer (Fig. 8.7) has one hydrogen bond acceptor site, thus it is logical that a descriptor that represents the hydrogen bond donor nature of the analyte may appear in the equation. The analyte descriptor MR is a combined measure of molecule size and polarizability. The descriptor MR is calculated from the refractive index, molecular weight and density of the analyte. As swelling in the polymer-carbon composite film is one mechanism of sensor response, it is logical that the molecular size of the analyte will appear in the equation describing the sensor response.

8.4.1.2   Approach II

The statistically most significant equation containing the descriptor H_{sorpt}, calculated using Approach II is:

$$\textit{Calculated activity } (A_1) = -0.008856\, H_{\text{sorpt}} + 0.118308\, HB_D^2$$
$$+ 0.000264\, MR^2. \tag{8.2}$$

The r^2 value of the calculated vs. experimental fit using Approach II for the training set is not significantly different than Approach I, as seen Fig. 8.6. On comparing statistically most significant equations of the QSAR sensor activity models for these approaches, (8.1) and (8.2); we see similarity in the descriptors that have appeared in the equations and as well as the functional form. There is quadratic dependence on the hydrogen bond donor (HB_D) and molar refractivity (MR) terms, and a linear dependence on the sensor response descriptor that appears in the selected equation, E_{pa} for Approach I and H_{sorpt} for Approach II.

8.4.1.3 Comparison of QSAR with Other Multivariate Sensor Response Approaches

The QSAR models for polyethylene oxide-carbon black show analyte descriptor terms MR and HB_D; similar terms also appear in the LSER approach [2, 39]. LSER is discussed in detail in Chap. 9 of this volume. In addition, the sensing film partition coefficient, K, is correlated in LSER with a linear combination of analyte solubility descriptors (solvation parameters). The regression coefficients obtained for the LSER models characterize the properties of the sensing film. The analyte property terms that appear in LSER are the excess molar refractivity, dipolarity/polarizability, hydrogen bond acidity and basicity parameters, and the gas/liquid partition coefficient. The LSER approach has been used for sorption studies of vapors in a polymer film only [39] and also for sorption in graphite/fullerene coatings [40]. The sensing film used in our studies is a polymer-carbon composite film; our earlier efforts [10] to use LSER to model response to analyte in polymer-carbon composites has resulted in poor correlation between calculated and measured sensor response. The QSAR approach used in this work is different from LSER, as it includes the polymer-analyte and carbon black-analyte interactions, as well as contributions from analyte-water interactions that represent experimentation conditions in the descriptor set.

8.4.1.4 Additional Considerations: QSAR Descriptors in Sensor Activity Model Development

Previous investigations have recognized that the partition coefficient of an analyte in a polymer correlates to polymer-carbon black sensor response [41, 42]. For gas phase detection of a target analyte molecule, the sensing film response is a function of the equilibrium partition coefficient, K, of the analyte molecule in the sensing film [39, 43, 44]. This equilibrium constant is defined as the ratio of the equilibrium concentrations of the analyte in the sensing film (C_s) to the bulk analyte concentration

(C_v) to which the sensing film is exposed. The concentration of analyte in the sensing film, C_s, is normally measured using piezoelectric techniques. The bulk analyte concentration in a carrier gas (air or nitrogen) depends on the vapor pressure of the target analyte at a given temperature. Evaluating partition coefficients by measuring experiments for mass uptake by the polymer-carbon composite film is time consuming and elaborate, so to include the partition coefficient functionality in our QSAR study, analyte vapor pressure [45] calculated at 300 K was added to the descriptor list. We have shown in our previous work [7] that, the A_1 coefficient for indole may be better predicted with the inclusion of vapor pressure in the molecular descriptor set.

8.4.2 QSAR Models for Sensor Arrays

The polymers used in the JPL ENose polymer-carbon composite sensing film array fall into different categories of chemical functionalities [9]. These include: hydrogen-bond acidic (HBA), hydrogen bond basic (HBB), dipolar and hydrogen bond, weak H-bond basic or acidic (MD-HB), and weakly dipolar with weak or no hydrogen-bond properties (WD).

The QSAR sensor activity models for selected polymer-carbon sensors using the combined descriptor set (analyte and sensor response) in the JPL ENose sensor array is summarized in Table 8.5 and the chemical structures of the polymer monomer units shown in Fig. 8.7. These equations in Table 8.5 are statistically most significant in a set of equations containing the polymer-analyte descriptor E_{pa}. A good fit is achieved for the training data set and test data as seen in Fig. 8.8. It can

Table 8.5 QSAR sensor activity equations for selected polymer-carbon sensing films of the JPL ENose array The polymer chemical functional classification is: hydrogen-bond acidic (HBA), hydrogen bond basic (HBB), moderately dipolar and weakly H-bond basic or acidic (MD-HB), and weakly dipolar with weak or no hydrogen-bond properties (WD)

Polymer (in polymer-carbon sensing films used in JPL ENose)	Polymer chemical functional classification	QSAR sensor activity equation	Statistical parameters
Polyethylene oxide	WD	$0.15207\,E_{pa}$ $+0.116727\,HB_D^2$ $+0.000241\,MR^2$	$r^2 = 0.86$ $F = 40.60$
Poly (styrene-co-maleic acid)	HBA	$-0.094\,E_{pa}$ $+0.102\,HB_D^2$ $+0.021\,\text{Dipole-}Y^2$ $-0.079\,\text{Dipole-}Y$	$r^2 = 0.92$ $F = 121.51$
Poly (epichlorohydrin-co-ethylene oxide)	MD-HB	$-0.02432\,E_{pa}^2$ $+0.130288\,E_{aa}^2$ $+2.00\text{E-06 VP}^2$	$r^2 = 0.99$ $F = 1350.77$
Poly (2-vinyl pyridine)	HBB	$-0.293916\,E_{aa}$ $+0.714156\,E_{pa}^2$ $-0.000341\,\text{VP}$ $-0.000171\,A_{pol}$	$r^2 = 0.70$ $F = 15.49$

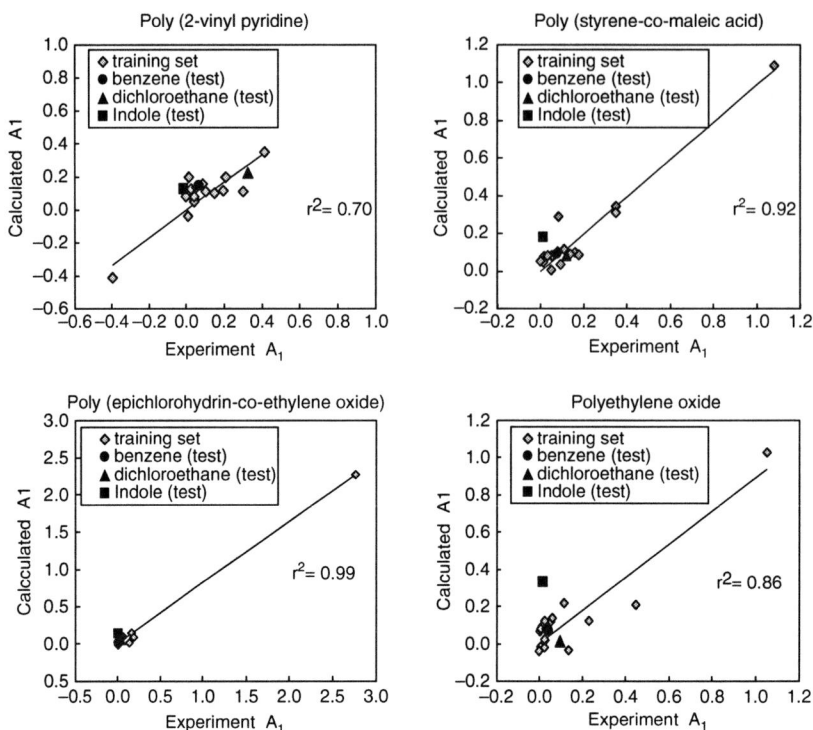

Fig. 8.8 QSAR modeled array response for 4 JPL ENose polymer-carbon sensing films. The plot shows QSAR calculated versus experimental sensor activity (coefficients A_1) for the training and test analyte set using combined descriptor set (analyte and sensor response). QSAR equations in Table 8.5 were used to determine calculated values for both the training and test analyte set. The $\mathbf{r}^2$ value is for the training data set and reflects the correlation between the calculated vs. the experimental values obtained

also be seen that the QSAR model shows good predictions of the test analyte activities, as seen from the close proximity of the test data points around the diagonal line.

The polyethylene oxide polymer described above fits in the category WD, weakly dipolar with weak or no hydrogen-bonding, based on its carbon backbone with oxygen linkages. Poly(styrene-co-maleic acid) is categorized as HBA, hydrogen bond acidic, based on the two carboxylic groups. As seen in the QSAR sensor activity equation of the poly(styrene-co-maleic acid)-carbon sensing film, along with the polymer-analyte interaction term (E_{pa}), the analyte properties that appear are the number of hydrogen donor sites (HB$_D$) and the components of the dipole moment. Poly(epichlorohydrin-co-ethylene oxide) is a copolymer that has ethylene oxide monomer units (same as polyethylene oxide) in combination with the epichlorohydrin monomer units, which contains a chlorine atom. The combination of these monomer units makes the polymer weakly dipolar. The QSAR sensor activity

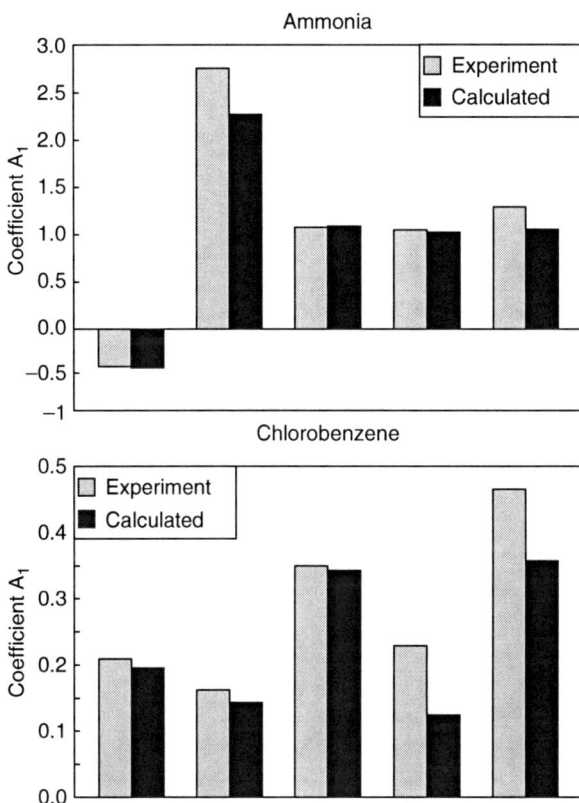

Fig. 8.9 QSAR modeled array response for the training analytes: ammonia and chlorobenzene, for the JPL ENose polymer-carbon composite sensors. Shown are a comparison of the calculated A_1 coefficients (using QSAR model) and experimental values for the sensors. From *left* to *right* are polymer-carbon composite sensors of: poly(2-vinyl pyridine), poly(styrene-co-maleic acid), poly(epichlorohydrin-co-ethylene oxide), polyethylene oxide and poly(4-vinyl pyridine)

model for poly(epichlorohydrin-co-ethylene oxide) contains terms related to analyte-analyte interactions (E_{aa}) and vapor pressure (VP) terms with the polymer-analyte interaction. Poly(2-vinyl pyridine), based on the nitrogen atom present in the side chain of the monomer unit, is categorized as HBB, hydrogen bond basic category. The term A_{pol} that appeares in the poly(2-vinyl pyridine) QSAR sensor activity equation is the sum of atomic polarizabilities that is calculated from molecular mechanics inputs.

As described in the earlier sections, the QSAR sensor response models in this work were developed by correlating experimental data of the training analyte set with molecular descriptors. Fig. 8.8 shows a fit of QSAR model calculated versus experimental A_1 coefficients values for analytes for the polymer-carbon composite

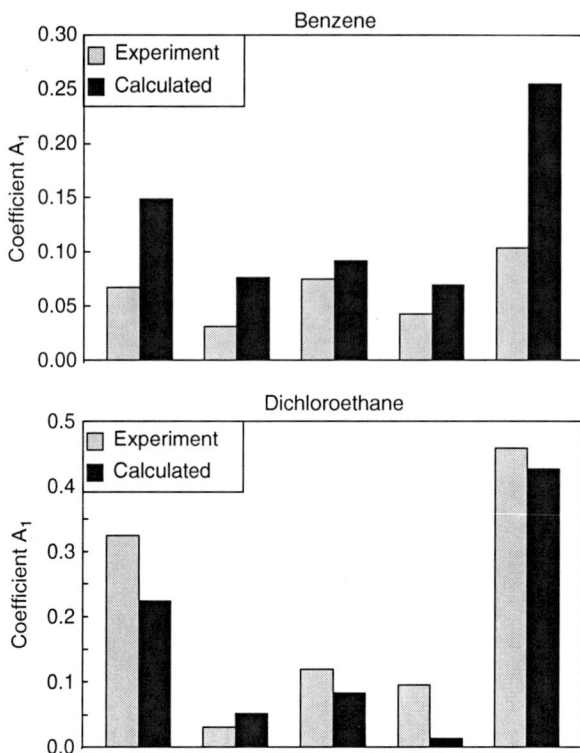

Fig. 8.10 QSAR modeled array response for test analytes: benzene and dichloroethane, for the JPL ENose polymer-carbon composite sensors. Shown are a comparison of the calculated A_1 coefficients (using QSAR model) and experimental values for the sensors. From left to right are polymer-carbon composite sensors of: poly(2-vinyl pyridine), poly(styrene-co-maleic acid), poly (epichlorohydrin-co-ethylene oxide), polyethylene oxide and poly(4-vinyl pyridine)

sensors, consisting of polymers: poly(2-vinyl pyridine), poly(styrene-co-maleic acid), poly(epichlorohydrin-co-ethylene oxide), and polyethylene oxide. A good fit is seen for all these sensors. The r^2 value corresponds to the fit the training analyte data set and the diagonal line fit corresponds to the condition where the calculated value is equal to the experimental value.

It would be interesting to consider some individual analytes from the training set and do a comparison between the calculated and experimental A_1 coefficients values for these sensors. The QSAR calculated and the experimental A_1 coefficients values for ammonia and chlorobenzene of the training set is shown in Fig. 8.9. The plot also includes data from an additional sensing film of poly(4-vinyl pyridine)-carbon composite, in-addition to the four polymer-carbon sensing films discussed above. Overall, there is a good match between these calculated and experimental values over the entire sensor array.

Next, to investigate is the prediction ability of the QSAR models for these polymer-carbon sensing films. As previously discussed, the test analytes are not a part of the training analyte set that is used in forming the QSAR models using GFA. Figure 8.8 shows how the activity of the test analytes is in close proximity of the diagonal line, which reflects a condition of the calculated coefficient value equal to the experimental value. We have plotted QSAR calculated and the experimental coefficients A_1 for test analytes benzene and dichloroethane in Fig. 8.10. It can be seen that a good prediction is observed for all the sensors which have different chemical functionalities. Furthermore, the above exercise suggests that we can use this prediction ability of QSAR sensor models, as a tool to predict the sensor response of untested analytes molecules to gauge the response of various sensors. This further suggests the applicability of the QSAR approach as a promising computational tool for sensing materials evaluation and selection [11].

8.5 Conclusions

The approach of developing validated sensor response models by correlating sensor response to analyte and sensor response descriptors using QSAR is a promising one. These validated QSAR sensor response models have shown good prediction capabilities. For a given sensing material, the validated model can be used to predict sensor responses to new analytes prior to experimentation. This can be expanded to other sensors in the array and an understanding of whether the existing set of sensors in an array will respond to the new analyte (s). Thus, QSAR sensor models can help in evaluation and selection, by allowing us to determine whether an analyte is likely to induce a weak or strong response in selected sensors. Subsequently, fewer experimental tests will need to be run on any given sensor array. The development of sensor response models for individual sensors in the array will provide a tool for generating virtual training sets for analytes that may not be easily tested, such as explosive or highly toxic or compounds.

In order to develop a model which will provide predictions of sensor response sufficient to use the calculated parameters in identification and quantification software it will be necessary to go beyond calculating the linear term of a quadratic equation. Further work on incorporating both terms is ongoing.

Acknowledgments This research was funded by the Advanced Environmental Monitoring and Control Program of NASA. This work was carried out at the Jet Propulsion Laboratory, California Institute of Technology under a contract with the National Aeronautics and Space Administration. The authors thank Prof. Anton Hopfinger and Dr. David Rogers for their very valuable suggestions and the Caltech Material and Process Simulation center (MSC) for use of computing facilities.

References

1. Eastman, M. P.; Hughes, R. C.; Yelton, G.; Ricco, A. J.; Patel, S. V.; Jenkins, M. W., Application of the solubility parameter concept to the design of chemiresistor arrays, *J. Electrochem. Soc.* **1999**, 146, 3907–3913
2. Grate, J. W.; Patrash, S. J.; Kaganove, S. N.; Abraham, M. H.; Wise, B. M.; Gallagher, N. B., Inverse least-squares modeling of vapor descriptors using polymer-coated surface acoustic wave sensor array responses, *Anal. Chem.* **2001**, 73, 5247–5259
3. Hierlemann, A.; Zellers E. T.; Ricco, A. J., Use of linear solvation energy relationships for modeling responses from polymer-coated acoustic-wave vapor sensors, *Anal. Chem.* **2001**, 73, 3458–3466
4. Nakamura, K.; Nakamoto T.; Moriizumi, T., Prediction of QCM gas sensor responses and calculation of electrostatic contribution to sensor responses using a computational chemistry method, *Mater. Sci. Eng. C* **2000**, 12, 3–7
5. Goddard III, W. A.; Cagin, T.; Blanco, M.; Vaidehi, N.; Dasgupta, S.; Floriano, W.; Belmares, M.; Kua, J.; Zamanakos, G.; Kashihara, S.; Iotov, M.; Gao, G. H., Strategies for multiscale modeling and simulation of organic materials: polymers and biopolymers, *Comput. Theo. Poly. Sci.* **2001**, 11, 329–343
6. Belmares, M.; Blanco, M.; Goddard III, W.A; Ross, R. B.; Caldwell, G.; Chou, S. -H.; Pham, J.; Olofson, P. M.; Thomas, C., Hildebrand and Hansen solubility parameters from molecular dynamics with applications to electronic nose polymer sensors, *J. Comp. Chem.* **2004**, 25, 1814–1826
7. Shevade, A. V.; Homer, M. L.; Taylor, C. J.; Zhou, H.; Manatt, K.; Jewell, A. D.; Kisor, A.; Yen, S. P. S.; Ryan, M. A., Correlating polymer-carbon composite sensor response with molecular descriptors, *J. Electrochem.* **2006**, 153, H209–H216
8. Ryan, M. A.; Zhou, H.; Buehler, M. G.; Manatt, K. S.; Mowrey, V. S.; Jackson, S. P.; Kisor, A. K.; Shevade, A.V.; Homer, M. L., Monitoring space shuttle air quality using the jet propulsion laboratory electronic nose, *IEEE Sens. J.* **2004**, 4, 337–347
9. Ryan, M. A.; Shevade, A. V.; Zhou H.; Homer, M. L., Polymer-carbon black composite sensors in an electronic nose for air-quality monitoring, *MRS Bull.* **2004**, 29, 714–719
10. Ryan, M. A.; Homer, M. L.; Zhou, H.; Manatt, K. S.; Manfreda, A., Toward a second generation electronic nose at JPL: sensing film optimization studies, In *Proceedings of the International Conference On Environmental Systems* **2001**, 2001–01–2308, Orlando, FL,
11. Ryan, M. A.; Homer, M. L.; Zhou, H.; Manatt, K.; Manfreda, A.; Kisor, A.; Shevade, A.; Yen, S. P. S., Expanding the analyte set of the JPL electronic nose to include inorganic species, *J Aerosp SAE Trans*, **2005**, 114, 225–232
12. Ryan, M. A.; Homer, M. L.; Zhou, H.; Manatt, K.; Manfreda, A.; Kisor, A.; Shevade, A.; Yen, S. P. S., Expanding the capabilities of the JPL Electronic nose for an international space station technology demonstration, In *Proceedings of the 36th International Conference on Environmental Systems*, Arlington, VA, **2006**
13. Shevade, A. V.; Ryan, M. A.; Taylor, C. J.; Homer, M. L.; Jewell, A. D.; Kisor, A. K.; Manatt, K. S.; Yen, S. -P. S., Development of the third generation JPL electronic nose for international space station technology demonstration, In *Proceedings of International Conference on Environmental Systems*, Chicago, IL, **2007**
14. Shevade, A. V.; Ryan, M. A.; Homer, M. L.; Kisor A. K.; Manatt, K. S., Off-gassing and particle release by heated polymeric materials, In *Proceedings of 38th International Conference on Environmental Systems*, San Francisco, CA, **2008**
15. Ryan, M. A.; Shevade, A. V.; Kisor, A. K.; Manatt, K. S.; Homer, M. L.; Lara, L. M.; Zhou, H., Ground validation of the third generation JPL electronic nose, In *Proceedings of 38th International Conference on Environmental Systems*, San Francisco CA, **2008**

16. Homer, M. L.; Lim, J. R.; Manatt, K.; Kisor, A.; Lara, L.; Jewell, A. D.; Shevade, A. V.; Ryan, M. A., Using temperature effects on polymer-composite sensor arrays to identify analytes, *Proc. IEEE Sens.* **2003**, 1, 144–147

17. Homer, M. L.; Lim, J. R.; Manatt, K.; Kisor, A.; Manfreda, A. M.; Lara, L.; Jewell, A. D.; Yen, S. -P. S.; Zhou, H.; Shevade, A. V.; Ryan, M.A., Temperature effects on polymer-carbon composite sensors: evaluating the role of polymer molecular weight and carbon loading, *Proc. IEEE Sens.* **2003**, 2, 877–881

18. Manfreda, A. M., Elucidating Humidity Dependence of the Jet Propulsion Laboratory's Electronic Nose Polymer-Carbon Composite Sensors, Research report submitted to the *California State Polytechnic University – Pomona* **2002**

19. Zhou, H., Homer, M. L., Shevade, A. V., Ryan, M. A., Nonlinear least-squares based method for identifying and quantifying single and mixed contaminants in air with an electronic nose, *Sensors* **2006**, 6, 1–18

20. Liu, J. Z.; Hopfinger, A. J., Identification of possible sources of nanotoxicity from carbon nanotubes inserted into membrane bilayers using membrane interaction quantitative structure-activity relationship analysis, *Chem. Res. Toxicol.* **2008**, 21, 459–466

21. Kurup, A.; Mekapati, S. B.; Garg R.; Hansch, C., HIV-1 protease inhibitors: A comparative QSAR analysis, *Current Med. Chem.* **2003**, 10, 1679–1688

22. Raymond, J. W.; Rogers, T. N.; Shonnard, D. R.; Kline, A., A review of structure-based biodegradation estimation methods, *J. Hazard. Mater.* **2001**, 84, 189–215

23. Perkins, R.; Fang, H.; Tong, W. D.; Welsh, W., Quantitative structure-activity relationship methods: Perspectives on drug discovery and toxicology, *Environ. Toxicol. Chem.* **2003**, 22, 1666–1679

24. Rogers, D.; Hopfinger, A. J., Application of genetic of function approximation to quantitative structure-activity relationships and quantitative structure-property relationships, *J. Chem. Inf. Comp. Sci.* **1994**, 34, 854–866

25. Brocchini, S.; James, K.; Tangpasuthadol, V.; Kohn, J., Structure-property correlations in a combinatorial library of degradable biomaterials, *J. Biomed. Mat. Res.* **1998**, 42, 66–75

26. Smith, J. R.; Seyda, A.; Weber, N.; Knight, D.; Abramson, S.; Kohn, J., Integration of combinatorial synthesis, rapid screening, and computational modeling in biomaterials development, *Macromol. Rapid Comm.* **2004**, 25, 127–140

27. Abramson, S. D.; Alexe, G.; Hammer, P. L.; Kohn, J., A computational approach to predicting cell growth on polymeric biomaterials, *J. Biomed. Mat. Res. Part A* **2005**, 73A, 116–124

28. Livingstone, D., Data Analysis for Chemists, Oxford University Press, New York **1995**

29. van Krevelen, D. W., Properties of Polymers: Their Correlation with Chemical Structure; their Numerical Estimation and Prediction from Group Contributions, Elsevier, New York, **1990**

30. Severin, E. J.; Lewis, N. S., Relationships among resonant frequency changes on a coated quartz crystal microbalance, thickness changes, and resistance responses of polymer-carbon black composite chemiresistors, *Anal. Chem.* **2000**, 72, 2008–2015

31. Blanco, M., Molecular silverware 1. General solutions to excluded volume constrained problems, J. Comput. Chem **1991**, 12, 237–247

32. Shevade, A. V.; Ryan, M. A.; Homer, M. L.; Manfreda, A. M.; Zhou, H.; Manatt, K., Molecular modeling of polymer composite-analyte interactions in electronic nose sensors, *Sens. Actuat. B: Chem.* **2003**, 93, 84–91

33. Cerius^2 v 4.2, Accelrys Inc., San Diego, CA

34. Berendsen, H. J. C, et al., Intermolecular Forces; Pullman, B., Ed.; Reidel, Dordrecht, Holland, **1981**, 331

35. Rappe, A. K.; Goddard, W.A., Charge equilibration for Molecular-dynamics simulations, *J. Phys. Chem.* **1991**, 95, 3358–3363

36. Mayo, S. L.; Olafson, B. D.; Goddard, W. A., Dreiding A. generic force-field for molecular simulations, *J. Phys. Chem.* **1990**, 94, 8897–8909

37. Steele, W. A., The Interaction of Gases with Solids Surfaces, Clarendon Press, Oxford, **1974**

38. Siperstein, F. R.; Myers, A. L., Mixed-gas adsorption, *AIChE J* **2001**, 47, 1141–1159

39. Grate, J. W.; Kaganove, S. N.; Bhethanabotla, V. R., Comparisons of polymer/gas partition coefficients calculated from responses of thickness shear mode and surface acoustic wave vapor sensors, *Anal. Chem.* **1998**, 70, 199–203

40. Grate, J. W.; Abraham, M. H.; Du, C. M.; Mcgill, R. A.; Shuely, W. J., Examination of vapor sorption by fullerene, fullerene-coated surface-acoustic-wave sensors, graphite, and low-polarity polymers using linear solvation energy relationships *Langmuir* **1995**, 11, 2125–2130

41. Severin, E. J.; Doleman, B. J.; Lewis, N. S., An investigation of the concentration dependence and response to analyte mixtures of carbon black/insulating organic polymer composite vapor detectors, *Anal. Chem.* **2000**, 72, 658–668

42. Doleman, B. J.; Severin, E. J.; Lewis, N. S., Trends in odor intensity for human and electronic noses: Relative roles of odorant vapor pressure vs. molecularly specific odorant binding, *Proc. Nat. Acad. Sci.* **1998**, 95, 5442–5447

43. Lonergan, M. C.; Severin, E. J.; Doleman, B. J.; Beaber, S. A.; Grubbs, R. H.; Lewis, N. S., Array-based vapor sensing using chemically sensitive, carbon black-polymer resistors, *Anal. Chem.* **1996**, 8, 2298–2312

44. Janghorbani, M.; Freund, H., Application of a piezoelectric quartz crystal as a partition detector-development of digital sensor, *Anal. Chem.* **1973**, 45, 325–332

45. National Institute of Standards and Technology (NIST)/TRC Vapor Pressure Database, version **2001**

Chapter 9
Design and Information Content of Arrays of Sorption-Based Vapor Sensors Using Solubility Interactions and Linear Solvation Energy Relationships

Jay W. Grate, Michael H. Abraham, and Barry M. Wise

Abstract The sorption of vapors by the selective polymeric layer on a chemical vapor sensor is described in detail and dissected into fundamental solubility interactions. The sorption process is modeled in terms of solvation parameters for vapor solubility properties and linear solvation energy relationships. The latter relationships model the log of the partition coefficient as the sum of terms related to specific types of interactions. The approaches are particularly applicable to the design and understanding of acoustic wave chemical vapor sensors such as those based on surface acoustic wave devices. It is shown how an understanding of solubility interactions informs the selection of polymers to obtain chemical diversity in sensor arrays and obtain the maximum amount of chemical information. The inherent dimensionality of the array data, as analyzed by principal components analysis, is consistent with this formulation. Furthermore, it is shown how new chemometric methods have been developed to extract the chemical information from array responses in terms of solvation parameters serving as descriptors of the detected vapor.

9.1 Introduction

A sorption-based sensor for vapor sensing, as defined in this chapter, consists of a microfabricated device with a thin layer of a selectively sorbent material on the surface, as shown in Fig. 9.1. The sensor's response arises from the quantity of

Jay W. Grate (✉)
Pacific Northwest National Laboratory, Chemistry and Materials Sciences Division, P.O.Box 999, WA 99353, Richland, USA
e-mail: jwgrate@pnl.gov

M.H. Abraham
Department of Chemistry, University College London, 20 Gordon Street, 1H 0AJ, London, WCIH OAJ, UK

B.M. Wise
Eigenvector Research, Inc., 3905 West Eaglerock Drive, Wenatchee, WA 98801, USA

M.A. Ryan et al. (eds.), *Computational Methods for Sensor Material Selection*, Integrated Analytical Systems, DOI 10.1007/978-0-387-73715-7_9, © Springer Science+Business Media, LLC 2009

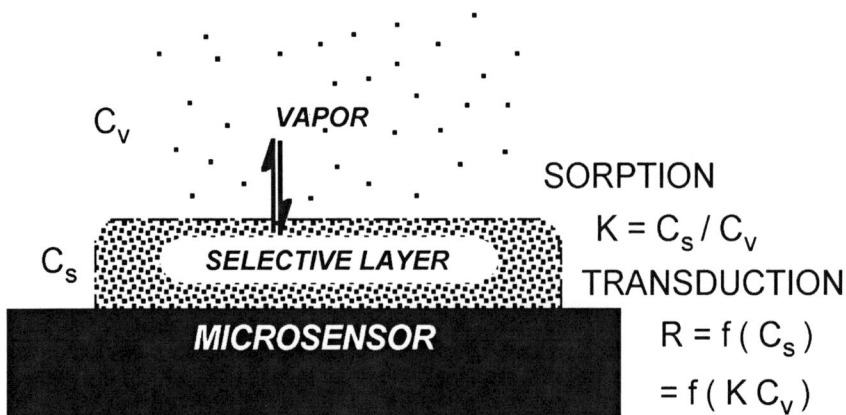

C_v VAPOR

SELECTIVE LAYER C_s

MICROSENSOR

SORPTION

$K = C_s / C_v$

TRANSDUCTION

$R = f(C_s)$

$= f(KC_v)$

Fig. 9.1 The absorption of vapor molecules into the bulk of a selective layer on the surface of a chemical microsensor

vapor absorbed in the selective layer and the method by which the device transduces this quantity of vapor into an analytical signal. The prototypical sorption sensor is a gravimetric acoustic wave device that transduces the amount of sorbed vapor as a mass into a change in resonant frequency. In this case, the sensor's signal is not dependent on specific analyte molecule properties except to the extent that they affect the amount of vapor absorbed.

The sorption of neutral vapor molecules into a selective layer entails transfer of vapor molecules from the gas phase where they have no interactions to a liquid or solid phase where new interactions occur. There are a defined number of fundamental interactions in this process.

The influences of these fundamental interactions on vapor absorption, by sorbents such as polymers, can be modeled systematically using Linear Solvation Energy Relationships (LSERs), as we shall describe below.

Because of the number and types of reversible interactions involved, and the great number of vapors with similar interactive properties, a sensor based on reversible absorption alone can never be perfectly selective. Accordingly, such sensors are typically used in arrays, where the collective responses of all the sensors give rise to a pattern that can be recognized using multivariate pattern recognition methods. On the other hand, the limited dimensionality, or rank, of the data from such arrays, and the fact that they can be modeled by LSERs, enables the extraction of chemical information from array responses as vapor solvation parameters, as we shall also describe below.

These approaches enable one to address basic issues with regard to the chemical information associated with arrays of sorption-based sensors, including: (1) what governs the selectivity of a sorbent layer on a sensor and how can you design or select a layer for a specific detection application; (2) what type of information and how much such information can be encoded in the responses of a sorption-based sensor array; (3) how you design a sensor array so that the multivariate response provide the most complete and selective information achievable; and (4) how you extract that information from the array data.

This chapter will be concerned with understanding the sorptive properties of selective layers, focusing on polymers, using solubility interactions and LSERs. Examples used in this chapter will come primarily from the use of polymer-coated surface acoustic wave (SAW) devices as sensor and sensor array platforms [1–6]. However, many of the concepts are relevant to any sensor using a selective layer for vapor absorption. Polymers are particularly useful as selective layers on sensors because they offer rapid reversible absorption for organic vapors; their selectivities are influenced by their chemical structures which can in turn be designed and synthesized, and can be readily applied to sensor surfaces as adherent thin films.

9.2   Vapor Sorption and Sensor Responses

The sorption of a vapor is characterized by the thermodynamic partition coefficient, K, which is the equilibrium distribution coefficient of the vapor between the gas phase at concentration C_v and the sorbent phase at concentration C_s, as given in (9.1).

$$K = C_s/C_v, \tag{9.1}$$

$$R = f(C_s), \tag{9.2}$$

$$R = f(C_v K). \tag{9.3}$$

The response, R, of a sorption-based chemical sensor is a function of the amount of vapor sorbed into the film, C_s, as indicated in Fig. 9.1 and (9.2). Since the partition coefficient relates C_s to C_v, the response can be related to the partition coefficient and the gas phase vapor concentration as shown in (9.3). If the function in (9.2) and (9.3) is a simple linear relationship, then these equations can be rewritten as (9.4) and (9.5) where S is a sensitivity factor.

$$R = SC_s, \tag{9.4}$$

$$R = SC_v K. \tag{9.5}$$

In these relationships, it is evident that the sensor response is dependent on the value of K; therefore interactions that promote sorption and raise the value of K will favorably influence the response of the sensor. The significance of K to vapor sorption and detection using coated acoustic wave sensors has been noted especially in the case of the quartz crystal microbalance (QCM) [7–9] and SAW sensors [10, 11].

Depending on the transduction mechanism of the sensor device, the sensitivity factor or function may depend only on the characteristics of the coated device, or it may also include factors that are specific to the particular vapor molecules that are absorbed. For an acoustic wave sensor acting as a purely gravimetric transducer, there are no vapor molecule specific sensitivity factors, and the response can be expressed as shown in (9.6) [11].

$$\Delta f_{\rm v} = \Delta f_{\rm s} C_{\rm v} K / \rho_{\rm s}. \tag{9.6}$$

The response is a frequency shift denoted by $\Delta f_{\rm v}$. The parameter $\Delta f_{\rm s}$ is a measure of the amount of polymer on the sensor surface given by the frequency shift due to the deposition of the film material onto the bare sensor. The parameter $\rho_{\rm s}$ is the density of the sorbent layer material. The quotient $\Delta f_{\rm s}/\rho_{\rm s}$ in (9.6) then represents the sensitivity factor S in (9.5); it only contains factors related to the characteristics of the coated device.

The QCM is a purely gravimetric sensor given sufficiently thin polymer films that move synchronously with the device surface [2, 3]. The SAW device can be a gravimetric device but polymer-coated SAW sensors often have an additional sensitivity to changes in the modulus of the polymer upon vapor sorption. The modulus change is related to the volume of the sorbed molecules through their influence on polymer film free volume [2, 12–18].

It is also possible to have a sorption-based sensor that responds to the amount of vapor as a volume rather than the amount as a mass. For example, a carbon particle/polymer composite can respond by this mechanism [12, 14, 19]. The response of a volume-transducing sensor can be expressed as given in (9.7)

$$R = v_{\rm v} S' C_{\rm v} K. \tag{9.7}$$

In this case the sensitivity factor S from (9.4) and (9.5) has been split into the sensitivity parameter S' and the vapor specific volume, $v_{\rm v}$. The latter factor is related to the volume fraction of vapor in the polymer/vapor solution, $\phi_{\rm v} = v_{\rm v} C_{\rm v} K$ and it represents a vapor molecule specific factor in the sensitivity [12, 14].

If a polymer-coated SAW sensor does respond to both the mass of sorbed vapor and its influence on polymer modulus via volume effects, then it acts as a mass-plus-volume-transducing device whose response can be expressed by (9.8). The first term has no vapor molecule-specific sensitivity factors while the second term has $v_{\rm v}$, which will vary from vapor to vapor.

$$\Delta f_{\rm v} = (\Delta f_{\rm s} C_{\rm v} K / \rho_{\rm s}) + v_{\rm v} S' C_{\rm v} K. \tag{9.8}$$

These relationships involving K relate to the steady state response of the sensor once the vapor phase concentration has equilibrated with the sorbent phase. In most sensor applications, it is desirable to reach this equilibration rapidly in order to obtain a fast sensor response. Accordingly it is important that vapors diffuse rapidly in the selective layer material. For polymers, the key parameter is the glass-to-rubber transition temperature, $T_{\rm g}$, which should be below the operating temperature of the sensor. Vapor diffusion is typically slow in glassy polymers below this transition temperature, but can be much more rapid above the transition temperature where the polymer is rubbery.

9.3   Solubility Interactions and LSERs

9.3.1   Solubility Interactions in Vapor Absorption

The absorption of a vapor from the gas phase entails the dissolution of the vapor into the bulk of the layer material, where the vapor molecule is the solute and the layer material is the solvent. Conceptually, the dissolution step can be broken down into the formation of a cavity within the material, occupation of that cavity by the solute, and the development of favorable interactions between the solute and the solvent. These interactions are, by definition, solubility interactions, including dispersion interactions, dipole-induced dipole interactions, dipole-dipole interactions, and hydrogen-bonding interactions [1]. These are the primary interactions that can occur between a neutral vapor molecule and an uncharged polymer material. The first three are often grouped together under the term van der Waals interactions.

Dispersion interactions are also known as London forces and could be characterized as instantaneous-dipole/instantaneous-dipole interactions. They are quite general and can occur in the absence of permanent dipoles, since momentary or "instantaneous" dipoles can exist due to the mobility of electrons around the nucleus. Dispersion interactions, for example, occur between a nonpolar alkane molecular solute and a nonpolar alkane solvent or simple aliphatic polymer. They also occur in the interactions between all other organic molecules and polymers, including those that may have additional heteroatoms and functional groups, and thus may participate in additional interactions at the same time.

Dipole induced-dipole interactions, also called induction interactions, occur between permanent dipoles and polarizable regions of a molecule or polymer. The nearby charge of the permanent dipole induces a dipole in the polarizable region, and hence the strength of this interaction depends on the strength of the perturbing dipole and the polarizability of the perturbed region. Dipole-dipole interactions are electrostatic interactions between the positively and negatively charged regions of permanent dipoles. The strength of the interaction depends on the dipole strengths and their orientations.

Hydrogen bonding interactions occur between the hydrogen atoms from the "hydrogen-bond acid" and basic heteroatoms of a "hydrogen-bond base". Hydrogen bonding acidity (A) and basicity (B) must be distinguished from proton transfer acidity and basicity. All references to acidity and basicity in this chapter refer to the former and not to proton transfer.

Proton transfer is a *reaction* between an acid and a base leading to the conjugate base and conjugate acid. Hydrogen bonding entails an *interaction* between a hydrogen-bond acid and a hydrogen-bond base without proton transfer. Although simple correlations between hydrogen bonding and proton transfer parameters can be made within chemical families, such correlations do not hold up across all different chemical families. There is no general relationship between proton transfer pKa values and hydrogen-bond acidity, for example [20–22]. Resonance

	A	B
Table 9.1 Parameters for hydrogen-bond acidity and basicity for selected solutes		
Hexane	0.00	0.00
Methanol	0.43	0.47
Phenol	0.60	0.30
Acetic acid	0.61	0.44
Urea	0.50	0.90
Triethylamine	0.00	0.79

stabilization of a conjugate base can be quite significant in influencing proton dissociation but is not so relevant to hydrogen-bonding interactions.

Table 9.1 provides solvation parameter values A and B for the hydrogen bonding acidity and basicity, respectively, for selected compounds as solutes. Although phenol and acetic acid have similar hydrogen bond acidities as indicated by their A values of 0.60 and 0.62 respectively, acetic acid is a stronger proton dissociation acid in water by five orders of magnitude. Urea is a stronger hydrogen bond base than triethylamine as indicated by their B values, but urea is actually a weaker proton acceptor base than triethylamine.

9.3.2 The LSER Model for Vapor Absorption

Solubility-dependent phenomena, such as partitioning of a vapor molecule between the gas phase and a sorbent polymer, can be systematically modeled in terms of solubility interactions using LSERs. These models are semi-empirical models expressing a measure of the solubility-dependent phenomenon as a linear combination of terms related to the fundamental interactions. LSERs have been successful in correlating a vast amount of solubility-dependent phenomena, often to the precision of the available data [21, 23–26]. The application of LSERs to the study of polymer-coated chemical sensors was introduced in 1988 [11] and has been described in detail in a number of articles and reviews [1, 3, 4, 22, 27, 28].

Abraham developed the general form of the LSER equation, and the scales of solvation parameters that are used to model vapor absorption. The LSER equation for vapor sorption is given in (9.9), where K is the partition coefficient as defined in (9.1) [1, 22, 24]. Recently the notation in (9.9a) has been simplified as shown in (9.9b) [26].

$$\log K = c + rR_2 + s\pi_2^H + a \sum \alpha_2^H + b \sum \beta_2^H + l \log L^{16}, \qquad (9.9a)$$

$$\log K = c + eE + sS + aA + bB + lL. \qquad (9.9b)$$

The independent variables in (9.9) are the solvation parameters serving as solute descriptors: E is the solute excess molar refractivity in units of $(cm^3 \, mol^{-1})/10$, S is the solute dipolarity/polarizability, A and B are the overall or summation hydrogen bond acidity and basicity, and L is the logarithm of the gas-hexadecane partition coefficient at 298 K.

Each term other than the constant contains a solute descriptor related to the vapor's solubility properties (E, S, A, B, L) and a coefficient (e, s, a, b, or l) representing the complementary solubility properties of the sorbent phase acting as the solvent (typically a polymer in the present context) [24, 26, 29]. Thus, $\log K$ is modeled as a linear combination of terms related to particular types of interactions, where $e\,E$ is a polarizability term, $s\,S$ is a dipolarity/polarizability term, $a\,A$ is a hydrogen-bonding term in which the vapor is the hydrogen-bond acid, $b\,B$ is a hydrogen-bonding term in which the vapor is the hydrogen-bond base, and $l\,L$ is a combined dispersion interaction and cavity term.

The solute descriptor E is a calculated excess molar refraction parameter that provides a quantitative measure of polarizable n and p electrons. The solute descriptor S measures a solute molecule's ability to stabilize a neighboring charge or dipole through dipole-dipole or dipole-induced dipole interactions. The hydrogen bonding parameters A and B measure effective hydrogen-bond acidity and basicity, respectively, of the solute molecule. The L parameter, mostly determined by gas-liquid chromatography, is a combined measure of exoergic dispersion interactions that increase L and the endoergic cost of creating a cavity in hexadecane leading to a decrease in L. All of these parameters are free energy related. The parameter scales were derived from measurements of complexation or partitioning equilibria [26, 30–34]. The E parameter can be obtained from an experimental refractive index of the liquid solute at 293 K, or can be obtained by an arithmetic calculation from group contributions [35]. A recent review sets out in detail the construction of the various solute parameters [26].

The l-coefficient is related to dispersion interactions and the cost of cavity formation in the solvent phase. The s-coefficient is related to the solvent phase dipolarity and polarizability. Similarly, the e-coefficient is related to polarizability. The a- and b-coefficients, being complementary to the vapor hydrogen-bond acidity and basicity, represent the solvent phase hydrogen-bond basicity and acidity, respectively. The coefficients and the constant c are obtained by regressing the measured partition coefficients of a series of diverse solute vapors against the known vapor solvation parameters using multiple linear regression. Typically the partition coefficients are determined by gas chromatographic measurements where the solutes are injected on a column where the sorbent material of interest is the stationary phase [36]. These partition coefficients represent the distribution coefficient with the vapor at infinite dilution.

LSER models have also been used to correlate the responses of polymer-coated acoustic wave sensors with the vapor solvation parameters. This approach is most rigorously correct if the sensor response is purely gravimetric and thus directly proportional to the partition coefficient as given in (9.6). For an acoustic wave sensor where the modulus contributes to the observed responses, as can be the case with polymer-coated SAW devices, the responses remain proportional to partition coefficients, however the proportionality is affected by an analyte specific sensitivity factor in the form of the vapor specific volume (see (9.8)) Nevertheless this is not a large perturbation and good correlations can be obtained. This approach was first shown by Zellers in 1993 for four polymers on SAW sensors [37]. In 2001, Grate et al. reported

LSER equations determined from SAW sensor responses for fourteen polymers that were "well-behaved" as sensor coatings and six less well-behaved materials, obtained from sensor responses to eighteen diverse vapors [27]. The first use of QCM devices to obtain data for the determination of LSER coefficients was reported in 2001 by Hierleman et al. [38] LSER modes were determined for six polysiloxanes.

9.3.3 Application to Polymeric Sensing Materials

The use of LSER models to understand polymer properties and the interactions involved in vapor absorption can be illustrated by examining some specific polymers. Figure 9.2 shows the structures of some polymers that have been

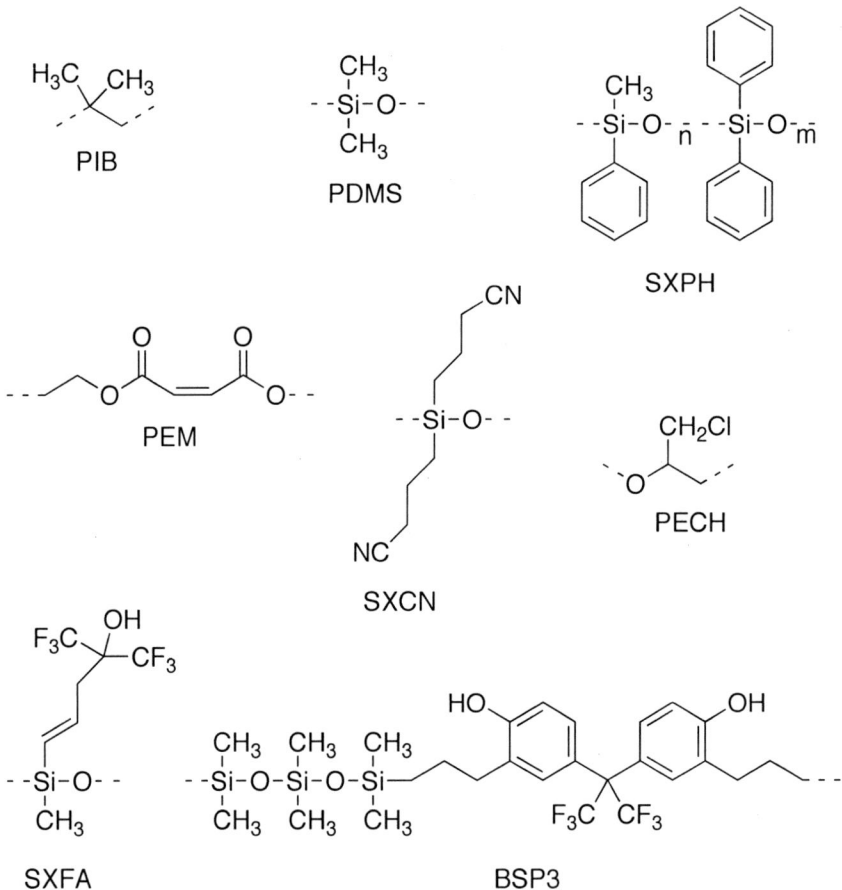

Fig. 9.2 The chemical structures of the repeat units of some diverse polymers used as sensing layers. The *top three* are low polarity or polarizable polymers. The *middle three* are basic and dipolar. The *lower two* polymers are strong hydrogen bond acids

Table 9.2 LSER coefficients for selected polymers

Polymer	Polarizability	Dipolarity/ polarizability	Hydrogen bond basicity	Hydrogen bond acidity	Dispersion/cavity
	e	s	a	b	l
PIB	−0.077	0.366	0.180	0.000	1.016
SXPH	0.177	1.287	0.556	0.440	0.885
SXCN	0.000	2.283	3.032	0.516	0.773
SXFA	−0.417	0.602	0.698	4.25	0.718
PEM	−1.032	2.754	4.226	0.000	0.865
PECH	0.096	1.628	1.450	0.707	0.831
PEI	0.495	1.516	7.018	0.000	0.770
SXPYR	−0.189	2.425	6.780	0.000	1.016
FPOL	−0.672	1.446	1.494	4.086	0.810

characterized by LSER methods or used as absorbent polymers on chemical vapor sensors (or both) [27, 36]. Table 9.2 provides the LSER coefficients for several polymers, as determined by using gas chromatographic methods to obtain the required partition coefficients [36]. The first four polymers in Table 9.2 illustrate structures with differing solubility properties as follows.

Poly(isobutylene) (PIB) is an aliphatic hydrocarbon polymer that can only interact by dispersion interactions. Such polymers are useful for absorbing aliphatic hydrocarbon vapors and are characterized by large l-coefficients. Another polymer with similar properties is poly(dimethylsiloxane) (PDMS).

The addition of aromatic groups to a polymer can increase the polarizability, as is the case with SXPH, a polysiloxane with 75% phenyl substituents and 25% methyl substituents. The gas chromatography phase OV-25 has the same composition [39]. Such materials are good for sensitivity to chlorinated hydrocarbons, and in combination with aliphatic materials (such as PIB), they help to distinguish between various low polarity vapors such as aliphatic hydrocarbons, aromatic hydrocarbons, and chlorinated hydrocarbons.

The nitrile groups of SXCN are highly dipolar and also basic, which can be seen in the high s-coefficient for dipolarity and the high a-coefficient for basicity. Another polymer with similar properties is poly(ethylene maleate)(PEM). Poly (epichlorohydrin)(PECH), which has also been a useful sensor material, has weaker dipolarity and basicity. Other basic polymers include polyethyleneimine (PEI) and a pyridine-substituted polysiloxane (SXPYR).

On the other hand, the fluoroalcohol groups of SXFA are strongly hydrogen bond acidic giving rise to a high b-coefficient for acidity [36, 40]. The latter polymer is excellent for the absorption of basic vapors, such as organophosphorus compounds, and has also been used in explosives detection [12, 27, 41–50]. A number of other polymers with strong hydrogen bond acidity such as BSP3, have been designed and synthesized [5, 6, 51, 52], and these were recently reviewed [53]. BSP3 has also been used in detection of organophosphorus compounds and in sensor arrays [5, 6, 27, 42, 51, 54–57]. Prior to the development of SXFA and BSP3, fluoropolyol (FPOL) was the sensing polymer most frequently used for its hydrogen-bond acidic properties [1, 5, 6, 11, 12, 18, 27, 36, 41–43, 48, 51, 58–70].

Table 9.3 Calculated interaction terms[a] for four vapors[b] on three polymers

Polymer	Vapor	Dipolarity/ polarizability	Hydrogen bonding	Hydrogen bonding	Dispersion/ cavity
		$(e\,\mathbf{E} + s\mathbf{S})$	$a\,\mathbf{A}$	$b\,\mathbf{B}$	$(c + l\,\mathbf{L})$
PIB	Hexane	0	–	–	1.94
	Et$_3$N	0.04	–	–	2.32
	DMF	0.45	–	–	2.45
	EtOH	0.13	0.07	–	0.74
SXCN	Hexane	0	–	–	0.43
	Et$_3$N	0.34	–	0.41	0.72
	DMF	2.99	–	0.38	0.82
	EtOH	0.96	1.12	0.25	–0.48
SXFA	Hexane	–	–	–	1.84
	Et$_3$N	0.05	–	3.36	2.10
	DMF	0.64	–	3.15	2.20
	EtOH	0.15	0.26	2.04	0.99

[a]Dashes indicate terms calculated to be zero

Given the solvation parameters for a vapor, and the LSER coefficients for a polymer, the interaction terms can be calculated. Examining interaction terms provides insights into the interactions that are most important in contributing to the log of the partition coefficient, and hence sensor selectivity and sensitivity.

Table 9.3 illustrates the calculation of interaction terms for three polymers and four vapors. Polarizability and dipolarity interactions can be taken as the sum of ($e\,\mathbf{E} + s\,\mathbf{S}$) where the $e\,\mathbf{E}$ term acts as a correction to the overall dipolarity/polarizability interaction indicated by $s\,\mathbf{S}$. The hydrogen bonding terms $a\,\mathbf{A}$ and $b\,\mathbf{B}$ represent hydrogen-bonding interactions where the polymer is a base in the first case and an acid in the second. Dispersion interactions favoring sorption can be difficult to separate from the cost of forming a cavity. These effects can be represented as the sum of the regression constant c and the $l\,\mathbf{L}$ term [71]. The interactions in Table 9.3 have been calculated according to these combinations.

With hexane, only dispersion interactions are possible with any of the polymers. Ethylamine is basic but only weakly dipolar, so one sees a strong hydrogen bonding interaction with acidic SXFA. Dimethylformamide is strongly dipolar and strongly basic. Accordingly, one sees a strong dipolar interaction with SXCN, and a strong hydrogen bonding interaction with acidic SXFA. Ethanol is the only hydrogen bond acidic vapor in this set, so one sees a significant hydrogen bonding interaction with basic SXCN polymer. As ethanol is also basic, it also hydrogen bonds with acidic SXFA.

In general, dispersion interactions are nearly always important in the sorption of vapors from condensable liquids. If the vapor has the solubility properties to interact by dipolar or hydrogen bonding interactions, then the polymer can be selected to set up such interactions. These polar interactions can make a significant contribution to the overall sorption and influence the selectivity of the sensor [22].

The discussion in this section has primarily addressed the use of LSERs to understand polymer properties and vapor absorption. LSERs can also be used for the purposes of prediction. Log K values can be predicted for known vapors on polymers whose LSER coefficients have been determined, according to (9.9), and then gravimetric acoustic wave sensor responses can be predicted according to (9.6). Such predictions have been used to assess the likely sensitivity and detection limits for a variety of toxic volatile organic compounds [41]. These results predicted that SAW sensor arrays should be effective at achieving detection at or below regulatory limits for many such compounds.

Many of the above topics have also been discussed in a review on gas chromatography stationary phases [25]. LSER coefficients are given for a wide variety of phases, including polymers, and are examined in terms of solute-phase interactions.

9.4   Chemical Diversity in Sensor Arrays

9.4.1   Chemical Diversity in Terms of Solubility Interactions

The purpose of any chemical analysis device or instrument is to determine chemical information about the sample. Using a sensor array, the collective response of the array must collect sufficient chemical information to distinguish responses from the analyte(s) of interest from responses due to potentially interfering vapors. Accordingly, the sensors in the array will ideally provide as much chemical information as possible about the vapor sample so as to enable successful discrimination. Given an array based on one type of transducer, such as a SAW array, the selection of coatings for the array determines the type of chemical information that the sensor responses will encode. Since the mechanism of the sensor response depends on the amount of vapor absorbed, and the mechanism of vapor sorption depends on fundamental solubility interactions, it follows that the chemical information encoded by the sensor responses are related to these solubility interactions.

To maximize the chemical information that such an array encodes, the array coatings must be selected so that all the relevant solubility interactions that a vapor molecule can participate in are set up in the array. For example, if the array lacked a hydrogen bond acidic coating, the array would collect no information about vapor basicity properties.

Thus the overall objective is to create an array where each sorbent material emphasizes a different solubility interaction or a combination of solubility interactions, while ensuring the whole array probes all the available interactions [1, 4, 22]. Ideally, the sensors would produce completely orthogonal responses, although this is not rigorously achievable using real materials and real absorption processes. Dispersion interactions, for example, will occur in all sorbent materials. Nevertheless, it is possible to maximize particular solubility interactions in a given material while minimizing others, in order to vary the selectivity of the materials within the array.

This strategy for varying the selectivities of the materials within a sorptive sensor array was originally set out in 1991 by Grate and Abraham [1]. A sensor array with diverse properties could be designed by including a hydrogen-bond acidic material with minimal basicity and modest dipolarity, a hydrogen-bond basic material with no acidity and minimal dipolarity, a dipolar material minimizing basicity and having no acidity and a nonpolar polarizable material. This type of array would thus include the full range of solubility properties discussed above, and when used in combination with pattern recognition methods, they provide good selectivity. Such an array could be called a chemically diverse array, and its responses would encode chemical information about vapor solubility properties.

For vapor sensing applications, arrays would also include one or more materials that have combinations of properties in order to maximize sensitivity to particular vapors of interest.

The benefits of including sensing materials with diverse properties, and maximizing these properties, can be further understood in terms of how the array responses spread out the data for different vapors in the 'feature' space. Given n sensors, feature space can be defined by n axes, each corresponding to the response of one of the sensors. The response to each vapor exposure can be plotted in this n-dimensional feature space. If several test vapors have similar properties, they will plot near each other in the feature space. If these vapors are different, and the sensor coatings can respond to these different properties, then they will plot out in different regions of the feature space, and they will be distinguishable. Thus a sensor array with the most chemically diverse set of coatings will best spread out a diverse set of vapors in the feature space, and thus facilitate discrimination. A sensor material that interacts strongly by a particular interaction will result in a sensor signal to an interacting vapor that plots farther out on the axis representing that sensor. Thus it is desirable not only to include all potential interactions in the array, but also to maximize them in particular materials, in order to best spread out the sensor responses in the feature space.

9.4.2   Dimensionality of Array Data

It is possible to create array systems with tens or hundreds of devices. Will an array with one hundred different sensors provide more types of chemical information than an array with ten different sensors?

Principal components analysis is a chemometric technique used to explore the variance in a data set. It determines a set of principal components that are completely orthogonal which can describe the variance in the data.

If n-dimensional feature space is defined by n-sensors that are cross-correlated, then the number of orthogonal principal components will be less than n. Many sensor array studies have taken large data sets and performed principal components analysis.

For example, Kowalski and Carey analyzed data from 27 coated QCM sensors exposed to 14 different vapors. Seven principal components could account for 95% of the variance in the data set. Thus, seven coatings could be selected with the goal to "span the space of all coatings using the fewest number of individual coatings while retaining analyte discrimination" [72].

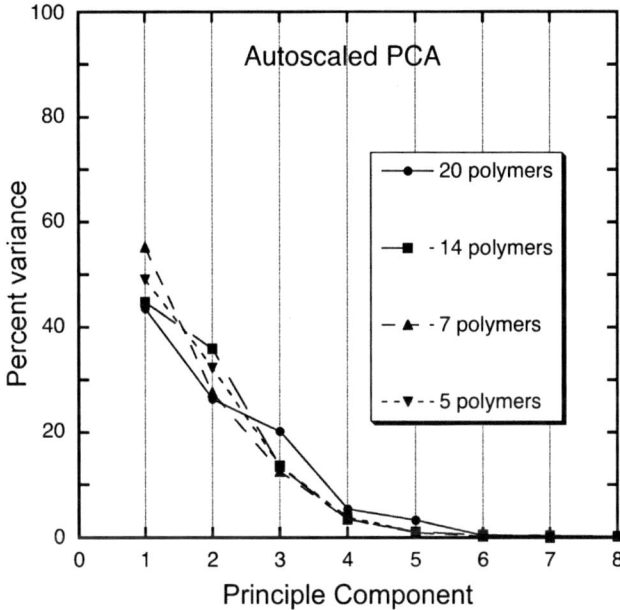

Fig. 9.3 Principal components plot from a data set of the responses of coated SAW sensors to eighteen organic vapors, pattern normalized and autoscaled, showing the effect of reducing the number of polymers on the spread of data in feature space, taking care to maintain chemical diversity in the array as the number of polymers is decreased

Figure 9.3 illustrates principal components analysis of data from twenty coated SAW sensors in response to eighteen chemically-diverse organic vapors [27]. Among the twenty coatings, fourteen were observed to be well-behaved sensors without anomalous response behaviors. In the analysis, the number of coatings was reduced from twenty, to fourteen, to seven, and finally to five, making certain to include a diverse set of coatings according to the criteria noted above. The results show that whether there are twenty coated sensors or just five diverse coated sensors, there are at most four or five orthogonal principal components.

Therefore, it follows that more coatings does not necessarily provide more types of chemical information. In addition, the dimensionality observed is consistent with the discussion of solubility interactions and LSERs above.

9.4.3   Polymer Selection Approaches

A coating for a chemical sensor in an array needs to meet several criteria. All the materials in the array should form thin films that result in sensors with rapid responses that are sensitive, reversible, and reproducible over time. The set of sensors for an array must provide a chemically diverse array. Given the large number of potential polymer coatings for sensors, methods are required for polymer selection. Many papers in the field of acoustic wave sensor arrays have described down selection of materials for arrays [42, 58, 60, 69, 72–86]. The effects of environmental factors such as temperature and humidity on sensor performance which can influence polymer selection have also been noted [18, 75, 87, 88]. Screening materials for stability and reproducibility as thin films on sensors is often a prerequisite to selection for a sensor array.

Several approaches can be used to select polymers for arrays and often some of them are used in combination. In one approach, a diverse set of polymers can be selected by considering how their structures will give rise to particular solubility properties, and by selecting a set that will give the full range of solubility properties described above.

Alternatively, a set of polymers could be characterized for their fundamental sorptive properties by LSERs. A set of fourteen polymers and sorbents was so characterized on the basis of gas chromatographic measurements [36], and another large set of polymers were modeled with LSERs using SAW sensor responses [27]. Then the LSER coefficients can be examined to select a set of polymers that span all the interactions represented in the LSER equation and maximize the solubility properties for particular interactions. In comparing coefficient values, it should be noted that the solvation parameters, and hence their corresponding coefficients, do not all scale with free energy equivalently. Free energy contributions can be calculated for particular solvent/solute pairs for comparison, as illustrated above in Table 9.3.

Additionally, the LSER coefficients for a large set of polymers can be examined using unsupervised learning methods to examine distances and clustering in feature space. Linear distances, and angular distances in pattern-normalized feature space, can be used to assess similarities and differences. Clustering of similar materials can be visualized from principal components plots and dendrograms [36]. Information from these approaches can be used to select dissimilar polymers to obtain chemical diversity.

Sensor response data can also be analyzed to select coatings. A diverse and large set of coatings can be used to generate a data set of vapor responses, using a diverse set of test vapors. These can be explored by unsupervised learning methods as just discussed, again using principal components plots and dendrograms for visualization. A set of coatings including one from each major group in a dendrogram, for example, can be selected to obtain a diverse array. This approach was described in the earliest papers on acoustic wave sensor arrays [4, 58, 60, 72].

Finally, polymer sets may be selected on the basis of the analytical purpose of the array, comparing different sets of coatings for their success at a particular analytical task the array has been designed to accomplish [4, 42, 60].

These various methods are not so different and typically they will all provide similar options for polymer selection and array design. Polymer-coated acoustic wave sensors and selection approaches have been reviewed previously [4]. A variety of array studies have recognized or used the principles of polymer solubility interactions to help guide selection of coatings for arrays [42, 74, 85, 89, 90].

9.5 Extracting Chemical Information from Sensor Arrays in Terms of Solvation Parameters

9.5.1 Information Content of Array Data

The responses of a gravimetric sorption-based sensor array are related to the sorption of vapors by the polymer coatings. Accordingly, the chemical information encoded by the array data is related to solubility interactions. The sorption of vapors and hence sensor responses can be modeled using LSERs. The original intent of applying LSERs to sorption-based sensors was to understand sorption and selectivity, and convert empirical sensor development to a predictive science. However, given that fact the data fit these models, which have a limited number of parameters, it follows that the information content of the array data have limited dimensionality. The limited dimensionality is confirmed in practice by PCA as described above. While the limited dimensionality represents a drawback in terms of selectivity, it can be an advantage in terms of being able to extract understandable information from the array data.

Consider that given vapor solvation parameters and polymer LSER coefficients, partition coefficients can be predicted, which can in turn be used to calculate sensor responses. If one knew the polymer parameters, and the partition coefficients or sensor responses, it should be possible to calculate the vapor solvation parameters. In this way, the responses of a well characterized array might be used to determine the properties of the detected vapor as solvation parameters, where these solvation parameters can then be regarded as descriptors [12–14, 27, 43].

Conventionally, sensor array data are processed using neural networks or statistical pattern recognition after a period of training, during which patterns to known vapors are determined. After training, the array can classify unknown samples that contain the known ones that were in the training. However, when confronted with an unknown one that was not in the training, the array may be unable to classify it or may misclassify it.

The use of array data to determine the descriptors of an unknown vapor has the potential to provide useful chemical information about samples even if they were not in the training set. This concept is shown in Fig. 9.4. The response vector, or pattern, is converted to descriptors in terms of the solvation parameters from LSERs. These descriptors could then be compared with the descriptors of known vapors in a data base to determine the unknown vapor's chemical identity or at least its chemical class.

Two approaches, to be described below, can be envisioned. In one approach, the response vector would be mathematically transformed into another vector containing all the descriptor values. In the second approach, models would be developed to convert a pattern vector to a single descriptor, and a set of such models would be developed to predict all the descriptors, one at a time.

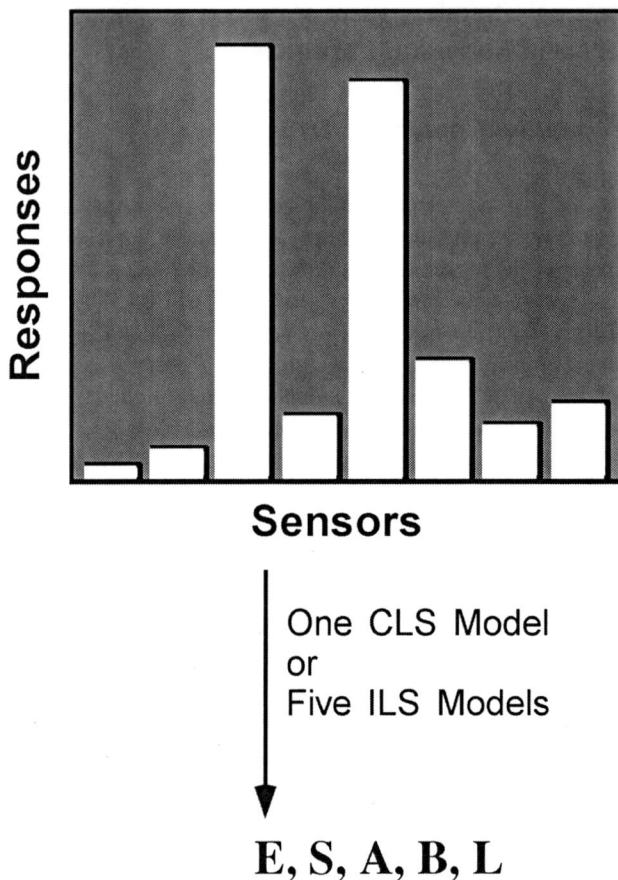

Fig. 9.4 The concept of converting a sensor array pattern vector to the five solvation parameters, using either a modified CLS approach or a set of ILS models

9.5.2 Classical Least Squares Approach

The transformation of the pattern vector into values for all the descriptors simultaneously can be carried out by a novel variant of classical least squares (CLS) methods. In conventional CLS approaches, as commonly used in absorbance spectroscopy, the response matrix R (samples by channels), containing the responses of the spectrometer, is modeled according to (9.10).

$$R = CS. \tag{9.10}$$

Here, C is a matrix of concentrations (samples by analytes). The matrix S contains the pure component spectra (analytes by channels) and if these are known, the concentrations C can be determined from R according to (9.11). (The superscript T denotes the transpose of a matrix and the superscript of -1 denotes the inverse of a matrix.)

$$C = RS^T (SS^T)^{-1}. \tag{9.11}$$

The LSER equation can be expressed in matrix algebra according to (9.12), which represents a collection of LSER relationships.

$$L = VP. \tag{9.12}$$

Matrix L, containing log K values, is related to matrix V containing the vapor solvation parameters, and matrix P containing the polymer parameters. Matrix V is defined so that each vector contains the five vapor solvation parameters and a vector of ones (number of vapors by six), and similarly P is defined to contain the polymer parameters including the constants (six by number of polymers).

Given an array of gravimetric sensors that respond according to (9.6), the responses can be expressed in matrix algebra according to (9.13).

$$R = C10^{(VP)}D^{-1}F. \tag{9.13}$$

Matrix R (vapors by polymers) contains the frequency shift values for the polymer-coated sensors in response to single component vapor samples. The matrix C (number of vapors by number of vapors) is a diagonal matrix of the concentrations of the vapors, and F (number of sensors by number of sensors) is a diagonal matrix of the Δf_s values of the sensors. Similarly, the matrix D (number of polymers by number of polymers) is a diagonal matrix of the densities of polymer materials on the sensors.

Given the responses, R, the polymer parameters, P, and the sensitivity factors, $D^{-1} F$, it is possible to solve for the vapor descriptors in V. However, the concentrations in C must also be known, which is unlikely since the concentration of an unknown vapor is not normally known.

Fortunately, the response model can be rearranged so that one solves for the log of the concentration simultaneously with determining the descriptor values [43]. Augmentation of the V matrix (to V_a) to contain the log of the vapor concentration in addition to the descriptors for each vapor, and augmentation of the P matrix (to P_a) with a vector of ones (see Table 9.1) yields the response model in (9.14).

$$R = 10^{(V_a P_a)}D^{-1}F. \tag{9.14}$$

Solutions for obtaining descriptors from array pattern vectors are then given by (9.15a) and (9.15b).

$$\{\log(RDF^{-1})\}P_a^T (P_a P_a^T)^{-1} = V_a, \tag{9.15a}$$

$$\{\log(rDF^{-1})\}P_a^T (P_a P_a^T)^{-1} = v_a. \tag{9.15b}$$

This solution in (9.15a) is shown for an entire matrix of responses, R, where V_a gives the values of the descriptors E, S, A, B, and L, as well as the log of the concentration. Alternatively, if the solution is expressed for a *single* sample as in (9.15b), the vector of descriptors within v_a is obtained from a single response vector, r.

Similar equations and solutions have been developed for volume-transducing sensors (see (9.7)) [14]. For mass-plus-volume-transducing sensors (see (9.8)) the mathematics are more complicated, but estimation and optimization approaches have been developed [12]. In general, studies have shown that these methods can be applied to mass-plus-volume transducing arrays such as polymer-coated SAW vapor sensor arrays in much the same way the approach is applied to purely mass-transducing sensor arrays. The quality of the results depends to some extent on the amount of variation in the extent to which modulus effects contribute to sensor sensitivities across the polymers in the array.

The potential for using these CLS approaches to classify array responses has been explored on large data sets using simulation methods [12, 14, 43]. Noise models were applied to synthetic sensor array data. The precision with which the solvation parameter descriptors can be obtained from array responses were determined as a function of the added noise. It was found that at 10% imposed noise, most descriptors were obtained back at near the precision to which they are known. Compared to the overall range of the descriptor values, the root mean square error of prediction even at 20% noise is rather small. These results for a simulation of a gravimetric sensor array are shown in Fig. 9.5.

One column is shown for each descriptor, with descriptor values on the left axis for all but the log L^{16} descriptor, which is referenced to the right axis. The black bars show the range of descriptor values. The small bars rising from zero on the

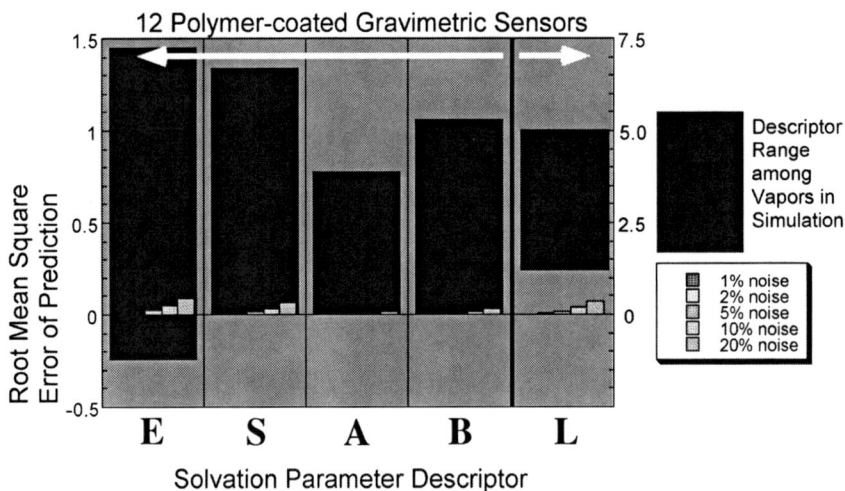

Fig. 9.5 Simulation results for obtaining descriptor values from array response data, comparing prediction errors at various noise levels to the overall descriptor value ranges

y-axis show the root mean square error of predictions for each descriptor at various imposed noise levels [12]. These bars are all clearly quite small compared to the overall range of descriptor values.

Once the descriptors for the unknown are obtained (found), they can be compared with tables of descriptor values for the known compounds. Using prediction errors to define an uncertainty around a descriptor value, matches can be determined in the data base where all five descriptors match, the found descriptors within this uncertainty, and are returned as matches to the unknown. It was observed in the simulations that at 10% noise, the vast majority of compounds were correctly matched, and where there were multiple possible matches, most were within the correct compound class [43].

The use of this modified CLS approach applies to array responses to single compounds. While conventional CLS returns the concentration of each vapor in a mixture, this modified approach treats the single detected vapor as a mixture of five "pure components" corresponding to the solvation parameters, yielding five descriptors for the vapor.

9.5.3 Inverse Least Squares Approach

The CLS approach just described determines the values of all the vapor descriptors simultaneously by converting the array pattern vector to a vector containing the descriptors and the log of the concentration. This approach requires that the interactive properties of the sorbent sensing layers be known and quantified as LSER-coefficients (polymer parameters), however the array need not have been trained on a specific vapor in order to determine its descriptor values.

Inverse least squares (ILS) methods can also be used to obtain vapor descriptors from array responses, in which case models are developed to determine each vapor descriptor individually [27, 43]. This approach does not require advance knowledge of polymer parameters, but it does require that an adequate calibration data set be created to derive the ILS models, which constitutes a form of training. Once ILS models are developed, however, an array could, in principle, be used to characterize an unknown vapor in terms of its descriptor values, even if the specific unknown vapor had not been in the training set used to develop the models.

The ILS method involves developing separate models for each vapor parameter, y, as a weighted sum of the responses according to (9.16).

$$y = Xb. \tag{9.16}$$

Here, X is the measured response and b is a vector of weights, generally determined by regression according to (9.17).

$$b = X^+y. \tag{9.17}$$

In (9.17), X^+ is the pseudo inverse of X, which is defined differently depending upon the type of regression to be used [91, 92]. In the case of polymer-coated acoustic wave sensors, y corresponds to one of the 5 vapor solvation parameters and X is the array response as the log of the measured responses, as suggested by (9.6) and (9.9). Thus, the sensor responses (predictor variables) are related to the vapor parameters (predicted variables).

Multiple linear regression (MLR) is the simplest ILS approach, where each descriptor is modeled as the linear combination of sensor responses. Principal components regression (PCR) and partial least squares (PLS) regression represent

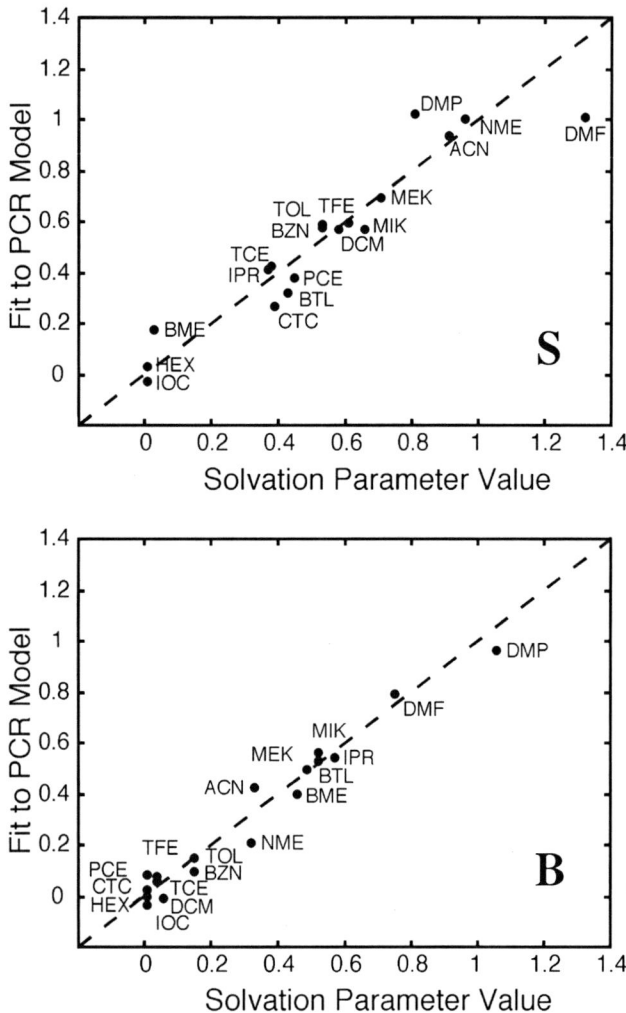

Fig. 9.6 Calibration results for two solvation parameters using six-factor PCR models to process SAW sensor array data into parameter values

additional ILS methods, which can be more effective at making predictive models. PCR and PLS model the parameter of interest as a linear combination of factors, where the factors are themselves linear combinations of sensor responses (i.e., principal components or latent variables). MLR maximizes correlation with the predicted variable (y), PCR captures maximum variance in the predictor variables (X), and PLS tries to do both by maximizing covariance.

In processing sensor array data to obtain descriptors, the objective is to be able to characterize the properties of the samples not in the training set, that is, to do prediction. MLR is generally not as effective in creating predictive models because it tends to overfit the data, modeling the noise as well as the meaningful chemical information. However, it does produce very good correlations, and the correlation of vapor descriptor values with array responses using MLR has been demonstrated successfully [27]. These correlations look very good compared to typical correlations in physical organic chemistry. The correlations developed by fitting using six factor PCR models are shown for two solvation parameters in Fig. 9.6.

In developing predictive models with PCR and PLS, selection of an appropriate number of factors is important; the factors should contain the chemical information while filtering out noise. Principal components analysis (PCA) and cross-validation techniques have been used in factor selection in analyzing a data set with twenty diverse polymers tested against 18 diverse vapors. For both PCR and PLS methods, the results supported the selection of six factors. Six factors were required, whereas more than six did not help. These results are consistent with models and studies described above. Principal components analysis (see Fig. 9.3) suggests that there are not more than five orthogonal components in pattern normalized data. There are 5 solvation parameters used to model vapor sorption, and concentration is also a separate factor in the log transformed data. Thus, the inherent dimensionality and structure of the experimental data were consistent with the models that were the basis of the approach.

9.6   Conclusions

This chapter has set out the basics of vapor sorption, modeling with LSERs, and the applications of this knowledge to acoustic wave sensor arrays. Although other sensor types may respond with different transduction mechanisms, and these transduction mechanisms may contain analyte specific sensitivity factors, the understanding of vapor sorption is still fundamental to understanding sensor response, and may actually be necessary for investigating transduction mechanisms. Therefore, the knowledge of vapor sorption as systematically set out in terms of solubility interactions and LSERs provides a foundation for understanding selectivity in vapor sensing.

In the field of acoustic wave sensors, the approach is particularly applicable, enabling understanding, prediction of sensor responses, and more recently, providing a new means to extract chemical information from the array data.

Acknowledgments JWG would especially like to acknowledge the many former coworkers and collaborators who played key roles in the development of polymer-coated acoustic wave sensors. He gratefully acknowledges funding from US Department of Energy, National Nuclear Security Administration, Office of Nonproliferation Research and Development (NA-22) for past funding and Laboratory Directed Research and Development funds of the US DOE, administered by the Pacific Northwest National Laboratory, for current funding. The William R. Wiley Environmental Molecular Sciences Laboratory, a US DOE scientific user facility operated for the DOE by PNNL is also acknowledged. The Pacific Northwest National Laboratory is a multiprogram national laboratory operated for the U.S. Department of Energy by Battelle Memorial Institute.

References

1. Grate, J. W.; Abraham, M. H., Solubility interactions and the selection of sorbent coating materials for chemical sensors and sensor arrays, *Sens. Actuators B* **1991**, 3, 85–111
2. Grate, J. W.; Martin, S. J.; White, R. M., Acoustic wave microsensors, *Anal. Chem.* **1993**, 65, 940A–948A, 987A–996A
3. Grate, J. W.; Frye, G. C., Acoustic Wave Sensors, In Sensors Update; Baltes, H.; Goepel, W.; Hesse, J., Eds.; VSH, Weinheim, **1996**, Vol. 2, 37–83
4. Grate, J. W., Acoustic wave microsensor arrays for vapor sensing, *Chem. Rev.* **2000**, 100, 2627–2648
5. Grate, J. W.; Kaganove, S. N.; Nelson, D. A., Carbosiloxane polymers for chemical sensors, *Chem. Innovations* **2000**, 30(11), 29–37
6. Grate, J. W.; Nelson, D. A., Sorptive polymeric materials and photopatterned films for gas phase chemical microsensors, *Proc IEEE* **2003**, 91, 881–889
7. Janghorbani, M.; Freund, H., Application of a piezoelectric quartz crystal as a partition detector: Development of a digital sensor, *Anal. Chem.* **1973**, 45, 325–332
8. Edmunds, T. E.; West, T. S., A quartz crystal piezoelectric device for monitoring organic gaseous pollutants, *Anal. Chim. Acta* **1980**, 117, 147–157
9. McCallum, J. J.; Fielden, P. R.; Volkan, M.; Alder, J. F., Detection of toluene diisocyanate with a coated quartz piezoelectric crystal, *Anal. Chim. Acta* **1984**, 162, 75–83
10. Snow, A.; Wohltjen, H., Poly(ethylene maleate)-cyclopentadiene: A model reactive polymer-vapor system for evaluation of a saw microsensor, *Anal. Chem.* **1984**, 56, 1411–1416
11. Grate, J. W.; Snow, A.; Ballantine, D. S.; Wohltjen, H.; Abraham, M. H.; McGill, R. A.; Sasson, P., Determination of partition coefficients from surface acoustic wave vapor sensor responses and correlation with gas-liquid chromatographic partition coefficients, *Anal. Chem.* **1988**, 60, 869–875
12. Grate, J. W.; Wise, B. M.; Gallagher, N. B., Classical least squares transformations of sensor array pattern vectors into vapor descriptors. Simulation of arrays of polymer-coated surface acoustic wave sensors with mass-plus-volume transduction mechanisms, *Anal. Chim. Acta* **2003**, 490, 169–184
13. Wise, B. M.; Gallagher, N. B.; Grate, J. W., Analysis of combined mass and volume transducing sensors arrays, *J. Chemometr* **2002**, 17, 463–469
14. Grate, J. W.; Wise, B. M., A method for chemometric classification of unknown vapors from the responses of an array of volume-transducing sensors, *Anal. Chem.* **2001**, 73, 2239–2244
15. Grate, J. W.; Zellers, E. T., The fractional free volume of the sorbed vapor in modeling the viscoelastic contribution to polymer-coated surface acoustic wave vapor sensor responses, *Anal. Chem.* **2000**, 72, 2861–2868
16. Grate, J. W.; Kaganove, S. N.; Bhethanabotla, V. R., Examination of mass and modulus contributions to thickness shear mode and surface acoustic wave vapour sensor responses using partition coefficients, *Faraday Discuss.* **1997**, 107, 259–283

17. Grate, J. W., Sensing glass transitions in thin polymer films on acoustic wave microsensors, In Assignment of the Glass Transition, ASTM STP 1249; Seylor, R. J., Ed.; ASTM, Philadelphia, **1994**, 153–164

18. Grate, J. W.; Klusty, M.; McGill, R. A.; Abraham, M. H.; Whiting, G.; Andonian-Haftvan, J., The predominant role of swelling-induced modulus changes of the sorbent phase in determining the responses of polymer-coated surface acoustic wave vapor sensors, *Anal. Chem.* **1992**, 64, 610–624

19. Severin, E. J.; Lewis, N. S., Relationships among resonant freqency changes on a coated quartz crystal microbalance, thickness changes, and resistance responses of polymer-carbon black composite chemiresistors, *Anal. Chem.* **2000**, 72, 2008–2015

20. Abraham, M. H.; Doherty, R. M.; Kamlet, M. J.; Taft, R. W., A new look at acids and bases, *Chem. Br.* **1986**, 22 551–554

21. Kamlet, M. J.; Doherty, R. M.; Abboud, J.-L. M.; Abraham, M. H.; Taft, R. W., Solubility: A new look, *CHEMTECH* **1986**, 16, 566–576

22. Grate, J. W.; Abraham, M. H.; McGill, R. A., Sorbent polymer coatings for chemical sensors and arrays, In Handbook of Biosensors: Medicine, Food, and the Environment; Kress-Rogers, E.; Nicklin, S., Eds.; CRC Press, Boca Raton, FL, **1996**, 593–612

23. Kamlet, M. J.; Taft, R. W., Linear solvation energy relationships. Local empirical rules - or fundamental laws of chemistry? A reply to the chemometricians, *Acta Chem. Scandinavica B* **1985**, 39, 611–628

24. Abraham, M. H., Scales of hydrogen-bonding: Their construction and application to physicochemical and biochemical processes, *Chem. Soc. Rev.* **1993**, 22, 73–83

25. Abraham, M. H.; Poole, C. F.; Poole, S. K., Classification of stationary phases and other materials by gas chromatography, *J. Chromatogr. A* **1999**, 842, 79–114

26. Abraham, M. H.; Ibrahim, A.; Zissimos, A. M., Determination of sets of solute descriptors from chromatographic measurements, *J. Chromatogr. A* **2004**, 1037, 29–47

27. Grate, J. W.; Patrash, S. J.; Kaganove, S. N.; Abraham, M. H.; Wise, B. M.; Gallagher, N. B., Inverse least-squares modeling of vapor descriptors using polymer-coated surface acoustic wave sensor array responses, *Anal. Chem.* **2001**, 73, 5247–5259

28. McGill, R. A.; Abraham, M. H.; Grate, J. W., Choosing polymer coatings for chemical sensors, *CHEMTECH* **1994**, 24, 27–37

29. Abraham, M. H.; Andonian-Haftvan, J.; Whiting, G.; Leo, A.; Taft, R. W., Hydrogen bonding. Part 34. The factors that influence the solubility of gases and vapours in water at 298 K, and a new method for its determination, *J. Chem. Soc. Perkin Trans.* **1994**, 2, 1777–1791

30. Abraham, M. H.; Grellier, P. L.; Prior, D. V.; Duce, P. P.; Morris, J. J.; Taylor, P. J., Hydrogen bonding. Part 7. A scale of solute hydrogen-bond acidity based on log K values for complexation in tetrachloromethane, *J. Chem. Soc. Perkin Trans. II* **1989**, 699–711

31. Abraham, M. H.; Grellier, P. L.; Prior, D. V.; Morris, J. J.; Taylor, P. J., Hydrogen bonding. Part 10. A scale of solute hydrogen-bond basicity using log K values for complexation in tetrachloromethane, *J. Chem. Soc. Perkin Trans.* **1990**, 2, 521–529

32. Abraham, M. H.; Whiting, G. S.; Doherty, R. M.; Shuely, W. J., Hydrogen bonding. XVI. A new solute solvation parameter, pi2h, from gas chromatographic data, *J. Chromatogr.* **1991**, 587, 213–228

33. Abraham, M. H.; Fuchs, R., Correlation and prediction of gas-liquid partition coefficients in hexadecane and olive oil, *J. Chem. Soc. Perkin Trans. II* **1988**, 523–527

34. Abraham, M. H.; Grellier, P. L.; McGill, R. A., Determinatin of olive oil-gas and hexadecane-gas partition coefficients, and caculation of the corresponding olive oil-water and hexadecane-water partition coefficients, *J. Chem. Soc., Perkin Trans. II* **1987**, 797–803

35. Abraham, M. H.; Whiting, G. S.; Doherty, R. M.; Shuely, W. J., Hydrogen bonding. Part 13. A new method for the characterisation of glc stationary phases - The laffort data set, *J. Chem. Soc. Perkin Trans. 2* **1990**, 1451–1460

36. Abraham, M. H.; Andonian-Haftvan, J.; Du, C. M.; Diart, V.; Whiting, G.; Grate, J. W.; McGill, R. A., Hydrogen bonding. XXIX. The characterisation of fourteen sorbent coatings

for chemical microsensors using a new solvation equation, *J. Chem. Soc. Perkin Trans.* **1995**, 2, 369–378

37. Patrash, S. J.; Zellers, E. T., Characterization of polymeric surface acoustic wave sensor coatings and semiempirical models of sensor responses to organic vapors, *Anal. Chem.* **1993**, 65, 2055–2066

38. Hierlemann, A.; Zellers, E. T.; Ricco, A. J., Use of linear solvation energy relationships for modeling responses from polymer-coated acoustic-wave vapor sensors, *Anal. Chem.* **2001**, 73, 3458–3466

39. Abraham, M. H.; Whiting, G. S.; Andonian-Haftvan, J.; Steed, J. W.; Grate, J. W., Hydrogen bonding XIX. The characterization of two poly(methylphenylsiloxane)s, *J. Chromatogr* **1991**, 588, 361–364

40. Grate, J. W., Siloxanes with strongly hydrogen bond donating functionalities, US Patent 5,756,631: **1998**

41. Grate, J. W.; Patrash, S. J.; Abraham, M. H., Method for estimating polymer-coated acoustic wave vapor sensor responses, *Anal. Chem.* **1995**, 67, 2162–2169

42. Grate, J. W.; Kaganove, S. N.; Patrash, S. J., Hydrogen-bond acidic polymers for surface acoustic wave vapor sensors and arrays, *Anal. Chem.* **1999**, 71, 1033–1040

43. Grate, J. W.; Wise, B. M.; Abraham, M. H., Method for unknown vapor characterization and classification using a multivariate sorption detector. Initial derivation and modeling based on polymer-coated acoustic wave sensor arrays and linear solvation energy relationships, *Anal. Chem.* **1999**, 71, 4544–4553

44. McGill, R. A.; Chung, R.; Chrisey, D. B.; Dorsey, P. C.; Matthews, P.; Pique, A.; Mlsna, T. E.; Stepnowski, J. L., Performance optimization of surface acoustic wave chemical sensors, *IEEE Trans. Ultrason. Ferroelec. Freq. Contr.* **1998**, 45, 1370–1379

45. Pinnaduwage, L. A.; Thundat, T.; Hawk, J. E.; Hedden, D. L.; Britt, R.; Houser, E. J.; Stepnowski, S.; McGill, R. A.; Bubb, D., Detection of 2,4-dinitrotoluene using microcantilever sensors, *Sens. Actuators B* **2004**, 99, 223–229

46. Patel, S. V.; Mlsna, T. E.; Fruhberger, B.; Klaassen, E.; Cemalovic, S.; Baselt, D. R., Chemicapacitive microsensors for volatile organic compound detection, *Sens. Actuators B* **2003**, *B*96, 541–553

47. Cunningham, B. T.; Kant, R.; Daly, C.; Weinberg, M. S.; Pepper, J. W.; Clapp, C.; Bousquet, R.; Hugh, B., Chemical vapor detection using microfabricated flexural plate silicon resonator arrays, *Proc. SPIE-Int. Soc. Opt. Eng.* **2000**, 4036, 151–162

48. Houser, E. J.; Mlsna, T. E.; Nguyen, V. K.; Chung, R.; Mowery, R. L.; Andrew McGill, R., Rational materials design of sorbent coatings for explosives: Applications with chemical sensors, *Talanta* **2001**, 54, 469–485

49. Mlsna, T. E.; Cemalovic, S.; Warburton, M.; Hobson, S. T.; Mlsna, D. A.; Patel, S. V., Chemicapacitive microsensors for chemical warfare agent and toxic industrial chemical detection, *Sens. Actuators B* **2006**, 116, 192–201

50. Patel, S. V.; Hobson, S. T.; Cemalovic, S.; Mlsna, T. E., Chemicapacitive microsensors for detection of explosives and tics, *Proc. SPIE-Int. Soc. Opt. Eng.* **2005**, 5986, 59860M/1–59860M/10

51. Grate, J. W.; Kaganove, S. N.; Patrash, S. J.; Craig, R.; Bliss, M., Hybrid organic/inorganic copolymers with strongly hydrogen-bond acidic properties for acoustic wave and optical sensors, *Chem. Mater.* **1997**, 9, 1201–1207

52. Grate, J. W.; Kaganove, S. N., Hydrosilylation: A versatile reaction for polymer synthesis, *Polymer News* **1999**, 24, 149–155

53. Grate, J. W., Hydrogen-bond acidic polymers for chemical vapor sensing, *Chem. Rev.* **2008**, 108, 726–745

54. Lu, C. J.; Zellers, E. T., Multi-adsorbent preconcentration/focusing module for portable-gc/microsensor-array analysis of complex vapor mixtures, *Analyst* **2002**, 127, 1061–1068

55. Lewis, P. R.; Manginell, R. P.; Adkins, D. R.; Kottenstette, R. J.; Wheeler, D. R.; Sokolowski, S. S.; Trudell, D. E.; Byrnes, J. E.; Okandan, M.; Bauer, J. M.; Manley, R. G.; Frye-Mason, G. C., Recent advancements in the gas-phase microchemlab, *IEEE Sensors J.* **2006**, 6, 784–795

56. Lu, C. J.; Whiting, J.; Sacks, R. D.; Zellers, E. T., Portable gas chromatograph with tunable retention and sensor array detection for determination of complex vapor mixtures, *Anal. Chem.* **2003**, 75, 1400–1409

57. Hsieh, M. D.; Zellers, E. T., Adaptation and evaluation of a personal electronic nose for selective multivapor analysis, *J. Occup. Environ. Hyg.* **2004**, 1, 149–160

58. Ballantine, D. S.; Rose, S. L.; Grate, J. W.; Wohltjen, H., Correlation of surface acoustic wave device coating responses with solubility properties and chemical structure using pattern recognition, *Anal. Chem.* **1986**, 58, 3058–3066

59. Wohltjen, H.; Snow, A. W.; Barger, W. R.; Ballantine, D. S., Trace chemical vapor detection using saw delay line oscillators, *IEEE Trans. Ultrason. Ferroelec. Freq. Contr.* **1987**, *UFFC-34*, 172–177

60. Rose-Pehrsson, S. L.; Grate, J. W.; Ballantine, D. S.; Jurs, P. C., Detection of hazardous vapors including mixtures using pattern recognition analysis of responses from surface acoustic wave devices, *Anal. Chem.* **1988**, 60, 2801–2811

61. Grate, J. W.; Klusty, M.; Barger, W. R.; Snow, A. W., Role of selective sorption in chemiresistor sensors for organophosphorous detection, *Anal. Chem.* **1990**, 62, 1927–1924

62. Fox, C. G.; Alder, J. F., Development of humidity correction algorithm for surface acoustic wave sensors. Part 1. Water adsorption isotherms on coated surface acoustic wave sensors, *Anal. Chim. Acta* **1991**, 248, 337–44

63. Snow, A. W.; Sprague, L. G.; Soulen, R. L.; Grate, J. W.; Wohltjen, H., Synthesis and evaluation of hexafluorodimethylcarbinol functionalized polymers as microsensor coatings, *J. Appl. Poly. Sci.* **1991**, 43, 1659–1671

64. Grate, J. W.; Rose-Pehrsson, S. L.; Venezky, D. L.; Klusty, M.; Wohltjen, H., Smart sensor system for trace organophosphorus and organosulfur vapor detection employing a temperature-controlled array of surface acoustic wave sensors, automated sample preconcentration, and pattern recognition, *Anal. Chem.* **1993**, 65, 1868–1881

65. Collins, G. E.; Rose-Pehrsson, S. L., Chemiluminescent chemical sensors for oxygen and nitrogen dioxide, *Anal. Chem.* **1995**, 67, 2224–30

66. Collins, G. E.; Buckley, L. J., Conductive polymer-coated fabrics for chemical sensing, *Synthetic Metals* **1996**, 78, 93–101

67. Grate, J. W.; Kaganove, S. N., Hybrid organic/inorganic copolymers films with strongly hydrogen-bond acidic properties for vapor sensing, *Polymer Preprints* **1997**, 38, 955

68. Zellers, E. T.; Park, J.; Tsu, T.; Groves, W. J., Establishing a limit of recognition for a vapor sensor array, *Anal. Chem.* **1998**, 70, 4191–4201

69. Park, J.; Groves, W. A.; Zellers, E. T., Vapor recognition with small arrays of polymer-coated microsensors. A comprehensive analysis, *Anal. Chem.* **1999**, 71, 3877–3886

70. Cai, Q.-Y.; Park, J.; Heldsinger, D.; Hsieh, M.-D.; Zellers, E. T., Vapor recognition with an integrated array of polymer-coated flexural plate wave sensors, *Sens. Actuators* **2000**, *B62*, 121–130

71. Martin, S. D.; Poole, C. F.; Abraham, M. H., Synthesis and gas chromatographic evaluation of a high temperature hydrogen bond acid stationary phase, *J. Chromatogr. A* **1998**, 805, 217–235

72. Carey, W. P.; Beebe, K. R.; Kowalski, B. R.; Illman, D. L.; Hirschfeld, T., Selection of adsorbates for chemical sensor arrays by pattern recognition, *Anal. Chem.* **1986**, 58, 149–153

73. Zellers, E. T.; Batterman, S. A.; Han, M.; Patrash, S. J., Optimal coating selection for the analysis of organic vapor mixtures with polymer-coated surface acoustic wave sensor arrays, *Anal. Chem.* **1995**, 67, 1092–1106

74. Rapp, M.; Boss, B.; Voigt, A.; Bemmeke, H.; Ache, H. J., Development of an analytical microsystem for organic gas detection based on surface acoustic wave resonators, *Fresenius J. Anal. Chem.* **1995**, 352, 699–704

75. Zellers, E. T.; Han, M., Effects of temperature and humidity on the performance of polymer-coated surface acoustic wave vapor sensor arrays, *Anal. Chem.* **1996**, 68, 2409–2418

76. Barie, N.; Rapp, M.; Ache, H. J., Uv crosslinked polysiloxanes as new coating materials for saw devices with high long-term stability, *Sens. Actuators, B* **1998**, *B46*, 97–103

77. Avila, F.; Myers, D. E.; Palmer, C., Correspondence analysis and adsorbate selection for chemical sensor arrays, *J. Chemom.* **1991**, 5, 455–65

78. Nakamoto, T.; Sasaki, S.; Fukuda, A.; Moriizumi, T., Selection method of sensing membranes in an odor sensing system, *Sens. Mater.* **1992**, 4, 111–119

79. Yokoyama, K.; Ebisawa, F., Detection and evaluation of fragrances by human reactions using a chemical sensor based on adsorbate detection, *Anal. Chem.* **1993**, 65, 673–7

80. Cao, Z.; Lin, H.-G.; Wang, B.-F.; Wang, K.-M.; Yu, R.-Q., Piezoelectric crystal sensor array used as an organic vapor sensing system, *Microchem. J.* **1995**, 52, 174–80

81. Eda, Y.; Takisawa, N.; Shirahama, K., Responses of polymer-coated piezoelectric crystals to organic vapors, *Sens. Mater.* **1995**, 7, 405–14

82. Cao, Z.; Lin, H.-G.; Wang, B.-F.; Chen, Z.-Z.; Ma, F.-L.; Wang, K.-M.; Yu, R.-Q., Discrimination of vapors of alcohols and beverage samples using piezoelectric crystal sensor array, *Anal. Lett.* **1995**, 28, 451–66

83. Deng, Z.; Stone, D. C.; Thompson, M., Selective detection of aroma components by acoustic wave sensors coated with conducting polymer films, *Analyst* **1996**, 121, 671–679

84. Cao, Z.; Lin, H.-G.; Wang, B.-F.; Xu, D.; Yu, R.-Q., A perfume odor-sensing system using an array of piezoelectric crystal sensors with plasticized pvc coatings, *Fresenius' J. Anal. Chem.* **1996**, 355, 194–199

85. Lau, K.-T.; Micklefield, J.; Slater, J. M., The optimisation of sorption sensor arrays for use in ambient conditions, *Sens. Actuators B* **1998**, *B*50, 69–79

86. Hoyt, A. E.; Ricco, A. J.; Bartholomew, J. W.; Osbourn, G. C., Saw sensors for the room-temperature measurement of co2 and relative humidity, *Anal. Chem.* **1998**, 70, 2137–2145

87. Liron, Z.; Greenblatt, J.; Frishman, G.; Gratziani, N., Temperature effect and chemical response of surface acoustic wave (saw) single-delay-line chemosensors, *Sens. Actuators B* **1993**, 12, 115–122

88. Hierlemann, A.; Wiemar, U.; Kraus, G.; Schweizer-Berberich, M.; Goepel, W., Polymer-based sensor arrays and multicomponent analysis for the detection of hazardous organic vapours in the environment, *Sens. Actuators B* **1995**, 26–27, 126–134

89. Hierlemann, A.; Wiemar, U.; Kraus, G.; Gauglitz, G.; Goepel, W., Environmental chemical sensing using quartz microbalance sensor arrays: Application of multicomponent analysis techniques, *Sens. Mater.* **1995**, 7, 179–189

90. Park, J.; Zhang, G.-Z.; Zellers, E. T., Personal monitoring instrument for the selective measurement of multiple organic vapors, *Am. Ind. Hyg. Assoc. J.* **2000**, 61, 192–204

91. Beebe, K. R.; Pell, R. J.; Seasholtz, M. B., Chemometrics: A practical guide; Wiley, New York, **1998**

92. Wise, B. M.; Gallagher, N. B., The process chemometrics approach to process monitoring and fault detection, *J. Process Control* **1996**, 6, 329–348

Part III
Designing Sensing Arrays

Chapter 10
A Statistical Approach to Materials Evaluation and Selection for Chemical Sensor Arrays

Baranidharan Raman, Douglas C. Meier, and Steve Semancik

Abstract We present a generic approach for designing sensor arrays for a given chemical sensing task. First, we present a correlation-based metric to systematically assess the analytical information obtained from the conductometric responses of chemiresistive films as a function of their operating temperatures and material composition. We illustrate how this measure can also be used to test the reproducibility of signals obtained from sensors of equal manufacture. Next, complementing the correlation-based analysis, we employ a statistical dimensionality-reduction algorithm to visualize the multivariate sensor response obtained from sensor arrays. We adapt this method to quantify the discriminability of chemical fingerprints. Finally, we show how to determine an optimal set of material compositions to be incorporated within an array for individual species' recognition when practical constraints/tradeoffs on fabrication are also considered. We validate our approach by designing a microsensor array for the task of recognizing a chemical hazard at sub-lethal concentrations in complex environments.

10.1 Introduction

How is a chemical sensor array designed for a specific application? Determining appropriate sensing materials with which to populate a sensor array is a key step that critically influences the device's performance. While an approach based on first principles may be applicable for tasks involving a few simple analytes in well-controlled background conditions, such studies become infeasible for broad-spectrum applications involving several complex targets in a wide variety of ambient conditions and in the possible presence of interfering gases. In this chapter,

B. Raman, D.C. Meier, and S. Semancik (✉)
Chemical Science and Technology Laboratory, National Institute of Standards and Technology, MD 20899, Gaithersburg, USA
e-mail: steves@nist.gov

M.A. Ryan et al. (eds.), *Computational Methods for Sensor Material Selection*, Integrated Analytical Systems, DOI 10.1007/978-0-387-73715-7_10, © Springer Science+Business Media, LLC 2009

we present a generic computational approach to this combinatorial problem. Our approach incorporates statistical methods for processing multivariate sensor responses to assess their similarity, test reproducibility, and to determine their analytical content. Furthermore, practical tradeoffs on manufacturability such as array size and the number of materials used, etc., can also be easily incorporated into our approach. We demonstrate this methodology for the illustrative problem of recognizing a target chemical hazard (ammonia, NH_3) at sub-lethal concentrations, under varying ambient conditions, and in the presence of interfering chemicals. To approximate relevant real-world conditions, NH_3 at immediate danger to life and health (IDLH; around 300 μmol/mol for NH_3) concentrations are introduced individually into synthetic air at three different relative humidities (10%, 30% and 70%) and/or infused with the vapors of any one of the following common products: bleach (Clorox [1]), interior house paint, window cleaner (Windex [1]) and floor stripper (ZEP [1]) (see Fig. 10.1). The challenge here is to evaluate different chemiresistive sensing materials, as well as their temperature-dependent responses, and to determine how to populate and operate a microsensor array for detecting and recognizing NH_3 in these well defined backgrounds and ambient conditions.

We note that a number of approaches have been proposed to optimize the composition of a chemical sensor array; readers are referred to [2–4] for a detailed review on this topic. In this chapter, we present specialized operational schemes to enhance the analytical content of sensor response, and adopt standard statistical algorithms to create qualitative and quantitative methods for designing and evaluating array-based solutions for a wide variety of detection problems.

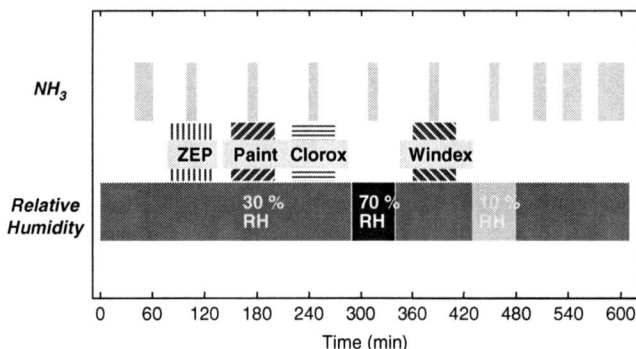

Fig. 10.1 Analyte delivery protocol used for data collection. The *solid continuous bar* in the *bottom row* indicates the humidity levels (10% RH, 30% RH and 70% RH) at different time periods during a cycle. Four common, commercially available interferences (*middle row*) each at 1% of their saturated vapor concentration were used to simulate realistic scenarios. The periods of NH_3 introduction are shown in the top row. The last three target analyte introductions were graded such that the concentrations gradually changed from zero to maximum and then returned in a similar fashion. Reprinted with permission from ref [21]. Copyright 2009 Elsevier

10.2   The Sensing Platform

A suitable platform to meet the analytical requirements of this task that provides considerable benefits with respect to its cost, speed, size and power consumption is the MEMS-based microhotplate array [5]. An optical micrograph of a single microhotplate array element is shown in Fig. 10.2a. Each array element contains three functional components (see layered schematic in Fig. 10.2b): a polycrystalline silicon resistor for heating, a chemically sensitive film, and interdigitated platinum electrodes that enable measurement of the conductance of the sensing film that is deposited onto the array element. The lateral dimensions of this miniaturized device are in the order of $\approx$100 μm, with a mass of $\approx$250 ng. The device has a thermal time constant of a few milliseconds, and an operating temperature range from ambient to 500°C. The localized temperature control offered by microhotplates, along with their fast heating/cooling characteristics, makes them ideal both for self-lithographic, thermally activated chemical vapor deposition (CVD) processes [6] and for device operation with temperature programming, wherein each element is independently cycled through multiple temperatures during sensor measurements [7]. Each microsensor array consists of a collection of individually addressable temperature-controlled elements as shown in Fig. 10.2c. A 40-pin dual in-line packaged device is shown in Fig. 10.2d.

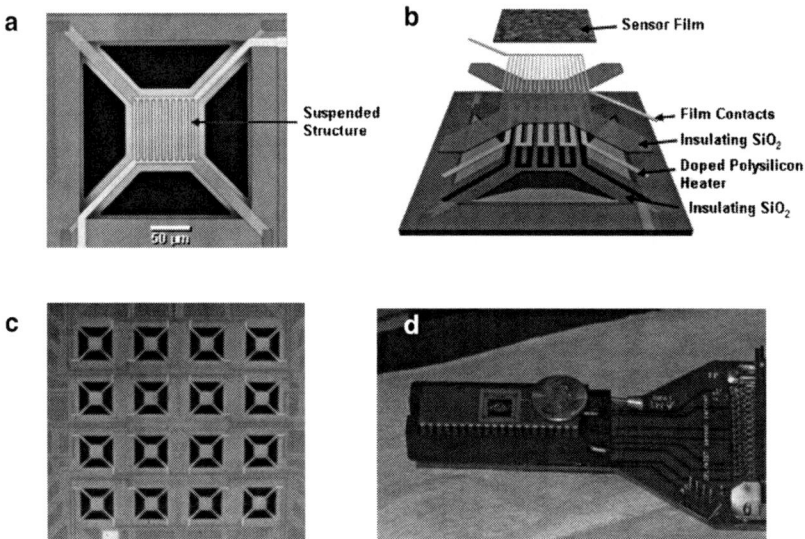

Fig. 10.2 Microsensor array platforms. (**a**) An optical microscopy image of a single microhotplate microsensor element. (**b**) A layered schematic showing the three primary components of the microsensor elements: polycrystalline silicon heater, interdigitated platinum electrodes, and metal oxide sensing film. (**c**) A microsensor array with 16 individually addressable, temperature-controlled elements. (**d**) A 40-pin dual in-line packaged microsensor device

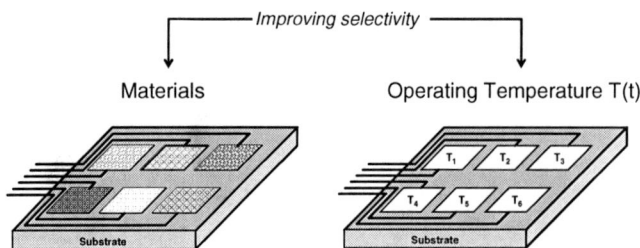

Fig. 10.3 Selectivity can be imparted through use of diverse materials each operated at a fixed operating parameter (*left panel*). A more sophisticated approach to enhance the selectivity of the array involves capturing the responses of each sensor element as a function of a chosen operating parameter (e.g. temperature)

Typically, selectivity is imparted to these sensor arrays by employing diverse sensing materials that ideally generate non-redundant information about the target chemical species (Fig. 10.3). However, for sensing tasks with complex analytical requirements, the selectivity of the device can be further enhanced by modulating an operating parameter of each sensor element (e.g. temperature of a chemiresistive sensor) and capturing the responses for a wide range of parameter values. For the ammonia detection and recognition problem, we will use four different materials and capture their responses at 32 different temperatures.

10.2.1 Sensing Materials

In general, elements in a sensor array can be populated with materials from different classes, including for example, metal oxides [8–13], conducting polymers [14], and others [2, 15]. Metal oxide semiconductor films were chosen as the chemically sensitive component in the devices used here, for a variety of reasons. They are known to undergo chemical interactions with gas species ranging from surface-mediated oxidation of analyte gases to charge transfer upon analyte chemisorption [8, 12, 15]. These interactions, as a result of electron transfers between adsorbed gases and a surface-depletion layer, cause a repeatable change in the electrical conductance of the film, thus yielding a measurable and recognizable signal [16]. These conductance changes have been shown to be temperature dependent, materials dependent, and most importantly, analyte dependent for compounds from the simple (e.g., CO) to the complex (e.g., chemical warfare agent molecules) [11, 16, 17]. The oxides are also robust materials capable of withstanding wide excursions of temperature, which we can use to enhance the analytical content of the signal stream.

For the purposes of broad-spectrum detection required by this task, we consider the following four semiconducting metal oxide films: tin (IV) oxide (SnO_2), tin (IV) oxide coated with titanium (IV) oxide (SnO_2/TiO_2), titanium (IV) oxide (TiO_2), and titanium (IV) oxide coated with ruthenium oxide (TiO_2/RuO_x). Tin oxide (SnO_2) has been a widely used chemiresistive transduction material in thick-film sensors [18], and has also been successfully employed in microhotplate arrays [19]. It provides sensitive detection responses for most hydrocarbons. Titanium oxide (TiO_2), while significantly more resistive than SnO_2, exhibits a larger signal range for water vapor [20]. Furthermore, the oxides of titanium and tin have different suboxide states available to them, providing for the possibility of different reaction pathways for incident analytes. It follows that these materials would be expected to produce different conductance signatures for different analytes. In this study, the electron-rich SnO_2 and RuO_x were combined with electron-poor TiO_2 films in order to increase the film baseline conductance and concomitantly modify the surface chemistry.

Each of these films was deposited onto the array using a CVD process. Briefly, vapor-phase precursors, tin(IV) nitrate anhydrous at 30°C, titanium (IV) 2-proxoide at 30°C, and triruthenium dodecacarbonyl at 65°C, were delivered sequentially at 7 standard cm^3/min (sccm). Each precursor was given a 10-min chamber equilibration period prior to film deposition. Furthermore, after a deposition cycle was completed, clean Ar was delivered through all delivery lines for 10 min prior to starting the next precursor, in an effort to reduce cross-contamination. After the initial equilibration period, the target microhotplate element(s) were heated to 375°C, decomposing the precursor to form localized thin solid film(s). Since the microhotplate elements are individually addressable, different materials can be deposited on a selected subset of the array. The deposition period for the SnO_2 and TiO_2 pure films was 20 s and 120 s, respectively. For the mixed metal oxide films, the two materials were deposited in series: SnO_2 for 5 s followed by TiO_2 for 60 s and TiO_2 for 60 s followed by Ru for 90 s. The deposited films were subsequently annealed in zero-grade dry air at 1 standard l/min (slm) for 30 min. A more detailed description of this CVD processes is available elsewhere [21]. Figure 10.4a shows the optical images of these different films, and the sensing film microstructures are shown as SEM micrographs in Fig. 10.4b.

10.2.2 Temperature Programmed Sensing

The metal oxide sensing films themselves, while robust, are only partially selective and will respond to many analytes. However, it can be expected that any given analyte/sensing material pair will generate distinguishing characteristics in its electrical conductance profile if one measures the effects of temperature-dependent variations in chemical interactions at the surface, which can/will involve other co-adsorbed species from the background. Hence, each sensing film within our microhotplate array is programmed to cycle through many temperatures in a pulsed mode such as that illustrated in Fig. 10.5, in order to capture a greater range information on temperature-dependent interactions between analytes and sensing films.

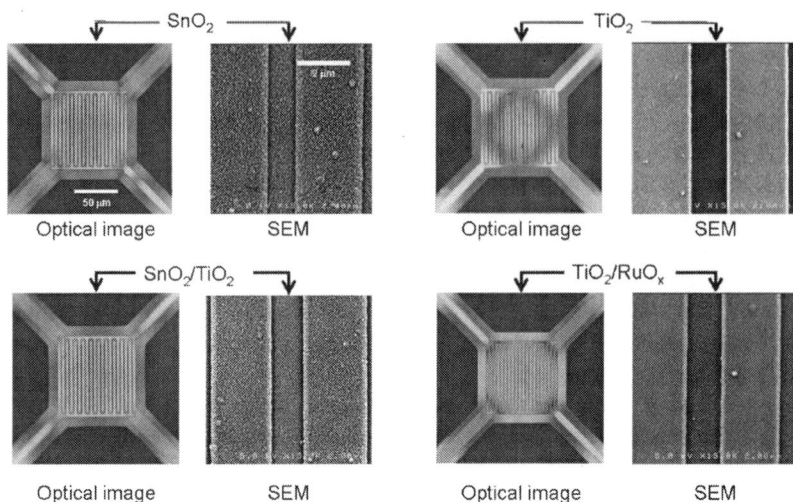

Fig. 10.4 Materials chosen for ammonia detection problem. (**a**) Optical microscopy images of the four films used in this study (size of scale bar applies to all panels): tin oxide (SnO_2), tin oxide coated with titanium oxide (SnO_2/TiO_2), titanium oxide (TiO_2) and titanium oxide coated with ruthenium oxide (TiO_2/RuO_x). Four copies of each material were used in arrays for this work. (**b**) Scanning electron microscopy images of the four films (scale bar applies to all panels). Reprinted with permission from ref [21]. Copyright 2009 Elsevier

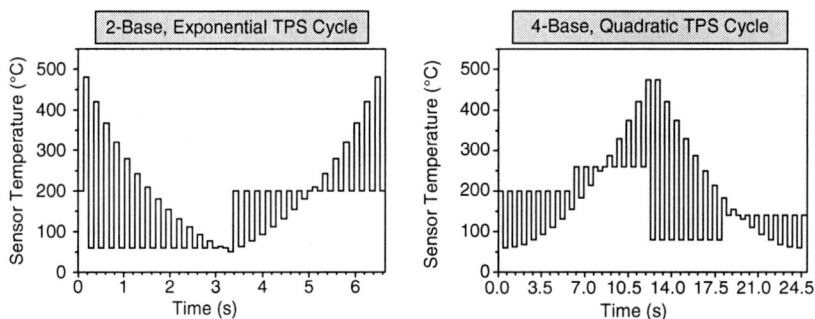

Fig. 10.5 Temperature Programmed Sensing. The temperature program used to operate the sensing elements toggles the temperature between multiple (left: 28; right: 32) ramp values that sample most of the temperature range of the device and different baseline temperature (left: 2; right: 4) values to allow "relaxation" toward some initial state prior to each ramp temperature. Moreover, different baselines also allow different film-analyte interactions (adsorption/desorption, decomposition, and reaction) at the sensing surface prior to the ramp measurements. The dwell time at each temperature (period of the pulse) was typically on the order of 200 ms. A conductance measurement was made at each base and ramp temperature, but only the ramp values were used for further analysis in this study

This mode of operation, which is expected to greatly enhance the analytical content of the microsensor signal over that of fixed-temperature sensing, is referred to as temperature-programmed sensing (TPS) [7, 11].

10.3 Analyte Delivery

A custom gas delivery manifold was used to separately deliver four analyte stream components: target-laden zero-grade dry air, interference-laden dry air, humid air, and balance dry air. This manifold features non-reactive tubing for each input stream connected to a central cell, where sample blending occurred. Metered concentrations of NH_3 were delivered from pre-blended commercial cylinders. Interference vapors were introduced by bubbling metered quantities of dry air through vessels containing the interference candidate, after which the saturated gas stream was delivered via dedicated lines to the central cell (thus interferences are reported in percent saturation). In order to reduce biases stemming from fractional evaporation of interference mixtures, the liquid sources were replaced regularly. Interference introduction was performed using this method for final concentrations of up to 1% saturation at room temperature. Humid air was generated by metering zero-grade dry air through a dew point generator. Test cell humidity could thus be varied between 0% relative humidity (RH) and 90% RH at 25°C using this apparatus. The water vapor concentrations at 10% RH, 30% RH, and 70% RH were 2.8 mmol/mol, 8.4 mmol/mol, and 19.6 mmol/mol, respectively. The balance dry air was adjusted such that at any point in time, the sum total flow rate of these four "single-component" streams into the central cell was 1 slm. A sensor cell that housed the microsensor array was placed downstream from the mixing manifold.

10.4 Statistical Methods for Sensor Material Selection

Capturing the conductance response of a variety of materials, each operated at different temperatures, generates a large data set from the sensor array for any analyte. Figure 10.6 illustrates the responses of a single tin oxide chemiresistor at the 32 ramp temperatures. The quadratic, 4-base TPS (refer to Fig. 10.5) was used in acquiring the data in Fig. 10.6. The top panel indicates the different analytes to which the sensor was exposed during the corresponding measurement (same as Fig. 10.1). The responses of the sensor with respect to the most recent baseline response are shown along the y-axis. Each trace indicates a single isotherm, i.e. response of SnO_2 at a specific temperatures. It is not clear from this multivariate sensor response (a) whether the different temperatures are generating orthogonal or similar information about the presence of ammonia, and (b) whether there is sufficient information to detect and identify all ammonia introductions that would be made in a testing sequence (refer to Fig. 10.1). In the remainder of this chapter, we will examine statistical methods that allow us to analyze such multivariate sensor responses.

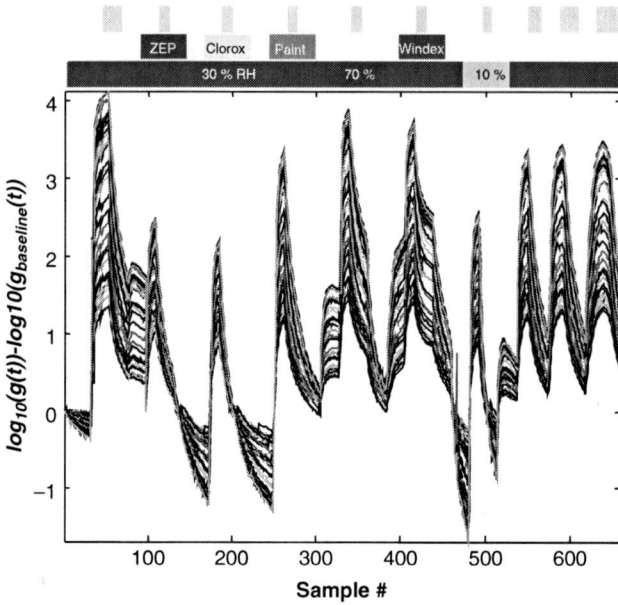

Fig. 10.6 Multivariate sensor responses. The responses of a single tin oxide chemiresistor at 32 different temperatures are shown. The top panel identifies the different measurement conditions over time (*x*-axis) to which the sensor was exposed during the measurement. The *y*-axis shows the response of the sensor registered with respect to the most recent baseline response. Each trace indicates a single isotherm, i.e. response of SnO_2 at a specific temperature

10.4.1 Assessment of Material Similarity

To assess similarity/orthogonality of conductometric responses generated by sensor materials and temperature programs, we present a measure based upon pair-wise correlation. Suppose x is a one-dimensional vector of conductance measurements made at each of the different training conditions using a material M_1 at temperature T_1 (equivalent to one isotherm in Fig. 10.6), and y is a one-dimensional vector of conductance measurements made using sensing material M_2 at temperature T_2, then the correlation between responses of M_1 at T_1 and M_2 at T_2 is calculated as

$$\frac{\text{Cov}(x, y)}{\sigma_x \sigma_y}, \qquad (10.1)$$

where $\text{Cov}(x,y)$ is the covariance of x and y, σ_x and σ_y are their respective standard deviations. A simple example illustrating the distribution of two simulated sensor responses for different degrees of correlations is provided in Fig. 10.7. It can be clearly seen that as correlation increases, the data converge along the diagonal i.e. given one sensor response the predictability of the other sensor behavior also

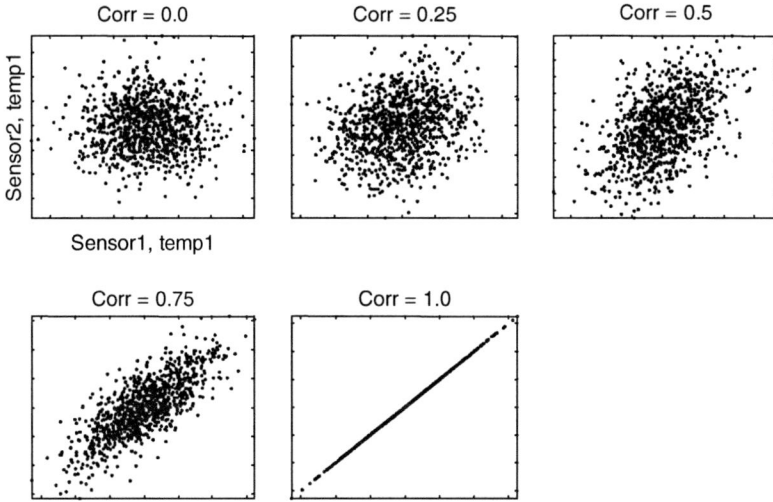

Fig. 10.7 Illustration of the distribution of two sensor responses occurring for different degrees of correlation between them

increases (indicates redundancy). An illustration of correlation analysis between SnO_2 isotherms at $60°C$ and $480°C$ is shown in Fig. 10.8.

10.4.1.1   Is Different Information Really Acquired at Different Temperatures?

In order to determine whether the TPS mode of operation generates additional information from each film composition, we compute, for each film type, the correlation between its responses to different conditions at different temperatures. A detailed plot illustrating the correlation coefficients for the four films employed in the illustrative NH_3 detection problem is shown in Fig. 10.9. Each pixel shows the correlation between two isotherms of the same material. Only the absolute values of the significant correlations ($P < 0.001$) are shown (the t-test is not valid for the diagonal elements, which represent self-correlations, as the denominator becomes zero). Lighter pixels indicate higher correlations and darker pixels represent lower correlations. The greater the correlation, the more similar or redundant is the information generated.

The correlation plots shown in Fig. 10.9 reveal that in the case of the TiO_2 films with and without RuO_x coating, the lower temperature responses correlate well amongst themselves, but not with those obtained at higher temperatures. Similarly, the high temperature responses correlate well only amongst themselves. This lack of correlation between the two temperature bands indicates that different analytical information is obtained from these bands. Interestingly, the TiO_2/RuO_x film shows a lack of correlation between the low-temperature features on the upward vs. downward portion of the temperature program, indicating a dependence on thermal history. In the case of SnO_2 films both with and without TiO_2, all temperature features appear

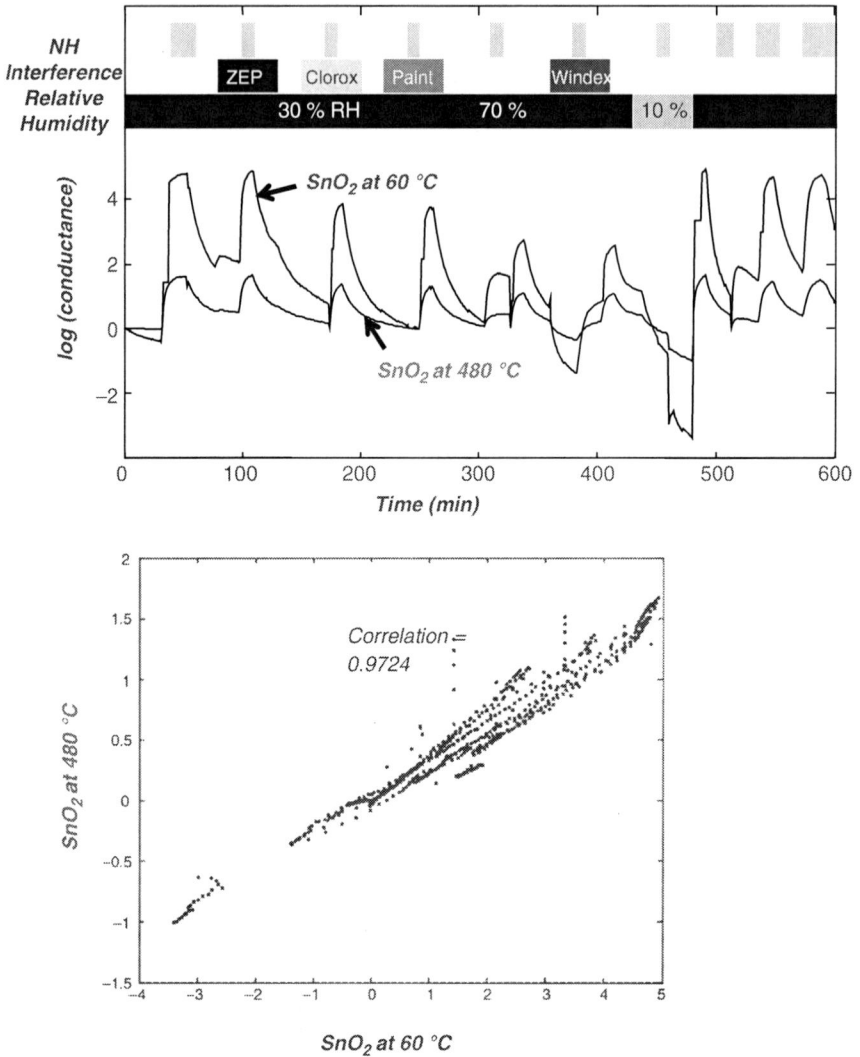

Fig. 10.8 Demonstration of correlation analysis. (*Top*) SnO$_2$ electrical conductance isotherms at 60°C and 480°C in response to different conditions indicated schematically at the top of the figure. (*Bottom*) The responses at 60°C vs. 480°C to different conditions lie along the diagonal, showing a greater degree of correlation between the two isotherms. Reprinted with permission from ref. [21]. Copyright 2009 Elsevier

to be well correlated (notice the change in the scale bar). However, as in the case of TiO$_2$ films, the high-temperature features and low-temperature features show a greater degree of correlation only amongst themselves. Hence, for the four sensing materials, different information is generated at low and high temperatures.

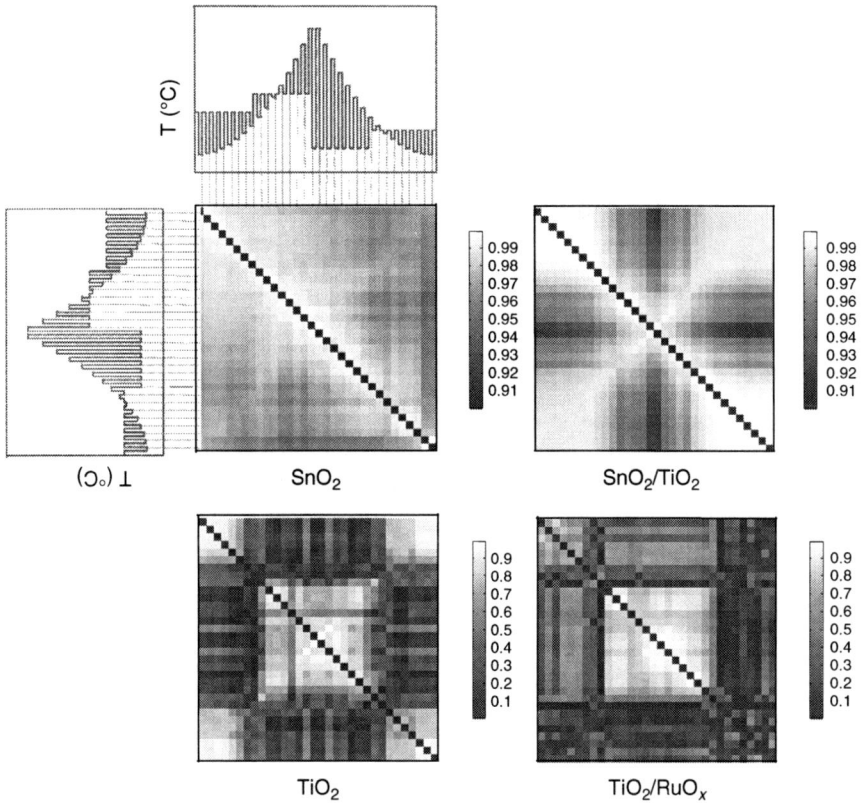

Fig. 10.9 Correlation analysis to assess similarity/orthogonality of material-temperature combinations. Intensity-coded representation of the correlation coefficients between conductometric responses of each of the four sensing materials at different temperatures. Note that light areas indicate high correlation and dark areas show lower degrees of correlation. Only the absolute values of correlations that are significant ($P < 0.001$) are shown. Reprinted with permission from ref. [21]. Copyright 2009 Elsevier

10.4.1.2 Is Different Information Generated by Different Film Types?

The self-correlations computed between temperature features within film types do not provide any insights into the similarity/orthogonality across multiple materials. To evaluate this, we compute cross-correlation across isotherms of different film types. Figure 10.10 shows the cross-correlation plots for a single copy of the four different metal oxides. The diagonal blocks show self-correlations and are essentially the same as in Fig. 10.9. We observe lower correlations between different film types, indicating that the four materials contribute non-redundant information towards NH_3 recognition.

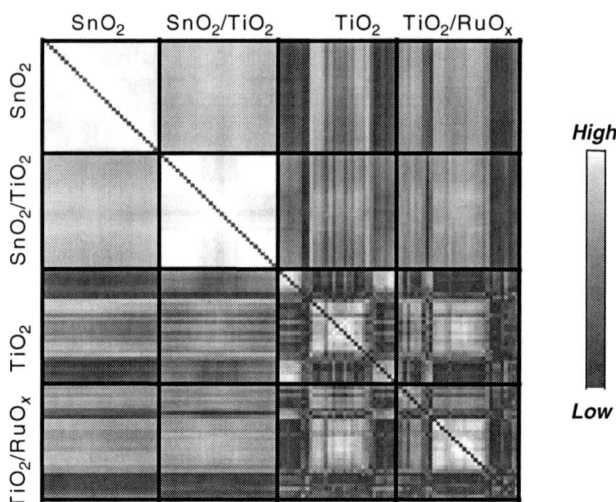

Fig. 10.10 Cross-correlation analysis. Lower correlation between responses of different film types at different temperatures reveals that the chosen sensing materials contribute non-redundant information for recognizing NH_3. Reprinted with permission from ref. [21]. Copyright 2009 Elsevier

10.4.2 Assessment of Material Reproducibility

To determine reproducibility of the chosen films, we fabricated four replicas of each film type in a 16-element microsensor array. Figure 10.11 shows cross-correlation across two copies of each material. A qualitative evaluation of the reproducibility of the sensing materials can be made by visually comparing the correlation patterns. Copies of a single film type (e.g., SnO_2 vs. SnO_2) show similar correlation patterns across temperature features, indicating that the films produced through the CVD process generate information that is highly reproducible between devices of equivalent manufacture. However, amongst the four film types used, the TiO_2/RuO_x films show the least-conserved cross-correlation patterns, especially at lower temperatures, indicating lower reproducibility compared to the other three film types used in these studies.

10.4.3 Dimensionality Reduction for Qualitative Analysis of Analytical Information

The correlation analysis uncovers relationships between conductometric responses of different materials at different operating temperatures, but does not provide insights into what information is contributed by each film type and whether the chosen materials and temperature programs provide sufficient analytical

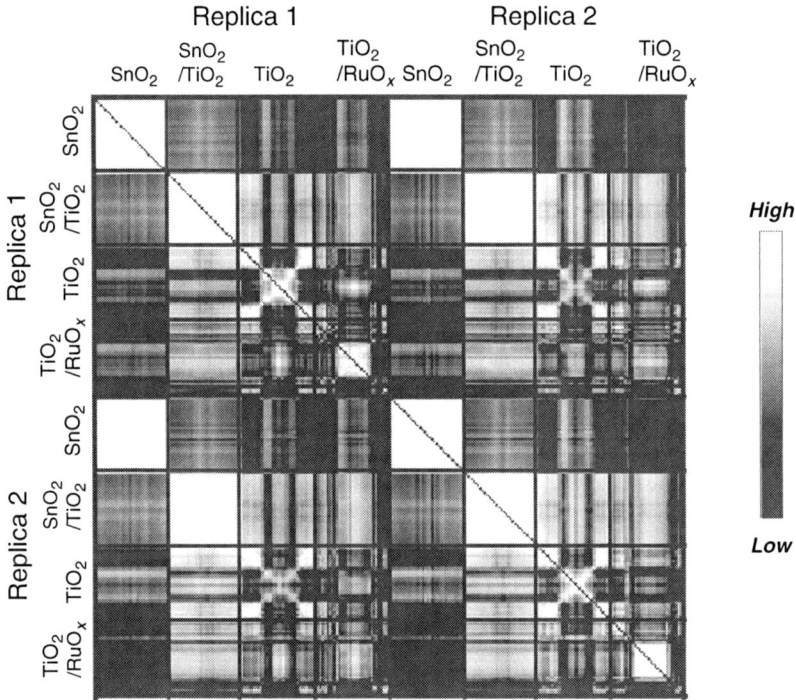

Fig. 10.11 Reproducibility analysis. Correlations across two copies of each sensing material deposited onto a 16- element array for the NH_3 detection problem. Reprinted with permission from ref. [21]. Copyright 2009 Elsevier

information to allow species recognition. In order to determine this, we visualize the multi-dimensional sensor array response using linear discriminant analysis (LDA) [22]. LDA is a dimensionality reduction technique that finds a few principal directions that maximize separation between classes and minimize variance within a single class. This is done by finding the eigenvectors of $S_W^{-1} S_B$ where S_W and S_B are the within-cluster and between-cluster scatter matrices, respectively, defined as follows:

$$S_W = \sum_{q=1}^{Q=8} \sum_{x \in \omega_q} (x - \mu_q)(x - \mu_q)^T, \tag{10.2}$$

$$S_B = \sum_{q=1}^{Q=8} (\mu_q - \mu)(\mu_q - \mu)^T, \tag{10.3}$$

$$\mu_q = \frac{1}{n_q} \sum_{x \in \omega_q} x \text{ and } \mu = \frac{1}{n} \sum_{\forall x} x, \tag{10.4}$$

where x is a linear projection of sensor response along $Q - 1$ linear discriminant axes, Q is the number of conditions (eight clusters corresponding to the following conditions: NH_3 at 30% RH, NH_3 at 10% RH, NH_3 at 70% RH, $NH_3 + ZEP$, $NH_3 +$ Clorox, $NH_3 + $ Windex, $NH_3 + $ Paint, and backgrounds without NH_3), μ_q and n_q are the mean vector and number of examples for condition q, respectively, n is the total number of examples in the dataset, and μ is the mean vector of the entire distribution. An illustration of the LDA dimensionality reduction where responses from two sensors are projected onto a single dimensional space is shown in Fig. 10.12.

Figures 10.13 and–10.16 show the scatter plot after dimensionality reduction of the multi-dimensional sensor response obtained by concatenating the sensor conductance values at the 32 ramp temperature of individual sensing films. Each three-dimensional object indicates one of the eight possible conditions of interest: NH_3 at 30% RH (■), NH_3 at 10% RH (◀), NH_3 at 70% RH (▶), $NH_3 + ZEP$ (◆), $NH_3 + $ Clorox (∗), $NH_3 + $ Windex (▲), $NH_3 + $ Paint (▽) and backgrounds without NH_3 (●; includes response to three humidity levels and four interferences without the target analyte). From the LDA plots it is clear that all of the chosen films can contribute, to varying degrees, to NH_3 detection and recognition. The ability to distinguish NH_3 clusters from the background improves as additional materials are included for this analysis (Figs. 10.17–10.19), and as the array size is systematically increased (Figs. 10.20–10.22).

10.5   Optimization of Array Configuration

To quantify the separability of the different analyte clusters and compare the films performance for this analytical problem, we derive a measure from Fisher's LDA [22] that can be defined as follows:

$$J = \frac{\text{trace}(S_B)}{\text{trace}(S_B) + \text{trace}(S_W)}, \tag{10.5}$$

where S_W and S_B are the within-cluster and between-cluster scatter matrices [see 10.2) and (10.3)], respectively. Being the ratio of the spread between classes relative to the spread within each class, the measure J increases monotonically as analyte clusters become increasingly more separable. Figure 10.23 compares 15 array configurations for NH_3 cluster separability alone. It is clear that the SnO_2 films (pure and doped with TiO_2) are better suited for fulfilling the analytical requirements of this problem. We can also observe that the separability increases as we add different materials (higher diversity – compare configurations 1, 5, 6, and 7) or add multiple copies of the same material (higher redundancy - compare configurations 1, 11, 12, and 13).

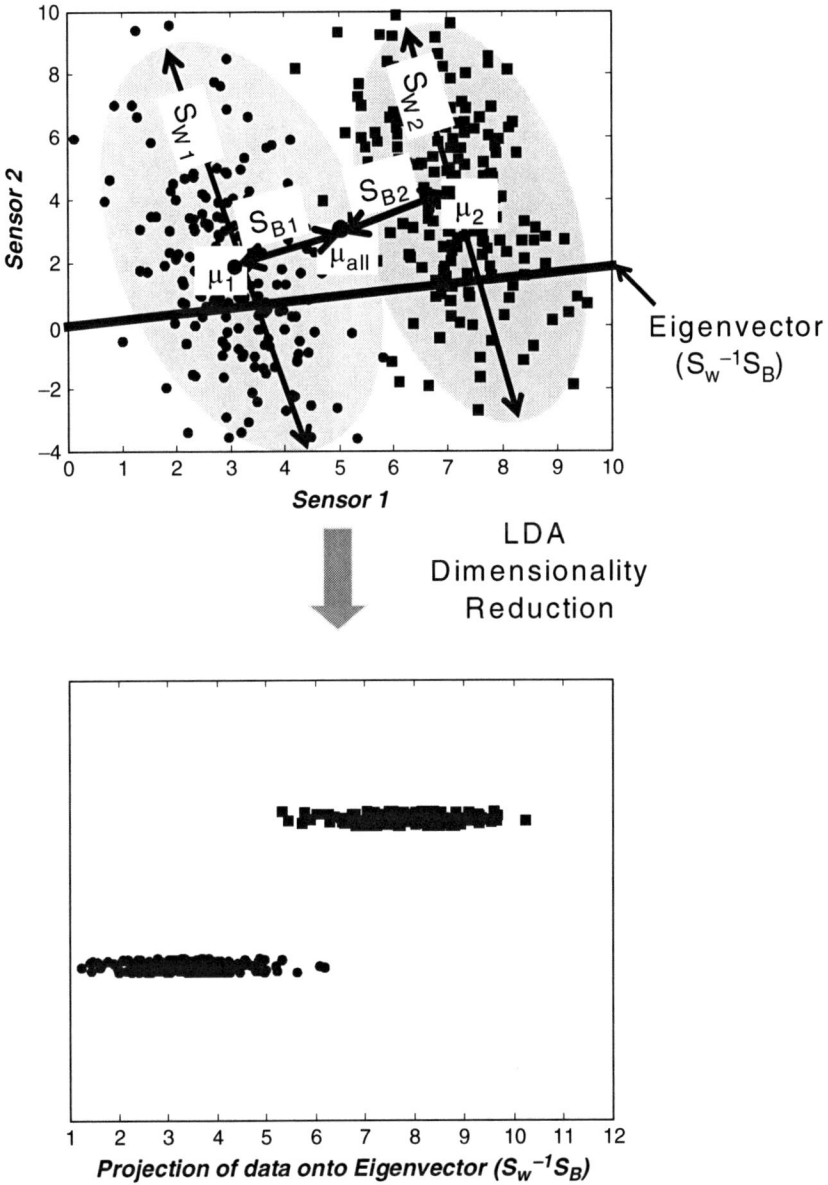

Fig. 10.12 Simulated responses from two sensors to two different conditions (*squares and circles*) are shown on the *top*. The two dimensional data is projected onto the eigenvector of the $S_W^{-1}S_B$ matrices to reduce dimensionality of the data (shown below)

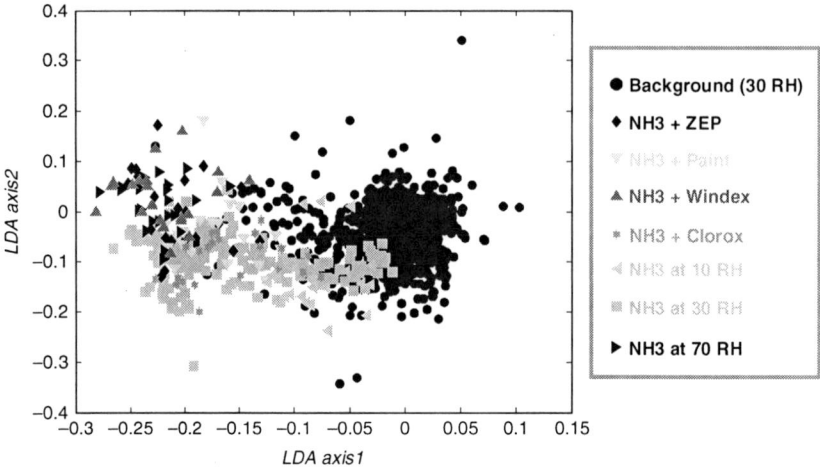

Fig. 10.13 Dimensionality reduction analysis of SnO_2 responses. Scatter-plot of the conducto-metric responses at different conditions (see legend) after dimensionality reduction using Fisher's linear discriminants

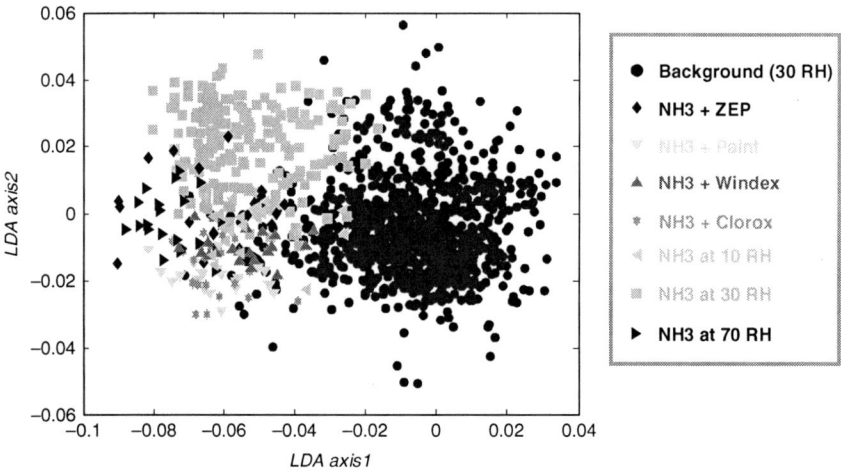

Fig. 10.14 Dimensionality reduction analysis of SnO_2/TiO_2 responses. Scatter-plot of the con-ductometric responses at different conditions (see legend) after dimensionality reduction using Fisher's linear discriminants

To determine the optimal material composition for the NH_3 detection problem, we define an objective function that takes into account the sufficiency of a solution (i.e. NH_3 cluster separability as defined in (10.5)), and a penalty term for incorpora-tion of practical constraints/tradeoffs on manufacturability. For discussions here,

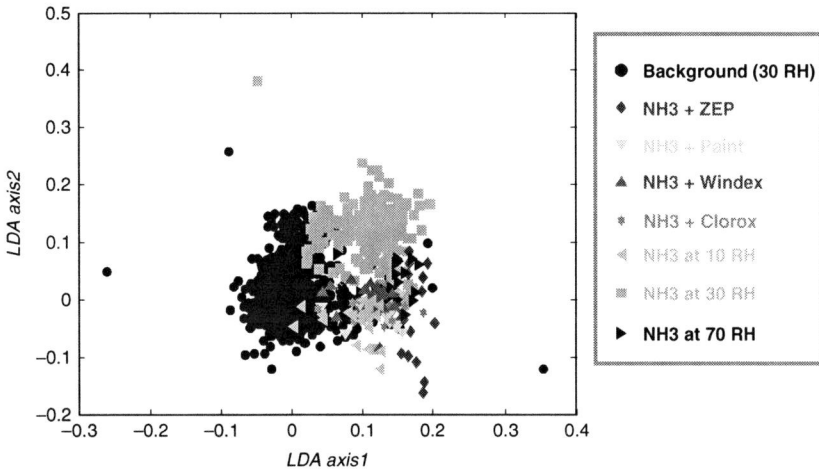

Fig. 10.15 Dimensionality reduction analysis of TiO_2 responses. Scatter-plot of the conductometric responses at different conditions (see legend) after dimensionality reduction using Fisher's linear discriminants

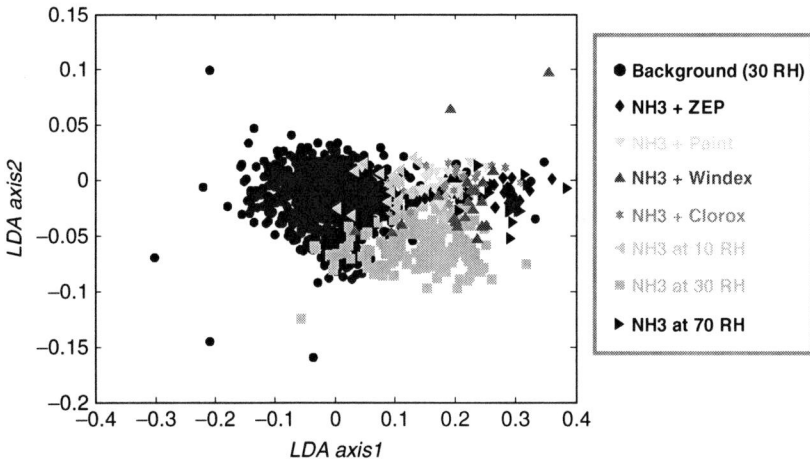

Fig. 10.16 Dimensionality reduction analysis of TiO_2/RuO_x responses. Scatter-plot of the conductometric responses at different conditions (see legend) after dimensionality reduction using Fisher's linear discriminants

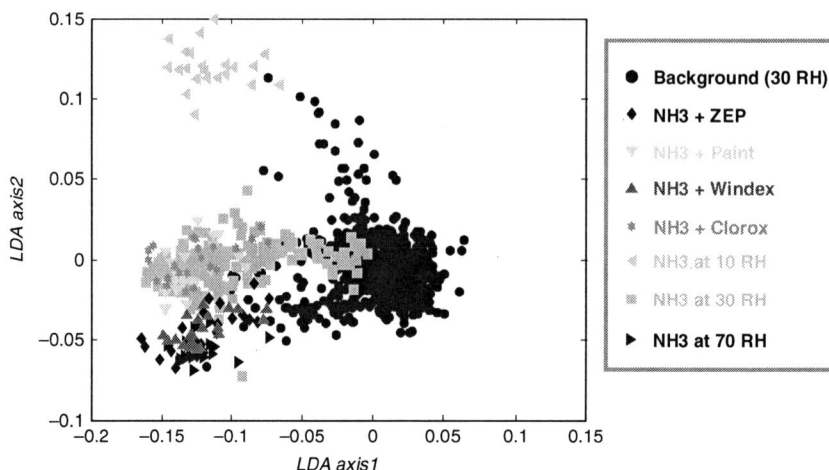

Fig. 10.17 Dimensionality reduction analysis of $SnO_2 + SnO_2/TiO_2$ responses. Scatter-plot of the conductometric responses at different conditions (see legend) after dimensionality reduction using Fisher's linear discriminants

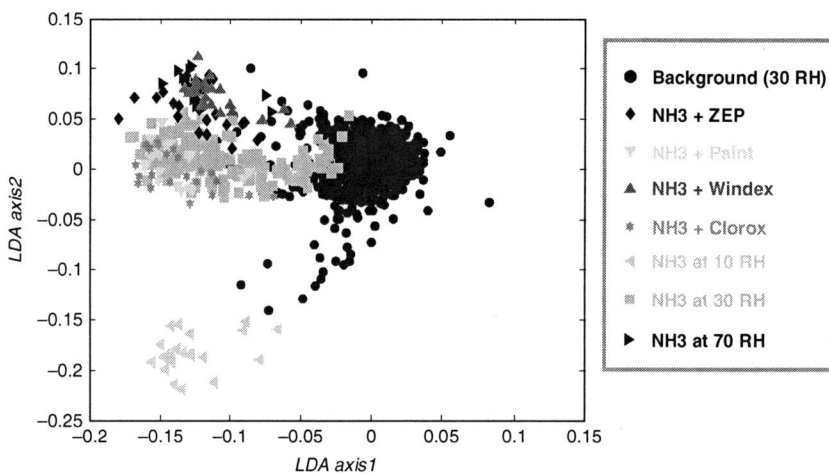

Fig. 10.18 Dimensionality reduction analysis of SnO_2+SnO_2/TiO_2+TiO_2 responses. Scatter-plot of the conductometric responses at different conditions (see legend) after dimensionality reduction using Fisher's linear discriminants

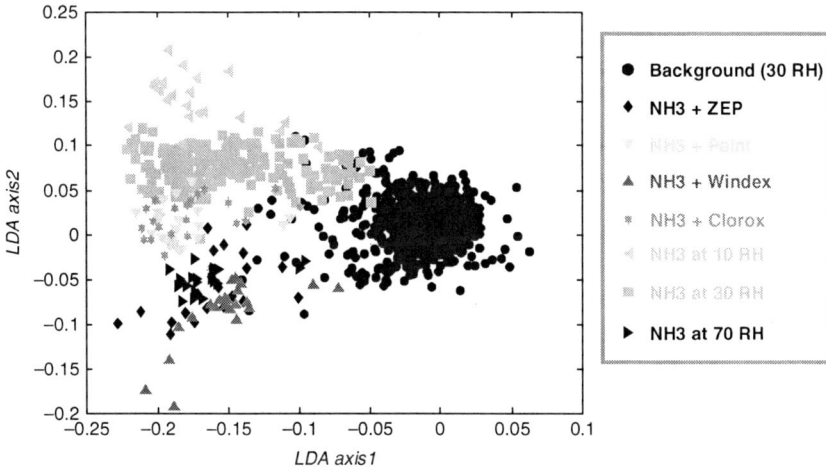

Fig. 10.19 Dimensionality reduction analysis of $SnO_2 + SnO_2/TiO_2 + TiO_2 + TiO_2/RuO_x$ responses. Scatter-plot of the conductometric responses at different conditions (see legend) after dimensionality reduction using Fisher's linear discriminants

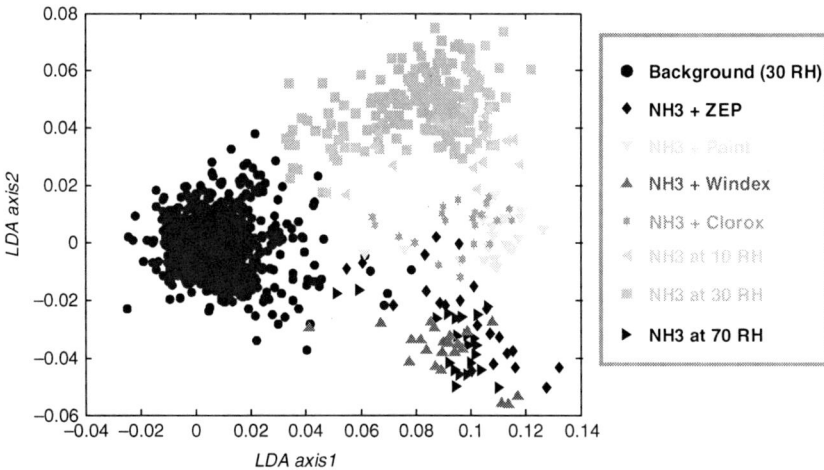

Fig. 10.20 Dimensionality reduction analysis of 2 copies of $SnO_2 + 2$ copies of $SnO_2/TiO_2 + 2$ copies of $TiO_2 + 2$ copies of TiO_2/RuO_x responses. Scatter-plot of the conductometric responses at different conditions (see legend) after dimensionality reduction using Fisher's linear discriminants

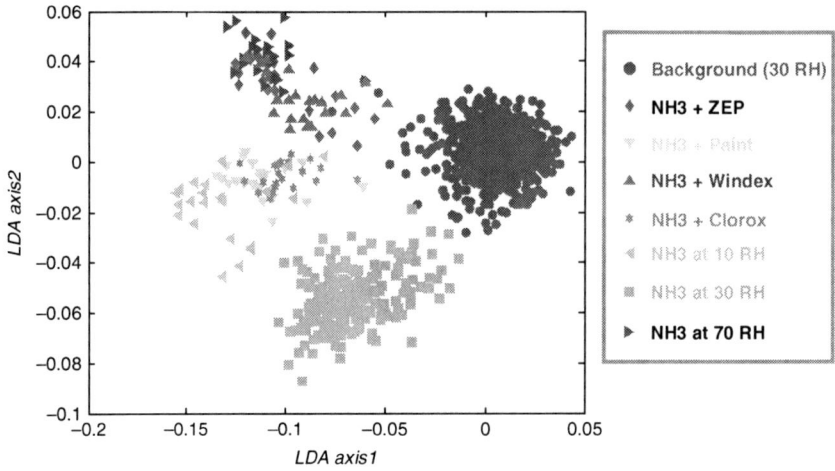

Fig. 10.21 Dimensionality reduction analysis of 3 copies of SnO_2 + 3 copies of SnO_2/TiO_2 + 3 copies of TiO_2 + 3 copies of TiO_2/RuO_x responses. Scatter-plot of the conductometric responses at different conditions (see legend) after dimensionality reduction using Fisher's linear discriminants

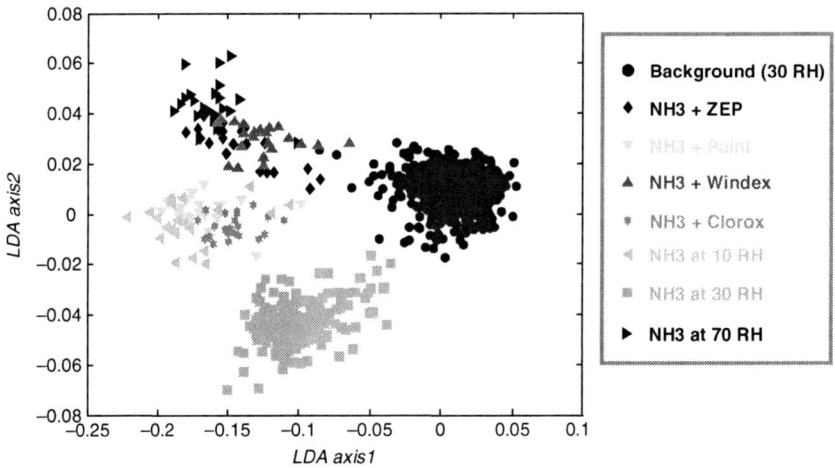

Fig. 10.22 Dimensionality reduction analysis of 4 copies of SnO_2 + 4 copies of SnO_2/TiO_2 + 4 copies of TiO_2 + 4 copies of TiO_2/RuO_x responses. Scatter-plot of the conductometric responses at different conditions (see legend) after dimensionality reduction using Fisher's linear discriminants

we will consider the following two resources independently: array size and the number of materials used.

Figure 10.24 shows the comparison of the same 15 configurations as in Fig. 10.23 with an extra penalty term for the array size:

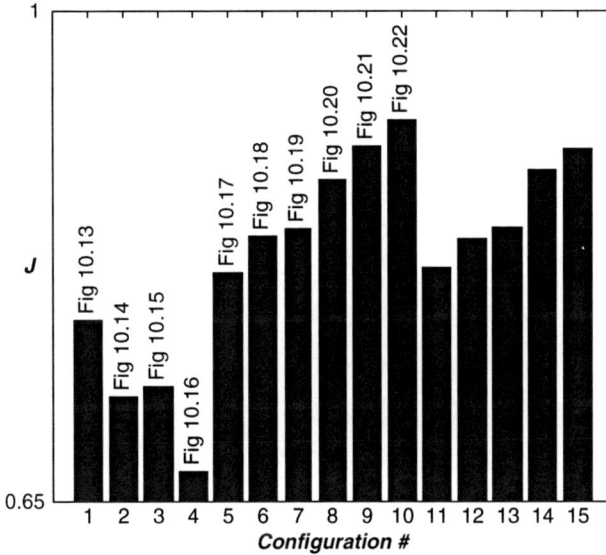

Fig. 10.23 Comparison of different combination of sensing materials for cluster separability (see 10.5): $1 - SnO_2$; $2 - SnO_2/TiO_2$; $3 - TiO_2$; $4 - SnO_2/RuO_x$; $5 - SnO_2 + SnO_2/TiO_2$; $6 - SnO_2 + SnO_2/TiO_2 + TiO_2$; $7 - SnO_2 + SnO_2/TiO_2 + TiO_2 + TiO_2/RuO_x$; $8 -$ two copies of $SnO_2 +$ two copies of $SnO_2/TiO_2 +$ two copies of $TiO_2 +$ two copies of TiO_2/RuO_x; $9 -$ three copies of $SnO_2 +$ three copies of $SnO_2/TiO_2 +$ three copies of $TiO_2 +$ four copies of TiO_2/RuO_x; $10 -$ four copies of $SnO_2 +$ four copies of $SnO_2/TiO_2 +$ four copies of $TiO_2 +$ four copies of TiO_2/RuO_x; $11 -$ two copies of SnO_2; $12 -$ three copies of SnO_2; $13 -$ four copies of SnO_2; $14 -$ four copies of $SnO_2 +$ four copies of SnO_2/TiO_2; $15 -$ four copies of $SnO_2 +$ four copies of $SnO_2/TiO_2 +$ four copies of TiO_2

$$J1 = \underbrace{\frac{\text{trace}(S_B)}{\text{trace}(S_B) + \text{trace}(S_W)}}_{NH_3 \text{ cluster separability}} - \underbrace{\gamma_1^* \text{ \# sensing elements}}_{\text{penalty term 1}}. \tag{10.6}$$

The penalty term imposes a constraint that each additional element added must improve the overall NH cluster separability by γ_1 ($=0.02$ in Fig. 10.24) in order to improve the objective function ($J1$). Due to this additional constraint, note that configuration # 10 (four copies of all four films), which provided maximum chemical hazard separability (see Fig. 10.23), is no longer the best solution. We find that the best array configuration with the fewest number of sensing elements to be a three-element array with a copy of the following sensing materials: SnO_2, SnO_2/TiO_2, and TiO_2.

Figure 10.25 shows the comparison of the same 15 configurations as in Figs. 10.23 and 10.24 but now with an extra penalty term for the number of sensing materials used:

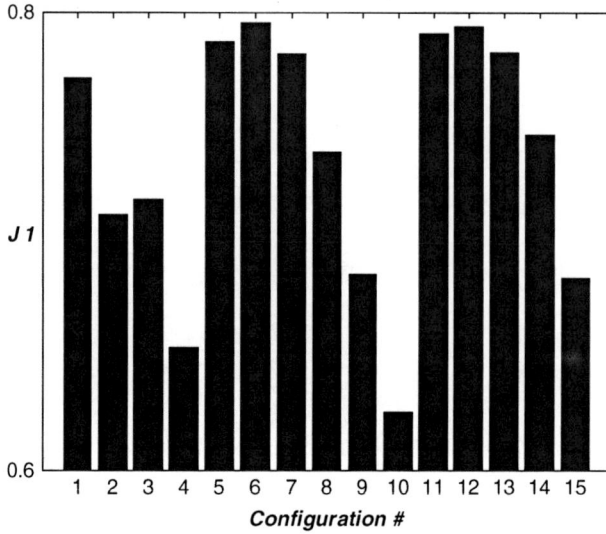

Fig. 10.24 Comparison of different combination of sensing materials for cluster separability with a penalty for the number of sensing elements used (see (10.6)): $1 - SnO_2$; $2 - SnO_2/TiO_2$; $3 - TiO_2$; $4 - SnO_2/RuO_x$; $5 - SnO_2 + SnO_2/TiO_2$; $6 - SnO_2 + SnO_2/TiO_2 + TiO_2$; $7 - SnO_2 + SnO_2/TiO_2 + TiO_2 + TiO_2/RuO_x$; $8 -$ two copies of $SnO_2 +$ two copies of $SnO_2/TiO_2 +$ two copies of $TiO_2 +$ two copies of TiO_2/RuO_x; $9 -$ three copies of $SnO_2 +$ three copies of $SnO_2/TiO_2 +$ three copies of $TiO_2 +$ four copies of TiO_2/RuO_x; $10 -$ four copies of $SnO_2 +$ four copies of $SnO_2/TiO_2 +$ four copies of $TiO_2 +$ four copies of TiO_2/RuO_x; $11 -$ two copies of SnO_2; $12 -$ three copies of SnO_2; $13 -$ four copies of SnO_2; $14 -$ four copies of $SnO_2 +$ four copies of SnO_2/TiO_2; $15 -$ four copies of $SnO_2 +$ four copies of $SnO_2/TiO_2 +$ four copies of TiO_2

$$J2 = \underbrace{\frac{\text{trace}\,(S_B)}{\text{trace}(S_B) + \text{trace}(S_W)}}_{NH_3 \text{ cluster separability}} - \underbrace{\gamma_2^* \,\#\, \text{sensing materials}}_{\text{penalty term 2}} . \qquad (10.7)$$

Note that the penalty term coefficient γ_2 ($=0.1$) was set at greater than γ_1. As array configurations that employed all four sensing materials in consideration were penalized more, we now find that the best array configuration employing the fewest sensing materials to be a four-element array with four copies of SnO_2 sensors alone.

10.6 Conclusions

We have presented statistical methods that provide a generalizable methodology for designing and evaluating array-based solutions for a specific detection problem. We demonstrated this approach and tuned a microsensor array for the problem of identifying NH_3 in the presence of interferences at fixed mixing-ratios and ambient condition changes by controlling the sensor material composition within the array and designing rapid temperature programs that enhance the analytical information

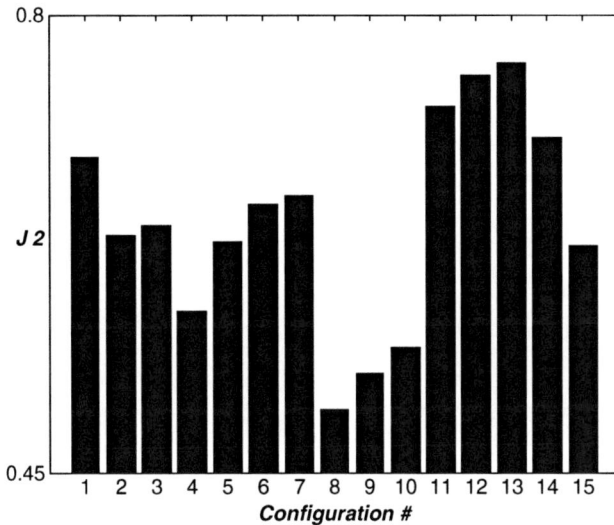

Fig. 10.25 Comparison of different combination of sensing materials for cluster separability with a penalty for the number of sensing materials used (see (10.7)): $1 - SnO_2$; $2 - SnO_2/TiO_2$; $3 - TiO_2$; $4 - SnO_2/RuO_x$; $5 - SnO_2 + SnO_2/TiO_2$; $6 - SnO_2 + SnO_2/TiO_2 + TiO_2$; $7 - SnO_2 + SnO_2/TiO_2 + TiO_2 + TiO_2/RuO_x$; $8 -$ two copies of $SnO_2 +$ two copies of $SnO_2/TiO_2 +$ two copies of $TiO_2 +$ two copies of TiO_2/RuO_x; $9 -$ three copies of $SnO_2 +$ three copies of $SnO_2/TiO_2 +$ three copies of $TiO_2 +$ four copies of TiO_2/RuO_x; $10 -$ four copies of $SnO_2 +$ four copies of $SnO_2/TiO_2 +$ four copies of $TiO_2 +$ four copies of TiO_2/RuO_x; $11 -$ two copies of SnO_2; $12 -$ three copies of SnO_2; $13 -$ four copies of SnO_2; $14 -$ four copies of $SnO_2 +$ four copies of SnO_2/TiO_2; $15 -$ four copies of $SnO_2 +$ four copies of $SnO_2/TiO_2 +$ four copies of TiO_2

obtained from each element in the array as a function of time. These advances are critical to the production of pre-programmed microsensors for non-invasive, real-time, multi-species recognition relevant to homeland security and other applications involving trace analyte detection in complex chemical cocktails.

Acknowledgments We acknowledge partial financial support of this project by the U.S. Department of Homeland Security, Science and Technology Directorate. BR was supported by a NIH(NIBIB)-NIST Joint Postdoctoral Associateship Award administered through the National Research Council. We thank Kurt Benkstein, Mike Carrier, Steve Fick, Jim Melvin, Wyatt Miller, Chip Montgomery, Casey Mungle, Jim Yost, Blaine Young, and Li Zhang for their valuable contributions to this project. We are grateful to Mark Stopfer for his helpful comments on an earlier version of this manuscript.

References

1. Mention of these and any other commercial products is strictly for provision of proper experimental definition, and does not constitute an endorsement by the National Institute of Standards and Technology
2. Hierlemann, A.; Gutierrez-Osuna, R., Higher-order chemical sensing, *Chem. Rev.* **2008**, 108, 563–613

3. Pearce, T. C.; Sanchez-Montanes, M., Chemical sensor array optimization: geometric and information theoretic approaches, In Handbook of Machine Olfaction: Electronic Nose Technology; Pearce, T. C.; Schiffman, S. S.; Nagle, H. T.; Gardner, J. W., Eds. Wiley-VCH, Weinheim, **2002**, 347–376.

4. Wilson, D.; Garrod, S.; Hoyt, S.; McKennoch, S.; Booksh, K. S., Array optimization and preprocessing techniques for chemical sensing microsystems, *Sens. Update* **2002**, 10, 77–106

5. Semancik, S.; Cavicchi, R. E.; Wheeler, M. C.; Tiffany, J. E.; Poirier, G. E.; Walton, R. M.; Suehle, J. S.; Panchapakesan, B.; Devoe, D. L., Microhotplate platforms for chemical sensor research, *Sens. Actuators B* **2001**, 77, 579–591

6. Semancik, S.; Cavicchi, R., Kinetically controlled chemical sensing using micromachined structures, *Acc. Chem. Res.* **1998**, 31, 279–287

7. Cavicchi, R. E.; Suehle, J. S.; Kreider, K. G.; Gaitan, M.; Chaparala, P., Fast temperature programmed sensing for micro-hotplate gas sensors, *IEEE Electron Device Lett.* **1995**, 16, 286–288

8. Batzill, M.; Diebold, U., The surface and materials science of tin oxide, *Sens. Actuators B* **1997**, 43, 45–51

9. Martinez, C. J.; Hockey, B.; Montgomery, C. B.; Semancik, S., Porous tin oxide nanostructured microspherers for sensor applications, *Langmuir* **2005**, 21, 7937–7944

10. Raman, B.; Hertz, J.; Benkstein, K.; Semancik, S., A bioinspired methodology for artificial olfaction, *Anal. Chem.* **2008**, 80, 8364–8371

11. Meier, D. C.; Taylor, C. J.; Cavicchi, R. E.; White, E.; Semancik, S.; Ellzy, M. W.; Sumpter, K. B., Chemical warfare agent detection using MEMS-compatible microsensor arrays, *IEEE Sens. J.* **2005**, 5, 712–725

12. Bârsan, N.; Weimar, U., Understanding the fundamental principles of metal oxide based gas sensors; the example of CO sensing with SnO_2 sensors in the presence of humidity, *J. Phys. Condens Matter* **2003**, 15, R813–R819

13. Tomchenko, A. A.; Harmer, G. P.; Marquis, B. T., Detection of chemical warfare agents using nanostructured metal-oxide sensors, *Sens. Actuators B* **2005**, 108, 41–55

14. Vaid, T. P.; Burl, M. C.; Lewis, N. S., Comparison of the performance of different discriminant algorithms in analyte discrimination tasks using an array of carbon black-polymer composite vapor detectors, *Anal. Chem.* **2001**, 73, 321–331

15. Albert, K. J.; Lewis, N.; Schauer, C.; Sotzing, G. A.; Stitzel, S. E.; Vaid, T. P.; Walt, D. R., Cross-reactive chemical sensor arrays, *Chem. Rev.* **2000**, 100, 2595–2626

16. Ding, J.; McAvoy, T. J.; Cavicchi, R. E.; Semancik, S., Surface state trapping models for SnO_2-based microhotplate sensors, *Sens. Actuators B* **2001**, 77, 597–613

17. Gaggiotti, G.; Galdikas, A.; Kačiulis, S.; Mattogno, G.; Šetkus, A., Temperature dependencies of sensitivity and surface chemical composition of SnO_x gas sensors, *Sens. Actuators B* **1995**, 24–25, 516–519

18. White, N.; Turner, J., Thick-film sensors: Past, present and future, *Meas. Sci. Technol.* **1997**, 8, 1–20

19. Panchapekesan, B.; Cavicchi, R.; Semancik, S.; DeVoe, D. L., Sensitivity, selectivity and stability of tin oxide nanostructures on large area arrays of microhotplates, *Nanotechnology* **2006**, 17, 415–425

20. Meier, D. C.; Semancik, S., Effects of Materials Chemistry on Conductometric Sensor Signals. In 2005 Materials Research Society Meeting Boston, **2005**

21. Raman, B.; Meier, D.; Evju, J.; Semancik, S., Designing and optimizing microsensor arrays for recognizing chemical hazards in complex environments, *Sens. Actuators B* **2009**, 137, 617–629

22. Duda, R. O.; Hart, P. E.; Stork, D. G., Pattern Classification, 2nd edn.; Wiley-Interscience, New York, **2000**, 115–121

Chapter 11
Statistical Methods for Selecting the Components of a Sensing Array

Margie L. Homer, Hanying Zhou, April D. Jewell, and Margaret A. Ryan

Abstract An electronic nose which uses an array of conductometric chemical sensors has been developed at the Jet Propulsion Laboratory; the JPL Electronic Nose is to be used as an event monitor in human habitat in a spacecraft. This sensor array is designed to identify and quantify 10–15 organic and inorganic species in air. The earlier generation/version JPL electronic noses consisted of 32 polymer-carbon black composite sensors; the target analytes included volatile organics as well as ammonia. This third generation electronic nose has a new suite of target analytes, and so, a new set of sensors was selected. In addition to volatile organic chemicals, the target analytes include the inorganic species: ammonia, sulfur dioxide and elemental mercury. The most recent array under development has 32 sensors; additional materials were selected in order to detect inorganic species and polymer-carbon black composite sensors were reevaluated. In the development of such a device, we must select sensors suitable for the detection of targeted analytes, and we must be able to evaluate both the sensors and the array response. This chapter will discuss the statistical tools and experimental criteria used to evaluate and select materials in the sensing array.

11.1 Introduction

The JPL Electronic Nose (ENose) is a fully operational system designed to fill the gap between an alarm which responds to the presence of chemical compounds with little or no ability to distinguish among them, and an analytical

M.L. Homer (✉), H. Zhou, and M.A. Ryan
Jet Propulsion Laboratory, California Institute of Technology, 4800 Oak Grove Drive, Pasadena, CA 91109, USA
e-mail: margie.l.homer@jpl.nasa.gov

A.D. Jewell
Tufts University Chemistry Department, 62 Talbot Avenue, Medford, MA 02155, USA

M.A. Ryan et al. (eds.), *Computational Methods for Sensor Material Selection*,
Integrated Analytical Systems,
DOI 10.1007/978-0-387-73715-7_11, © Springer Science+Business Media, LLC 2009

instrument that is able to distinguish all the compounds present but with no real-time or continuous event monitoring ability. The specific analysis scenario targeted for this development is one of leaks or spills of specific compounds. It has been shown in the analysis of samples taken from space shuttle flights and the International Space Station (ISS) that, in general, the air is kept clean by the air revitalization system and that the contaminants are present at levels significantly lower than the Spacecraft Maximum Allowable Concentrations (SMACs) [1]. The JPL ENose has therefore been developed to detect targeted chemical species released suddenly into the breathing environment; the sensing system in the JPL ENose is a chemical sensing array made up of 32 semi-selective conductometric sensors.

Two generations of the JPL ENose have previously been shown to be able to detect, identify and quantify a variety of organic analytes, as well as the inorganic species, ammonia and hydrazine [2–4]. The first generation ENose was tested in the laboratory and subsequently on Space Shuttle flight STS-95 in 1998 for 6 days, and was shown to detect, identify and quantify nine volatile organic compounds as well as ammonia and water against a breathing-air background [2, 3]. The second generation ENose was tested in the laboratory and shown to detect, identify and quantify 20 volatile organic compounds as well as ammonia and water against an air background [4]. In these two generations of the JPL ENose, the 32 sensors in the array were all polymer–carbon composite conductometric sensors.

The third generation ENose has been developed for demonstration and testing on board the ISS. For this application, the target analytes include eight organic and three inorganic species; the species which are new to the JPL ENose are elemental mercury and sulfur dioxide [5, 6]. As SO_2 is toxic at parts-per-million (ppm) level concentrations (the OSHA Time-Weighted Permissible Exposure Limit is 5 ppm), and elemental mercury is toxic at parts-per-billion (ppb) levels, a system monitoring cabin air quality should be able to rapidly identify and quantify these contaminants at these concentrations so that appropriate measures may be taken to protect the health of the crew.

In developing the third generation ENose, we elected to use the same system platform as the second generation device and to design and optimize a sensing array which would be sensitive to both the organic and inorganic analytes. Key to this design was the decision to maintain the platform, which meant maintaining 32 sensor elements; thus, we elected to study the existing array for its response to organic analytes and then to determine which sensors could be replaced, either with other polymer–carbon composite sensors or with other materials to provide adequate response to inorganic analytes. Designing an array to include the detection of inorganic analytes included the possibility of using new sensor platforms for the new sensors as well as developing novel sensing films for the new target analytes. Once selected, the new sensor array was evaluated for performance. The final test of array suitability was the application of the Levenberg-Marquardt Non-Linear Least Squares (LMNLS) identification and quantification algorithm developed for the earlier generation ENoses [7].

Table 11.1 Analytes of interest, with target concentration ranges at 1 atm

Analyte	Target Conc (ppm)	Quantification range (ppm)
Ammonia	5	1.6–15
Mercury	0.01	0.003–0.03
Sulfur Dioxide	1	0.3–3
Acetone	270	90–810
Dichloromethane	10	3–30
Ethanol	500	166–1,500
Octafluoropropane	20	6–60
Methanol	10	3–30
2-Propanol	100	30–300
Toluene	16	5–50

Table 11.1 shows the targeted analytes, which include volatile organic compounds as well as inorganic species, and concentration ranges of interest. We have previously reported the successful detection of SO_2 for concentrations as low as 0.2 ppm (200 ppb) [5] and of elemental mercury at concentrations as low as 2 ppb [6] by selection of specific materials for the task. The ability of ENose to detect ammonia in the single ppm to sub-ppm range has been shown in the previous generations [2, 3].

Selection and optimization of the ENose sensing array to detect the new set of analytes was an iterative process which included a series of evaluations and analyses of the sensors as individual elements and of the array as a whole. The first round of analysis determined whether the existing array was adequate for the set of analytes under consideration. Both experimental and statistical analysis of the array determined that the array was not adequate for the detection of the third generation set of analytes.

Based on analysis of the existing data and modeling, as described in Chaps. 3 and 8 of this volume, a preliminary set of sensors was selected for evaluation. This chapter will focus on the evaluation of this preliminary set of sensors using statistical evaluation of sensor responses and sensor reproducibility. The final selection of sensors for the array includes the consideration of statistical analysis as well as more practical considerations such as ease of sensor fabrication and sensor reproducibility. We also present results from the final evaluation of the array through the use of the LMNLS identification and quantification algorithm.

11.2   Initial Sensor and Array Evaluation

11.2.1   Experimental Methods

The analysis of sensor arrays involves fabricating sensor arrays for testing and exposing the arrays to a set of target analytes at the concentration ranges of interest.

Details for fabricating polymer-carbon composite sensors are described elsewhere [2, 3]. The ENose lab at JPL maintains three calibrated gas handling systems for exposing sensors to measured concentrations of organic compounds, SO_2, ammonia and mercury in humidified air. These systems combine computer controlled gas delivery and dilution in air with data acquisition. All systems include control of background humidity and sensors can be exposed to analytes of interest in humidity ranges of 0–90% RH at 20–25°C [8]. As the function of the JPL ENose is to be used in crew habitat, all testing is done in air at or near 1 atm., in a temperature range 20–30°C.

11.2.2 Selection of Sensors for Preliminary Evaluation

The initial selection of sensors for evaluation was based on several criteria along with the analysis of sensors and sensor development from the two earlier generations of ENoses. In the first two generations of ENose, the arrays were selected by consideration of polymer type and ligands, and how the polymers were predicted to respond to analytes based on bonding. Polymer ligand types were classified as hydrogen bond acidic (HBA), hydrogen bond basic (HBB), dipolar and hydrogen bond basic (DBB), moderately dipolar (MD) and weakly dipolar (WD). Ligand types were matched with analytes, and a distribution of sensor materials in appropriate classifications were selected. Arrays were selected based on experimental data developed in the laboratory, using a combination of statistical and experimental techniques. In the third generation, initial materials were based on materials used in the first two generations; in addition, consideration was given to the concept of "like dissolves like," where, in looking for a polymer that would respond to octafluoropropane, we evaluated halogenated polymers. Finally, materials to detect SO_2 and Hg were selected based on literature reports and on the results of modeling sensor-analyte interaction, as described in Chap. 3 of this volume.

11.2.3 Initial Sensor Analysis

For evaluation, we took data for polymer-carbon composite sensor response developed in the Generation 1 and Generation 2 efforts as well as polymers added based on preliminary analysis as described above. After these rounds of preliminary evaluation, 16 sensor formulations were selected for further testing and four sets of sensors were fabricated. The polymers used to make these sensors are shown in Table 11.2. Each set included 32 sensors. Two array sets, the "organic sets", were tested only for response to nine compounds, acetone, ammonia, dichloromethane, ethanol, formaldehyde, octafluoropropane, methanol, 2-propanol and toluene, at a range of concentrations. Two additional sets were used, one for testing only SO_2 and one for testing only mercury. The organic, SO_2, and mercury sets were not

Table 11.2 Polymers used in third generation preliminary analysis

Polymer	Abbreviation	Classification
Poly(4-vinyl phenol)	A	HBA
Methyl vinyl ether/maleic acid 50/50	C22	HBA
Poly(styrene-co-maleic acid)	PScMA	HBA
Polyamide resin	C38	HBB
Poly(N-vinyl pyrrolidone)	F	HBB
Vinyl alcohol/vinyl butyral, 20/80	C90	DBB
Ethyl cellulose	EC1	DBB
Poly(2,4,6-tribromostyrene)	C71	MD
Poly(vinyl acetate)	E	MD
Poly(caprolactone)	E15	MD
Soluble polyimide, Matrimid 5218	Mat5218	MD
Poly(epichlorohydrin-co-ethylene oxide)	PEC/EO	MD
Poly(vinylbenzyl chloride)	PVBC	MD
Styrene/isoprene, 14/86 block copolymer	C88	WD
Ethylene-propylene diene terpolymer	EPDT	WD
Polyethylene oxide (MW 100,000)	PEO100/Q	WD

cross-exposed in order to prevent potential poisoning to any given sensor set. Testing the sensors against the analytes allows the development of response curves for each sensor to each analyte. Development of response curves is critical to identification and quantification of the analyte, and has been discussed in detail in description of the LMNLS algorithm [7].

Figure 11.1 shows the sensor response of ethyl cellulose to ethanol as well as the curve for sensor response across a range of concentrations. Sensor response is expressed as normalized change in resistance, or dR/R_0 where, in these plots, R_0 is sensor resistance at time = zero. Ethyl cellulose is one of the most responsive sensors and responds strongly to all of the analytes except ammonia and octafluoropropane; the figure shows strong, repeated response to ethanol; the same sensor showed no significant response to ammonia. The response curve shows a linear relationship between magnitude of response and ethanol concentration for this polymer-carbon composite sensor.

These types of data, sensor resistance change as a function of analyte concentration and the response curve, can be used to make simple evaluations of each sensor response and determine how many sensors are responding to each analyte, but cannot be used to determine how well the array will perform in identifying and quantifying analytes and in distinguishing one from another.

When we looked at the response curves, we saw that for most of the organic analytes, either many sensors responded or few sensors responded. Acetone, ethanol, 2-propanol and toluene all had many sensors that responded. In addition, we could see that the sensors that responded to toluene were the same sensors that responded to 2-propanol; this could pose a selectivity or identity problem. Ammonia had a moderate number of responders. Dichloromethane, octafluoropropane and methanol had very few responders. When we examined the polymers individually we could see that EPDT responded to 2-propanol but not to ethanol or methanol so it could play a role in distinguishing alcohols. Matrimid responded to

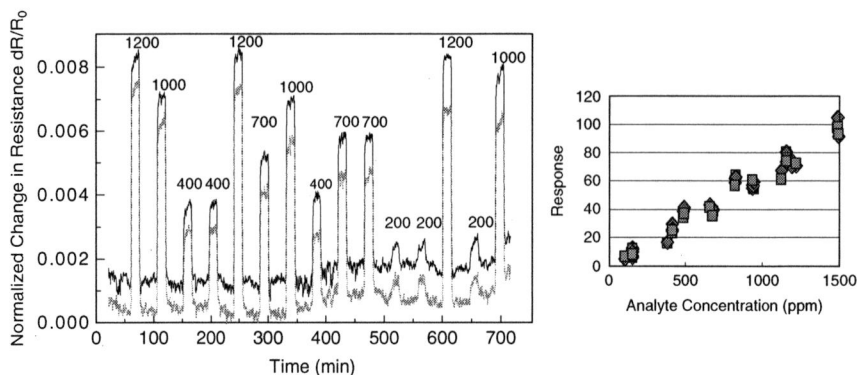

Fig. 11.1 Sensor response for two ethyl cellulose polymer–carbon composite sensors to a range of concentrations of ethanol. The numbers above responses in the left plot correspond to the concentration of ethanol delivered (ppm)

Table 11.3 Summary of simple analysis of polymers

Polymer (abbrev.)	Preliminary Analysis
PScMA	1. Toluene response is not consistently reproducible
	2. Reproducible response for acetone, ammonia, ethanol and propanol
C38	Responds to all organic analytes except octafluoropropane and dichloromethane
C90	Sensors reproducible
EC1	1. Responds to dichloromethane and somewhat to octafluoropropane
	2. No response to ammonia or methanol
Matrimid 5218	1. Responds to dichloromethane and acetone
	2. Responds to ethanol, but not propanol or methanol
	3. Easy to fabricate reproducible sensors
C88	Responds to dichloromethane and all others except methanol and octafluoropropane
EPDT	1. Responds to 2-propanol but not ethanol or methanol
	2. Good for toluene

dichloromethane and acetone; in addition it responded to ethanol but not 2-propanol or methanol, and so could also play a role in alcohol selectivity. Additional preliminary analysis of the polymer responses is summarized in Table 11.3. Clearly, application of a statistical method of array evaluation was necessary.

11.3 Statistical Array Analysis

To select better and more suitable sensor elements for an array designed to detect the targeted analytes, statistical analyses were performed on the available data from first and second generation sensor sets to guide selection of the sensor set described in Sect. 11.2. We evaluated individual sensor performance in each sensor set in terms of selectivity, reliability, and sensitivity with regard to analytes in Table 11.1. We then scored each sensor's "goodness" by these metrics individually and overall. This provided good insight and a statistical basis for selecting sensor material from each sensor set. Other methods such as (hierarchical) cluster analysis (CA) and principal component analysis (PCA) were also performed but did not add significant insight into sensor performance, and so results from these analyses were not used in sensor selection.

11.3.1   Selectivity and Diversity

Selectivity is the ability of the array to distinguish one analyte from all the others. This is naturally one of the most important criteria in selecting a sensing array. Quantification of selectivity relies on calculating the relative distance between array fingerprints for pairs of analytes. An array fingerprint or signature is a graphical representation of the response of the entire array to an individual analyte. Array response to a single analyte, as shown in Fig. 11.2, is a typical representation of a fingerprint.

Exposing the sensors to each analyte at a range of concentrations yields the individual response curves for each sensor to each analyte; the array fingerprint for each analyte is constructed by selecting a response magnitude in the middle of the concentration range from the response curve and showing that as the single sensor response to an analyte in a histogram.

Figure 11.2 shows the normalized fingerprints of the seven analytes used in optimizing for response to organic compounds. The fingerprints alone do not, however, tell us whether we will be able to distinguish one analyte from another, or how reliable the sensors are. Statistical analysis of the array begins with examining cross-analyte fingerprint distance. This distance sums the differences between fitted fingerprints of mth and nth analytes, over 32 sensors, normalized by the mean of their fingerprints.

Cross-analyte distance is defined as

$$\Delta S_{mn} = \frac{1}{K} \sum_{i}^{K} |X(i,m) - X(i,n)|, \tag{11.1}$$

where $X(i,m)$ is the ith sensor's normalized resistance change for the mth gas and the summation is over K sensors used [7].

In principal, a small value for ΔS_{mn} implies poor distinguishability between analytes, and a large value for ΔS_{mn} implies good distinguishability. This was not always the case when the LMNLS algorithm was applied to array response to analytes; distances alone did not adequately predict whether the array, in combination with our analysis software, could distinguish similar analytes [6, 7].

Thus, an alternative approach to calculation of selectivity was adopted, where we started by considering both array response, or selectivity, and individual sensor response to all analytes, or diversity.

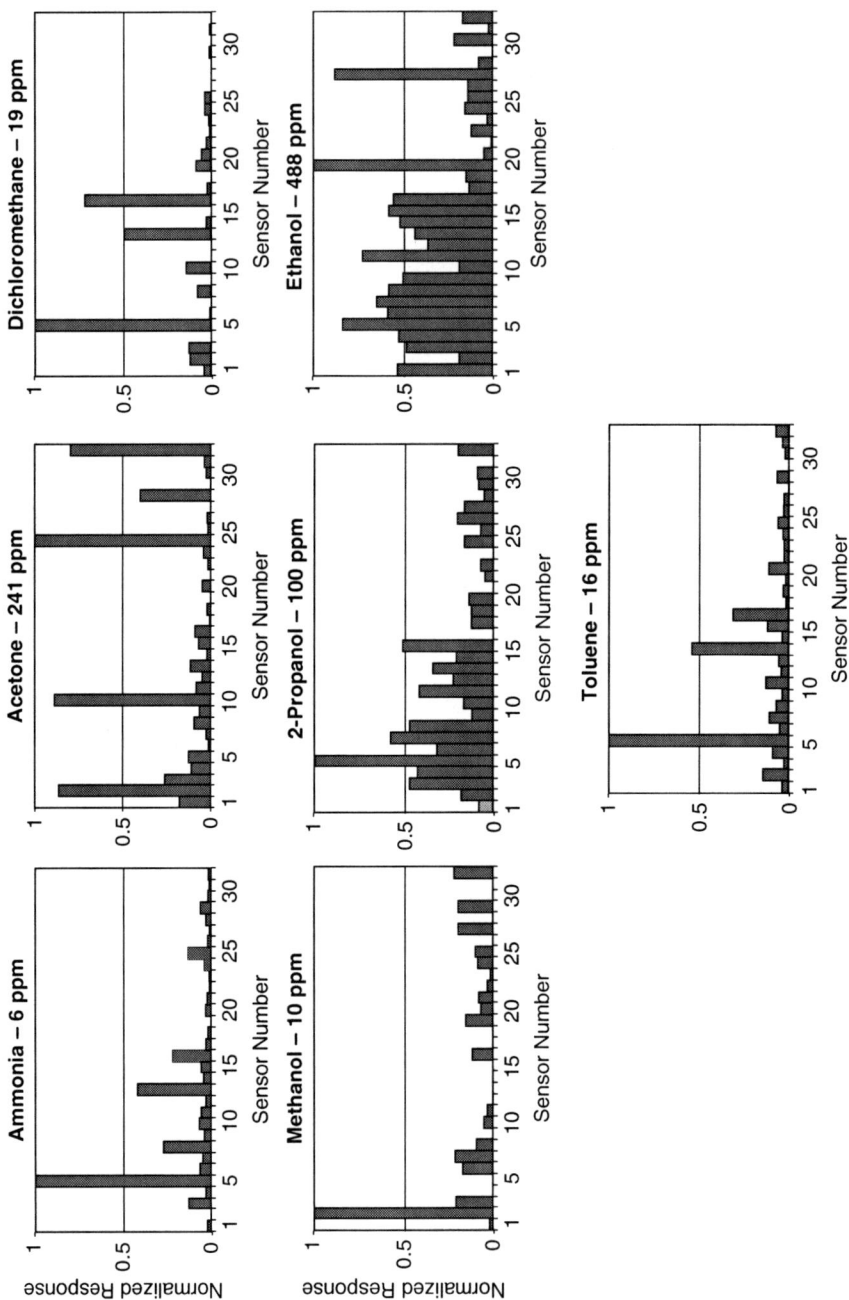

Fig. 11.2 Fingerprints of analytes used in optimizing ENose polymer–carbon composite sensor array

In Selectivity Method 1, overall normalized analyte response X, found by dividing the individual intensities by the maximum of absolute intensities across all sensors for each analyte, was used to calculate a selectivity:

$$\text{Selectivity } 1 = \sum_{m,n} |X(i,m) - X(i,n)| / (X(i,m) + X(i,n)). \tag{11.2}$$

Here the summation is over all the possible $N(N-1)/2$ pairs of analyte combinations (for N target analytes) for a given sensor s. X is the normalized analyte response.

In Selectivity Method 2, comparing the normalized responses to the analytes across a sensor can also give us a measure of an individual sensor's role in the array diversity.

$$\text{Selectivity } 2 = X(i,:)_{\max} - X(i,:)_{\min} \tag{11.3}$$

where $X(i,:)_{\max}$ is the response of sensor i to the analyte it responds to most strongly, and $X(i,:)_{\min}$ is the response of sensor i to the analyte it responds to most weakly. A large selectivity is preferred since it indicates a large range in response magnitude for the analytes to which the sensor responds. A smaller selectivity indicates that the sensor responds similarly to all the analytes and will not contribute to distinguishability. Figure 11.3 shows this measure for the tested array of 32 polymer–carbon composite sensors against seven analytes.

The two methods, Selectivity 1 and Selectivity 2, generally agree with each other; the second method usually results in larger differences in number, and therefore is better for visualization or visual inspection.

Diversity is another approach to determine whether an array is likely to show selectivity toward analytes. We determine diversity by calculating the standard deviation of a sensor's response strength distribution across N analytes; this approach ensures that selectivities as calculated in (11.2) and (11.3) are not dominated by one or two very large separations:

$$\text{Diversity} = \text{std}(\text{sort}(X(i,:))), \tag{11.4}$$

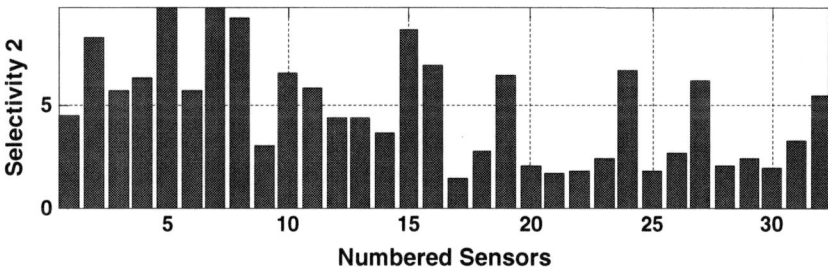

Fig. 11.3 Selectivity 2 across preliminary array for seven analytes as calculated by (11.3)

where a given sensor's response to N analytes is first sorted in the magnitude order before calculating the standard deviation.

It is generally desirable to select sensors whose responses to analytes are more diverse. Figure 11.4 shows the diversity of 32 sensors across nine analytes; the numbers in each plot represent the sensor number followed by the diversity as calculated by (11.4). The bars represent a normalized response to each analyte, but are sorted from the weakest response to the strongest response so the analyte order is not the same for each sensor. For example, in Fig. 11.4, sensors 1, 2, 10 and 11 show a triangular shape for sorted distribution of sensor response, showing good diversity. Sensors 16, 17 and 21 show a flattened shape, where sensor response is mostly the same, and so this sensor shows poor diversity. Similarly, sensor 31 has poor diversity because there is one dominant analyte response, and the remaining responses are similar. For individual sensors, it would be ideal to have sorted (minimum to maximum) response to analytes as seen in Sensors 10 and 11. Sensing materials for which the standard deviation shows that the large separation is dominated by one or two strong responses, as in Sensor 31, may be eliminated from the array.

This diversity is, however, by itself insufficient to select sensors for an array, as this does not determine that there is adequate distinguishability among analytes. Although it might be unlikely, it is possible that each sensor has a similar magnitude response to each analyte, making for poor distinguishability, so diversity must be taken together with other measures in drawing conclusions on the suitability of an array.

11.3.2   Reliability/Variation

A further consideration in selecting elements in an array is *Reliability*, or the ability of sensors and the array to repeat a response to the same stimulus over time. Reliability is a measure of the individual sensor scatter, and is expressed as the inverse of Reliability, or Variation. Although, in principle, selectivity can tell how distinct a fingerprint for one analyte is from another, it alone is often not sufficient to ensure good sensor material selection. As we search possible sensor materials to detect new analytes or analytes at very low concentration ranges, it is apparent that Reliability and Sensitivity can be major limiting factors in overall performance in detecting and identifying target analytes.

We define the Variation or scatter as the inverse of reliability for a given sensor, as the relative difference between actual vs. fitted analyte responses, where fitted response is based on the response curve shown in Fig. 11.1 and used in constructing the LMNLS algorithm for identification and quantification of analytes:

$$\text{Variation}\left(\text{Reliability}^{-1}\right) = \sum_n \sum_j \left|X(i,n) - x_j(i,n)\right| / X(i,n). \qquad (11.5)$$

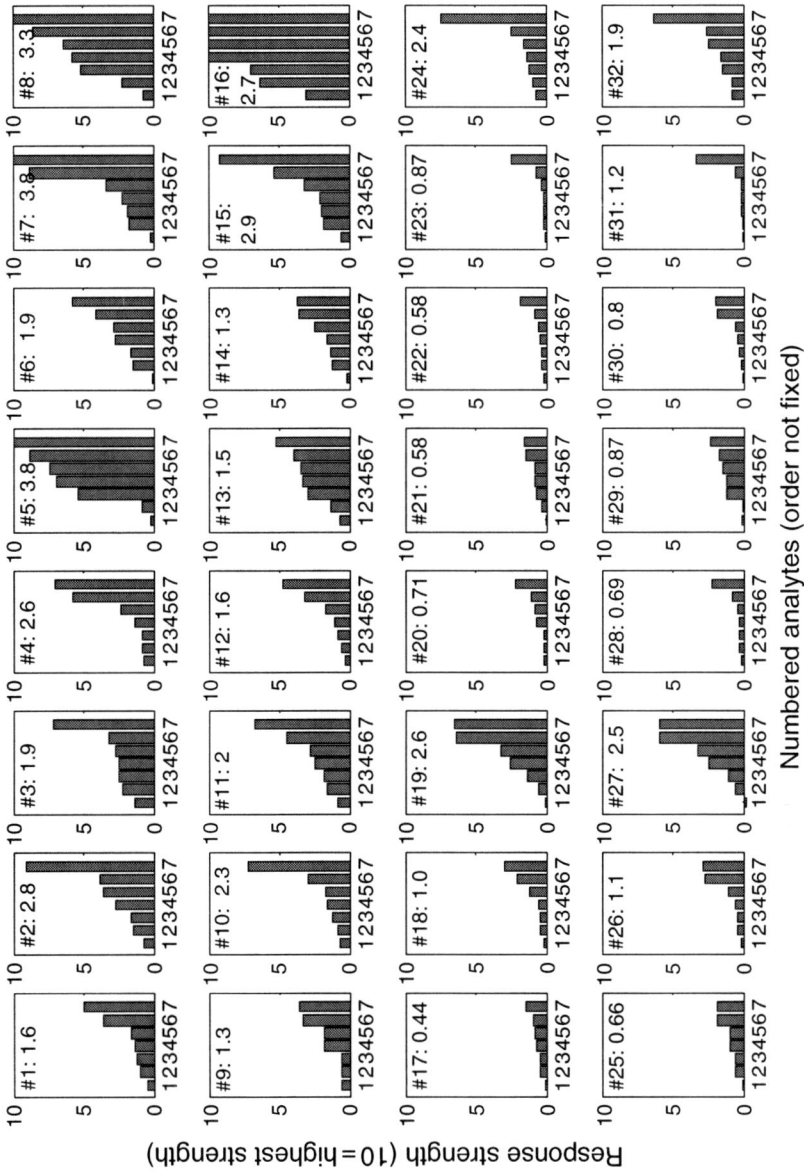

Fig. 11.4 Sensor diversity – sensor response strength distribution among seven analytes. Each subplot shows a sensor's relative response to seven analytes sorted as minimum to maximum response magnitude (analytes not in fixed order)

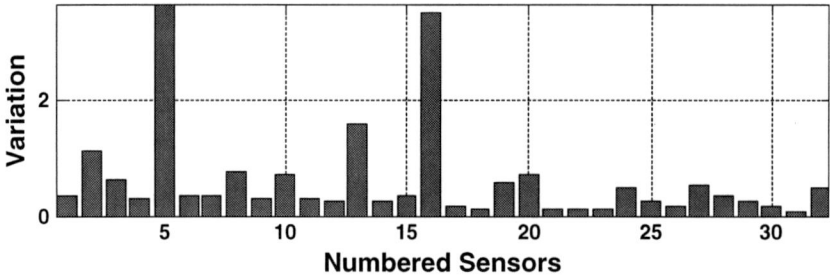

Fig. 11.5 Variation of 32 sensors' response to seven analytes as calculated by (11.5)

In (11.5) x_m is the mth actual normalized response of an analyte, the first summation is over M test curves collected for an individual analyte, and the second summation is over all N analytes.

Figure 11.5 shows a plot of calculated Variation for the 32 element polymer–carbon composite array against seven analytes. We can see that Sensors 5, 13 and 16 have particularly large Variations; these sensors are noisy and thus do not repeat response to a stimulus reliably. In selecting sensors for their Reliability, the smallest possible relative Variation is desired. The large Diversity shown in Sensors 5 and 13 (see Fig. 11.4) is likely caused by sensor variation and not by true diversity, whereas the large diversity in Sensors 1, 2, 10 and 11 is likely to be true diversity, because the Variation is small. The large Variation seen for Sensor 16 is reflected in the poor Diversity but large response seen in Fig. 11.4. Sensors with a Variation less than 1 are preferred.

11.3.3 Sensitivity

Sensitivity of a sensor is a measure of the magnitude of response of that sensor to the stimulus of the analyte set. The sensitivity of a sensor is important, particularly as the JPL ENose is challenged to detect several analytes that are difficult to detect or are expected to appear at very low concentration ranges.

We define the sensitivity as the mean of normalized response strength:

$$\text{Sensitivity} = \sum_n |X(s,n)| \qquad (11.6)$$

with the summation over all analytes for a given sensor s.

Figure 11.6 shows a plot of Sensitivity for the 32 element polymer–carbon composite array, summed across all analytes.

We can see that Sensors 5 and 16 have large measures of Sensitivity, but, as with Diversity, this large magnitude can be attributed to the large variation in response and is not likely to be caused by a high sensitivity. Larger Sensitivities are

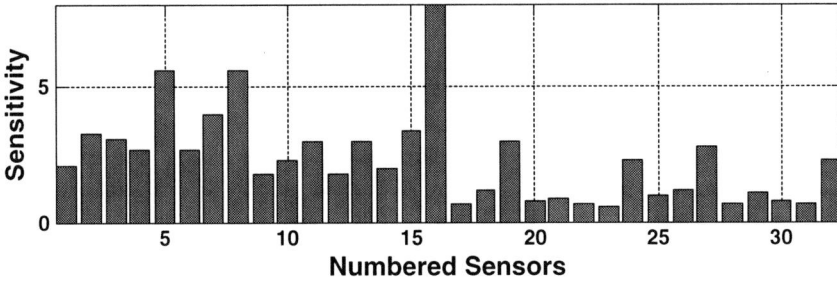

Fig. 11.6 Sensitivity of 32 sensors' response to seven analytes as calculated by (11.6)

preferred; there is no fixed value above which sensitivity must be acceptable for an array; as a general rule, we prefer the Sensitivity to be greater than 1.

11.3.4 Analysis

When performed on individual analyte level, the evaluation of these metrics can give us insight into which sensors are contributing to or undermining the identification of specific analytes.

Using these statistical analysis tools, we evaluated all three sensor array sets we have: first generation, second generation, and pre-third generation sensor data with regard to the selected target analytes. The statistical measures described above were taken on the pre-third generation set, from which we ranked the sensors in order of best to worst by the three measures Diversity, Reliability and Sensitivity based on their responses to seven analytes used for evaluation. Figures 11.3–11.6 show each sensor's performance in terms of reliability, selectivity, and sensitivity with regard to these target analytes. The next step will be to rank sensors from the best to the worst in terms of these individual metrics, then to calculate how suitable each sensor is for the array to detect the analytes under consideration

	Better sensors							Poorer sensors				
Selectivity	7	5	8	15	2	16	...	28	20	25	22	21	17
Reliability	31	23	22	21	18	30	...	20	6	2	13	16	5
Sensitivity	16	8	5	7	15	2	...	20	22	31	28	17	23

To score a sensor's overall goodness, G, or suitability for inclusion in an array, we first discarded sensors 5, 16, 13 and 2 for high variation (poor reliability), 17 for poor selectivity and 23, 28, 31, 22, 20, 30 and 21 for poor sensitivity. We then used the following combination of the above metrics on the remaining 20 sensors:

$$G = (\text{Selectivity}/\text{Variation}) * \text{Sensitivity} \qquad (11.7)$$

As a result, we ranked the sensors in order of suitability for an array to detect the seven test analytes in the following order:

Better sensors																Poorer sensors		
7	15	8	11	4	6	27	19	24	18	1	14	3	32	12	10	9	26	29	25

Based on all of these analysis approaches of both the individual sensors and the entire array, we selected eight sensor materials for further testing. The best sensor materials, corresponding to the ranking above, were poly(4-vinyl phenol), poly (styrene-co-maleic acid), polyamide resin, 20/80 vinyl alcohol/vinyl butyral, ethyl cellulose, poly(caprolactone), Matrimid 5218, and ethylene-propylene diene terpolymer. Materials that were removed from the array include poly(N-vinyl pyrrolidone), poly (epichlorohydrin-co-ethylene oxide), polyethylene oxide, poly(vinylbenzyl chloride), poly(2,4,6-tribromostyrene), and 50/50 methyl vinyl ether/maleic acid. Two polymer materials, poly(vinyl acetate) and styrene/isoprene, 14/86 ABA block polymer were in the middle region, and ultimately were retained in the array.

11.4 Selection of Sensors for Inorganic Analytes

In order to detect SO_2 and elemental mercury it was necessary to develop additional sensors to add in to the array. Materials to detect SO_2 and Hg were selected based on literature reports and on the results of modeling sensor–analyte interaction, as described in Chap. 3 of this volume. Since this sensor development has been discussed in other papers as well as in Chap. 3, the topic is covered briefly here.

11.4.1 Selection of SO_2 Sensors

Prior work on SO_2 detection has focused on fluorescence, chemiluminescence, or electrochemical/amperometric methods, none of which are compatible with the existing JPL ENose platform, where all of the sensors are chemiresistors [9]. Reversible SO_2 sorption by polymers has been demonstrated [10], so it was determined that this approach was likely to work for SO_2 detection and to be compatible with the JPL ENose.

In order to screen candidate materials efficiently, experiments were carried out using microarrays with interdigitated electrodes and embedded heaters. This approach also assisted in the optimization of sensing parameters such as operating and regeneration temperatures, and carbon-black loading. Polymers that were under investigation are derivatives of linear and cross-linked poly-4-vinyl pyridine and vinyl benzyl chloride functionalized with various free-amine containing substituents [5]. After exposure to SO_2 only, two polymers were selected and made into polymer-carbon black composite sensors. These two polymers are both poly-4-vinyl pyridine derivatives with a quaternary and a primary amine [5]; these polymers were predicted to respond to SO_2 by quantum mechanical modeling (see Chap. 3), and were found to respond strongly in laboratory testing. These materials were not included in optimization experiments as described in Sect. 11.3, as the sensors selected in Sect. 11.3 had very small responses to SO_2. In contrast, the strong responses of two poly-4-vinyl pyridine derivatives, called EYN2 and EYN7, resulted in high selectivity of these materials to this one analyte. Thus, these two poly-4-vinyl pyridine derivatives were included in the JPL ENose Third Generation sensor array as selective sensors.

11.4.2 Development of Mercury Sensors

A literature review of the development of sensors for elemental mercury detection showed that work had focused primarily on thin films of gold films and other noble metals to form a metal-mercury amalgam, and reading the change in resistance with amalgam formation [6, 9]. In addition, palladium chloride with tetrahydroxyethyl-ethylenediamine (THEED) has been used in one study of Hg sensing [11].

For this array, we considered and tested several sensing materials for Hg detection, including gold films, treated gold films, sintered palladium chloride ($PdCl_2$) thick films, polymer-carbon composite thick films, and thin gold films on polymer-carbon composites. All materials were tested in flowing, humidified air at 20–25°C. Relative humidity in all the tests was ~30%.

The material selected for this array was thick sintered $PdCl_2$ films [6]. These sensing films showed good sensitivity and reproducibility of response to Hg concentrations of 2–10 ppb at 23°C in humidified air. $PdCl_2$ sensor response magnitude does not increase above 10 ppb Hg [6]. These sensing films also show partial regeneration under mild conditions, temperatures <40°C, and good repeatability of response for concentrations 10 ppb and under [6]. As with the polymer sensors selected for detection of SO_2, these $PdCl_2$ sensors had very small response to other analytes and so could be used as selective sensors for Hg. They have moderate response to water and ammonia, but are not needed to distinguish these analytes from other targeted analytes.

11.5 Final Selection of Array

The final selection of the array was an iterative process. This included further statistical analysis of the array after exposure to all analytes of interest. Where there were problems with reproducibility or distinguishability, additional sensors were tested and analyzed. Different carbon loadings for certain polymers were considered, and additional halogenated polymers were tested. As sensors were swapped in and out, additional statistical analyses were performed. The final consideration was, of course, performance of the data analysis program.

The sensing materials selected for the final array are shown in Table 11.4. As can be seen in the table, some materials are in the array in duplicate, for redundant sensors, and some are in the array with two or more different carbon loadings. Many polymers show different behavior with different carbon loadings, and so are considered to be separate sensors.

Table 11.4 Sensing materials used in the 3rd Generation JPL ENose

Sensor Number	Polymer	Carbon/Polymer weight (%)
1	Poly(4-vinylphenol)	15
2	Poly(4-vinylphenol)	20
3	Poly(4-vinylphenol)	20
4	Poly(2,4,6-tribromostyrene)	10
5	Poly(2,4,6-tribromostyrene)	10
6	Poly(2,4,6-tribromostyrene)	15
7	Poly(styrene-co-maleic acid)	15
8	Poly(styrene-co-maleic acid)	15
9	Poly (2,2,2-trifluoroethyl methacrylate)	15
10	Poly (2,2,2-trifluoroethyl methacrylate)	20
11	Poly(t-butylaminoethyl methacrylate)	20
12	Poly(t-butylaminoethyl methacrylate)	20
13	Poly (ethylene-co-acrylic acid)	10
14	Polystyrene	12
15	Matrimid 5218	12
16	Ethyl cellulose	12
17	Polyamide resin	15
18	Styrene/isoprene, 14/86 block polymer	15
19	Vinyl alcohol/vinyl butyral, 20/80	12
20	Ethylene-propylene diene terpolymer	15
21	Poly(4-vinylphenol)	15
22	Poly(4-vinylphenol)	20
23	poly(t-butylaminoethyl methacrylate)	20
24	poly(t-butylaminoethyl methacrylate)	20
25	EYN2	15
26	EYN2	15
27	EYN7	15
28	EYN7	15
29	Carbon nanotubes	NA
30	Carbon nanotubes	NA
31	Sintered $PdCl_2$	NA
32	Sintered $PdCl_2$	NA

11.6 Evaluation of Final Array

The ENose was tested for response and ability to identify and quantify ten chemical species, shown in Table 11.1. A quantification is considered to be correct if the concentration value reported by the ENose data analysis software is $\pm 50\%$ of the independently measured concentration of the species. The error range of $\pm 50\%$ in quantification is based on the approximate range with which the SMACs are set; toxicities are not well known for most chemical species, and SMACs are set very conservatively.

Success rate in analyte identification and quantification is the number of correctly identified and quantified deliveries of analyte, minus the number of false negatives and false positives, divided by the number of deliveries of the analyte. A false negative is failure to detect the delivery and presence of a targeted chemical species. A false positive is a report of the presence of a chemical species when it was not present.

Misidentification of a targeted species, for example identifying propanol as methanol, or identifying a confounder as a targeted species, is considered to be a false positive.

11.6.1 Training Sets

As the ENose is an array-based chemical sensor device, before it can be used as an air quality monitor, training sets must be acquired. Based on the training sets, the patterns of array response to targeted analytes under specified conditions are included in the data analysis algorithm.

Data from the ENose are recorded for each individual sensor as resistance versus time. As the ENose is designed to function as an event monitor, the data are analyzed as change in resistance vs. time. Individual sensor resistances are recorded simultaneously, with a point being taken every 20 s. While it would be possible to take data more or less frequently than three times a minute, this data rate has been established as an optimum rate to show fairly rapid changes in the environment without overwhelming computer memory with data. Our data analysis approach defines an "event" as a change in the composition of the environment which lasts longer than 10 min, or 30 points at the standard data rate, in part because events of duration shorter than ten minutes cannot practically be addressed or mitigated using either breathing apparatus or clean-up techniques. The data analysis algorithm needs about ten points ($\sim$ 3 min) to establish that resistance has changed significantly. Based on the data rate and needs of the data analysis algorithm, training sets are established using vapor deliveries, or events, of 30–45 min duration.

The data analysis algorithm is a Levenberg–Marquardt non–linear least squares fitting approach to deconvolution of change in resistance across the sensing array into identification and quantification of the analyte causing response in the sensors. The analysis approach has been discussed in detail previously [7].

Training sets were established for the ten analytes in Table 11.1 as well as formaldehyde. Tier 1 analytes are ammonia, sulfur dioxide and mercury. Tier 2 analytes are the remaining analytes in Table 11.1. Formadehyde is the Tier 3 analyte and was a goal not a requirement for this project. The environmental conditions for the training sets vary only in water content. As there is temperature control in the sensing chamber in the ENose, the environmental temperature does not influence the temperature at which analytes are detected, identified and quantified. The relative humidity of the environment will be altered if the temperature of the sensing chamber is different from the temperature of the environment, so for training sets, the humidity is regulated as ppm water. Training sets were made in a background of filtered house air with water concentrations of 5,000, 10,000, 15,000 and 20,000 ppm. These concentrations correspond roughly to 20%, 40%, 60% and 80% relative humidity at 21°C, and cover the specified range of humidities for the specified range of temperatures.

In designing training sets, the range of analyte concentration to which sensors are exposed is set at 1/3–3 times the target concentration. This range is divided into 10–12 concentrations, and the sensor array exposed to those concentrations at each water content. A total of 1,599 different analyte exposures were made to establish the training sets: 325 for Tier 1 compounds and 1,274 for Tiers 2 & 3. The exposures were divided approximately equally among four humidity levels, although there was a larger number at the two lowest humidities in both sets, Tier 1 and Tiers 2 & 3. Formaldehyde, the one Tier 3 compound, had the fewest exposures as it was a goal rather than a requirement in this program.

Training sets give insight into the conditions under which the ENose operates best, which conditions cause difficulties in identification and quantification, and allow calculation of performance as accuracy of identification and quantification, number of false negatives and number of false positives using a large number of trials.

Training sets can be used to judge the accuracy of the identification and quantification algorithm even though they are used to establish the coefficients. Accuracy of the algorithm has been computed using half the data to establish the algorithm and coefficients and the other half to test the application, and using all the data for both functions. The statistical difference in results is insignificant. In the success rates for identification and quantification shown below, all the data are used for both functions.

11.6.2   Success Rates

Success rates for analyte detection, identification and quantification at a range of environmental conditions have been discussed in detail [12]. In training set data, success in detection signifies that the analyte was detected, identified correctly, and quantified within $\pm 50\%$ of the measured delivered concentration of the analyte. An "event" is defined as a change in environment caused by the presence of a targeted species. A false positive is detection of an "event," where there was no event, or the misidentification of an event (e.g. identification of toluene as methanol.) A false negative is failure to identify that an event has happened, with no question of identification or quantification.

Accuracy in identification of analytes was best at nominal humidity conditions, approximately 10,000 ppm water (30–40% RH). The overall success rate for identification and quantification is better when the water content of the air is not at its highest; for Tier 1 species, the overall success rate at nominal temperature and humidity is 93%, and for Tier 2 the overall success rate at nominal conditions is 85%.

At conditions of 50–60% relative humidity with environmental temperature above 21–22°C, the increased water absorbed by polymer based sensors lowers the sensor responses to other vapors. In addition, high water content in polymer based sensors can result in dissolution of some Tier 2 compounds, particularly oxygen-containing compounds, which will result in a different type of capture of analyte molecules in the sensing film. At high humidity, overall success rate for Tier 1 species is 82% and to 74% for Tier 2 species. Sensors for mercury are inorganic, and so are not significantly affected by humidity; the fall in success with humidity for Tier 1 species is caused by a slight fall in success with ammonia and a significant fall in success for SO_2.

11.6.3 Conclusion

In a total of 1599 exposures, the overall success rate for Tier 1 species was 89%, and for Tiers 2 and 3, 80%, over all humidity and pressure conditions. Weighting the success rates for number of chemical species in each category, the overall success rate for identification and quantification of delivered species in training sets was 83% over all conditions.

Success rates at nominal conditions, with an environmental temperature of 21–22°C, with a sensing chamber temperature of 25°C and about 40% RH, the success rate for Tier 1 species was 93%, and for Tiers 2 & 3 was 85%. The overall success rate for all species under nominal conditions was 87%. This level of success rate is a final assessment of whether the array has been optimized for this set of analytes. A success rate approaching 90% is very good, although some further improvement could be possible. Future work will consider whether it is possible to optimize a sensor set over all environmental conditions, so that success rate does not vary with water content of the background air.

This optimization work did not specifically consider sensor lifetime, although it has been found that polymer-carbon composite sensors, not including those specifically selected for SO_2 detection, have a lifetime of more than 2 years under frequent operation.

Acknowledgments The work discussed here was carried out at the Jet Propulsion Laboratory, California Institute of Technology, under contract with the National Aeronautics and Space Administration (NASA). Work was sponsored by Advanced Environmental Monitoring and Control Project of the Exploration Systems Mission Directorate.

References

1. Nicagossian, A. E.; Hunton, C. L.; Pool, S. L., Space Physiology and Medicine, Lea and Febiger, Philadelphia, PA, **1994**
2. Ryan, M. A.; Homer, M. L.; Zhou, H.; Manatt, K. S.; Ryan, V. S.; Jackson, S. P., Operation of an electronic nose aboard the space shuttle and directions for research for a second generation device, In Proceedings of International Conference on Environmental Systems, **2000**, 00ICES-259, Toulouse, France
3. Ryan, M. A.; Zhou, H.; Buehler, M. G.; Manatt, K. S.; Mowrey, V. S.; Jackson, S. P.; Kisor, A. K.; Shevade, A. V.; Homer, M. L., Monitoring space shuttle air quality using the jet propulsion laboratory electronic nose, *IEEE Sens. J.* **2004**, 4, 337–347
4. Ryan, M. A.; Homer, M. L.; Zhou, H.; Manatt, K.; Manfreda, A., Toward a second generation electronic nose at JPL: sensing film optimization studies, In Proceedings of International Conference on Environmental Systems, **2001**, 2001–01–2308, Orlando, FL
5. Ryan, M. A.; Homer, M. L.; Zhou, H.; Manatt, K.; Manfreda, A.; Kisor, A.; Shevade, A.; Yen, S. P. S., Expanding the capabilities of the JPL electronic nose for an international space station technology demonstration, *J. Aerosp. SAE Trans.* **2006**, 2006–01, 225–210
6. Shevade, A.V.; Ryan, M. A.; Taylor, C. J.; Homer, M. L.; Jewell, A. D.; Kisor, A. K.; Manatt, K. S.; Yen, S. -P. S., Development of the third generation JPL electronic nose for international space station technology demonstration, In Proceedings of International Conference on Environmental Systems, **2007**, 2007–01–3149, Chicago IL, USA

7. Zhou, H.; Homer, M. L.; Shevade, A. V.; Ryan, M. A., Nonlinear least-squares based method for identifying and quantifying single and mixed contaminants in air with an electronic nose, *Sensors* **2006**, 6, 1–18

8. Nix, M. B.; Homer, M. L.; Kisor, A. K.; Soler, J.; Torres, J.; Manatt, K.; Jewell, A.; Ryan, M. A, Sniffing out problems for humans in space, *IEEE Potentials* **2007**, 26, 18–24

9. Ryan, M. A.; Homer, M. L.; Zhou, H.; Manatt, K.; Manfreda, A.; Kisor, A.; Shevade, A.; Yen, S. P. S., Expanding the analyte set of the JPL electronic nose to include inorganic species, *J. Aerosp. SAE Trans.* **2005**, 2005–01–2880, 225

10. Diaf A.; Garcia J.I.; Beckman, E.J.,Thermally reversible polymeric sorbents for acid gases – CO_2, SO_2, and NOx, *J. App. Polym. Sci.* **1994**, 53, 857–875

11. Ruys, D. P.; Andrade, J. F.; Guimaraes, O. M., Hg detection in air using a coated piezoelectric sensor, *Anal. Chim. Acta* **2000**, 404, 95–100

12. Shevade, A. V.; Ryan, M. A.; Homer, M. L.; Kisor A. K.; Manatt, K. S., Off-gassing and particle release by heated polymeric materials, In Proceedings of 38th International Conference on Environmental Systems **2008**, 2008–01–2090, San Francisco CA, USA

Chapter 12
Hybrid Arrays for Chemical Sensing

Kirsten E. Kramer, Susan L. Rose-Pehrsson, Kevin J. Johnson,
and Christian P. Minor

Abstract In recent years, multisensory approaches to environment monitoring for chemical detection as well as other forms of situational awareness have become increasingly popular. A hybrid sensor is a multimodal system that incorporates several sensing elements and thus produces data that are multivariate in nature and may be significantly increased in complexity compared to data provided by single-sensor systems. Though a hybrid sensor is itself an array, hybrid sensors are often organized into more complex sensing systems through an assortment of network topologies. Part of the reason for the shift to hybrid sensors is due to advancements in sensor technology and computational power available for processing larger amounts of data. There is also ample evidence to support the claim that a multivariate analytical approach is generally superior to univariate measurements because it provides additional redundant and complementary information (Hall, D. L.; Linas, J., Eds., Handbook of Multisensor Data Fusion, CRC, Boca Raton, FL, 2001). However, the benefits of a multisensory approach are not automatically achieved. Interpretation of data from hybrid arrays of sensors requires the analyst to develop an application-specific methodology to optimally fuse the disparate sources of data generated by the hybrid array into useful information characterizing the sample or environment being observed. Consequently, multivariate data analysis techniques such as those employed in the field of chemometrics have become more important in analyzing sensor array data. Depending on the nature of the acquired data, a number of chemometric algorithms may prove useful in the analysis and

Kirsten E. Kramer (✉)
Cognis Corporation, Cincinnati Innovation Concept Center, 4900 Este Avenue, Building 53 Cincinnati, OH 45232-1419 (513) 482-2242
e-mail: kirsten.kramer@cognis.com

Susan L. Rose-Pehrsson and Kevin J. Johnson
Naval Research Laboratory, Chemistry Division, Code 6181, Washington, DC 20375-5342, USA,
e-mail: rosepehrsson@nrl.navy.mil

Christian P. Minor
Nova Research, Inc., Alexandria, VA, USA

M.A. Ryan et al. (eds.), *Computational Methods for Sensor Material Selection*,   265
Integrated Analytical Systems,
DOI 10.1007/978-0-387-73715-7_12, © Springer Science+Business Media, LLC 2009

interpretation of data from hybrid sensor arrays. It is important to note, however, that the challenges posed by the analysis of hybrid sensor array data are not unique to the field of chemical sensing. Applications in electrical and process engineering, remote sensing, medicine, and of course, artificial intelligence and robotics, all share the same essential data fusion challenges. The design of a hybrid sensor array should draw on this extended body of knowledge. In this chapter, various techniques for data preprocessing, feature extraction, feature selection, and modeling of sensor data will be introduced and illustrated with data fusion approaches that have been implemented in applications involving data from hybrid arrays. The example systems discussed in this chapter involve the development of prototype sensor networks for damage control event detection aboard US Navy vessels and the development of analysis algorithms to combine multiple sensing techniques for enhanced remote detection of unexploded ordnance (UXO) in both ground surveys and wide area assessments.

12.1 Introduction

Current trends in chemical sensing are geared toward the use of sensor arrays, particularly for gas-phase analysis and head-space analysis of samples containing volatile compounds [1–3]. Construction of a device comprised of an array of sensing elements is most commonly motivated by the desire to detect and classify multiple chemical species with a single measurement [4]. Vapor sensing has widespread applications in environmental monitoring [5, 6], food analysis [7, 8], as well as clinical [9] and industrial settings. There is much interest in producing miniaturized, portable, cost-effective gas sensing devices that may serve as a substitute for bulkier, more expensive analytical instrumentation [10].

The most common elements of gas sensor arrays are based on metal-oxide semiconductors (MOS) [11, 12], metal-oxide semiconducting field effect transistors (MOSFETs) [13], acoustic wave sensors [14–17], polymer-coated sensors [18–21], and optical sensors [22]. In recent years, much research has been devoted to the use of arrays based on the concept of mammalian olfactory sensing whereby a collection of semiselective sensors are used to produce a unique signature or "fingerprint" for different compounds of interest [23–28]. Many of these electronic nose or *enose* (EN) sensor arrays are commercially available and have found use in a variety of applications. Based on a similar principle, the tasting mechanism, etongue (ET) arrays have been constructed for liquid sampling [29]. The challenge in designing an effective sensor is in choosing the elements of the array to produce unique response profiles for each class of species under study and also to demonstrate consistency over time. If the variance in the response between identical sensors (i.e., sensor-to-sensor reproducibility) is higher than the variances that distinguish one chemical fingerprint from another, then the sensors will not have the required selectivity or reliability for multisensing [30]. Although extensive research has been performed in the area of materials and coatings for gas sensors [31, 32],

EN sensor arrays often lack the ability to discriminate between similar compounds, making them unsuitable for many types of analyses [5, 33, 34].

If an array of semiselective sensors is not powerful enough for the analysis at hand, one option is to investigate the use of hybrid sensing: in other words, the use of an array that employs more than one kind of transducer, sensing principal, or sampling procedure that acts to increase or expand the information content of the data. Hybrid sensing schemes have yet to gain widespread usage in chemical sensing applications; however, a number of sensor array reviews [3, 24, 35] and publications [36–40] have characterized these approaches as cutting-edge techniques that may gain more acceptance in the near future. Although the integration of more than one kind of transducer may increase the complexity of the hardware, the benefits of utilizing different transducers might well be worth the additional efforts in the fabrication process.

In this chapter, the use of a hybrid array will be discussed as a way of increasing the analytical potential of a sensor. The choice of sensing elements in a hybrid array is not ruled by the cost or convenience of the manufacturing process, but by the aim of maximizing the analytical capability of the system so that challenging sensing tasks may be undertaken. There are numerous EN and ET sensors that employ transducers identical in nature, but few that integrate more than one type of transducer or response value. Combining more than one type of transducer, instrument, or sampling parameter acts to increase the orthogonality (or uniqueness) of the sensing mechanisms utilized by the array by increasing the likelihood that the profiles or fingerprints produced by each compound will also be more unique in character. In many reports involving the use of a hybrid sensor, the concept of orthogonality is discussed in the introductory or theory section. The experimental results can be used to gauge whether the theoretical benefits of hybrid sensing are proven to work in practice and whether the increased hardware complexity is worth the effort of a particular analysis.

An approach to data integration called *multisensor data fusion* [37] will be discussed, as the terminology, data analysis architecture, theoretical premises, and other such aspects largely overlap with the concept of hybrid sensing. For some hybrid arrays, the fusion of data may be optimal after different levels of processing or combined at different stages in the data analysis – issues that are commonly addressed in data fusion approaches. Multi-element sensory data may be fused at a "low abstraction level" whereby the responses are given relatively equal weight in the analysis and can be combined side-by-side in their raw (or standardized) format. A higher level of abstraction may be necessary for more complicated hybrid arrays or when the analysis involves a more challenging classification such as mixture analysis. A high abstraction level may perform an analysis [such as principal component analysis (PCA)] of each system individually, then merge the most relevant pieces of information; these pieces of information may be features extracted from the data, known properties of the data, or the outputs of a more sophisticated (quantitative/ qualitative) analysis of the data. Fusion may occur at a variety of levels in a tiered analysis, hierarchical network, or tree-based method. Most EN systems reported in the literature use a low abstraction level, although examples of those using higher abstraction levels will be discussed in this chapter.

In the first part of this chapter, examples from the literature will be given in order to illustrate that hybrid sensors are not only attractive from a theoretical standpoint, but are also being successfully implemented in practice. Although the survey does not claim to be an exhaustive review of hybrid sensing, it attempts to highlight some preliminary studies wherein hybrid EN schemes have been employed. The results of most studies indicate that a collection of hybrid sensory data is much superior to an array employing only one type of transducer or measurement parameter. In the latter sections of this chapter, we will present three studies conducted at the Naval Research Laboratory in the areas of fire/hazardous event detection aboard naval vessels and the detection of unexploded ordnance material beneath the ground. These studies employ hybrid sensor arrays, chemometric data analysis, and multisensor data fusion techniques. The first section will discuss the development and initial testing of a multicriteria sensor array for fire detection, the second project involves a "volume sensor" that employs machine vision, spectral sensors and acoustic signatures for fire/hazardous event detection, and the final study is the detection of unexploded ordnance (UXO) materials using a combination of airborne magnetometer data and image analysis.

12.2   Hybrid Arrays Involving Low Level Data Fusion

The following section gives examples of hybrid sensing whereby the data are fused at a relatively low abstraction level. Common configurations are arrays of colocated sensing elements, which vary in terms of transduction mode.

12.2.1   Hybrid Array Composed of MOS-MOSFET-CO_2 Sensors

Perhaps one of the first publications reporting a hybrid sensor array that served as an enose was by Winquist et al. [38] An array of ten MOSFETs, four doped tin dioxide (SnO_2, also referred to as Taguchi or TGS) sensors, and a carbon dioxide (CO_2) detector based on infrared (IR) absorption was used to detect meat quality in ground beef and pork samples. The MOSFET sensors were chosen because of their response to hydrogen, hydrogen sulfide, amines, and alcohols, while the TGS sensors were useful for detecting saturated hydrocarbons, alcohols, and humidity. Microbial growth causes the evolution of CO_2, so this sensor was also useful in the array. For two types of classification algorithms, artificial neural network (ANN) and abductory induction mechanism, all 15 sensors provided good predictions of meat type as well as storage time. When the carbon dioxide sensor was omitted, the prediction of storage time was less successful. The researchers demonstrated the value of using a diverse array of sensors tailored to their particular application.

This hybrid array was also used to analyze paper quality [39]. In this study, only four of the 15 sensors (two MOSFET and two TGS) were necessary for distinguishing five classes of paper when principal component analysis and cluster analysis were used. This design was also used by Börjesson et al. for the classification of grain samples [40]. In this study, the researchers found that responses within each sensor class (either MOSFET or TGS) were highly correlated, while essentially no correlation existed between the two classes of transducers. Thus, the between-group sensing capabilities were much higher and results were much improved when both classes of sensors were used in the analysis rather than only one class. The researchers concluded that the different classes of sensors were responding to different groups of volatile species, and speculated that an even wider variety of sensing mechanisms may improve the performance of the array.

This hybrid array was also tested for monitoring two biopharmaceutical processes, the production of human growth hormone from *Escherichia coli* and the human factor VIII from hamster cells [41]. The human growth hormone production was monitored for 33 h and was categorized into three separate stages using the PCA. Results improved when an additional sensor that measured dissolved oxygen was combined with the PCA data, demonstrating the benefits of fusing different types of data. The human factor VIII was categorized in four different production stages and a reduced set of seven elements (the CO_2 sensor, five MOSFET, and one SnO_2) gave comparable results to the full set of 15 sensors. The fact that the subset contained each type of transducer indicated that the hybrid collection of data was valuable, while the sensors of the same transducers may have produced redundant responses.

This same array was also used for the on-line monitoring of yeast production [42] where fusion of the sensor array data with parameters such as reactor volume and aeration rate enabled the prediction of cell mass and ethanol concentration using ANN. In another study, this array was used to monitor a batch cultivation [43] and the researchers found that a reduced sensor subset of three MOSFET, one TGS, and the CO_2 detector was useful for predicting ethanol concentration. The reduction in the number of variables from 15 to 5 was much more amenable to the ANN algorithm and since the four TGS sensors produced similar response patterns, only one was used in the final analysis. Similar to the other studies mentioned above, the researchers found that the most useful subset comprised at least one sensor from each of the three transducer types, indicating that a richer profile was achieved using a between-group set of transducers data rather than a within-group.

A similar sensor array composed of nine MOSFET, four TGS, one IR CO_2 detector, and one electrochemical oxygen (O_2) detector, was used for the classification of bacteria [44]. The O_2 detector was added because both CO_2 and O_2 are important indicators of bacterial growth, as stated by the authors. In this study it was found that only four of the features, two TGS, the CO_2 and the O_2, were necessary for classification.

12.2.2 Hybrid Modular Sensing System

Mitrovics et al. developed a hybrid MOdular SEnsing System (MOSES) which combined tin oxide gas sensors, polymer-coated quartz microbalances (QMBs), electrochemical cells, and metal-oxide semiconductor field effect (MOSFET) sensors into one device [45]. In the prototype, each type of sensing mechanism was composed of an eight-element array that was housed in a separate module, providing flexibility in the sampling process, maintenance, and sensor configuration. The researchers recommend improving performance of vapor sensor arrays by increasing the feature space of the array in one of three ways (1) increasing the number of sensors in a single transducer array, (2) using different transducers to measure various properties of each sensing material, or (3) modulating the measurement conditions (i.e., temperature, voltage, frequency, or reference gas delivery) [46]. The researchers advocate compiling features that are more orthogonal or independent in nature, a task that is not always achievable using the first approach (use of only one transducing principle). The MOSES sensor contains separate modules of different classes of transducers that can be modulated and optimized for the specific application of interest.

In preliminary studies, samples of volatile organic compounds (VOCs), plastic materials, textiles, coffees, olive oil, and tobacco were tested, demonstrating the usefulness of the hybrid array in a wide variety of applications [47]. The combination of measurements from the different transducers was particularly useful for separating coffee brands, where it was necessary to use responses from two SnO_2 elements, five MOSFET elements, three electrochemical cells, seven QMB sensors, as well as a temperature, and a humidity sensor. Analysis was carried out using PCA and ANN.

The MOSES design was also tested for the identification of ozone using an array of conductivity sensors based on indium oxide (In_2O_3), SnO_2 sensors, QMBs (referred to in this publication as transversal shear mode resonators, TMSRs), and a module with electrochemical sensors [48]. Although only the sensors based on indium oxides showed correlation with ozone concentration, the full array was useful for discriminating between the interfering gases tested. In another publication [49], the researchers used PCA plots to show that a hybrid array consisting of eight MOS and eight QMB can better separate homologous aldehydes in vegetable oil as well as textile materials used in the automotive industry than an array composed of only one transducer type. This array also demonstrated good reproducibility over several months in studies involving the analysis of packing materials used by the food industry [50, 51].

In another study, an array containing seven QMBs, eight SnO_2 sensors, and four electrochemical cells, (called the *Moses II*), was tested to detect contraband food products for potential use at US border points [52, 53]. The researchers found that as few as two sensing elements (one QMB and one electrochemical cell) produced similar or better results compared to the full set of 19 sensors. Similar to other researchers, the authors found that a reduced sensor set that included only those relevant for the analysis was superior to using the full set of sensors. The authors also reported that use of the hybrid array gave better results than restricting the inputs to any single sensor class.

12.2.3   Other Hybrid Sensor Arrays

As mentioned previously, sensor arrays based on one type of transducer often yield response profiles that are too similar in nature to act as unique "fingerprints" for separating classes of compounds. In a recent study, analysis of honey was performed using an array of 10 MOSFET and 12 MOS and it was found that only three sensors (two MOSFET and one MOS) were needed for classification using ANN [54]. In this case, the output from only one of the MOSFETS was just as effective as the array of 10.

In another experiment, a hybrid EN composed of nine MOS and ten MOSFET sensors were used to characterize car seat foam materials [55]. A plot of the first two principal components revealed that the two kinds of transducers were separated along the two (orthogonal) axes and the authors noted that the two types of sensors brought independent information to the response data.

12.2.4   Conclusions

The consensus of the studies highlighted in this section was that the incorporation of multiple transducers was valuable and often at times necessary for successful analysis. Data representing the full range of the transducers consistently produced superior results compared to subsets of a single type. Feature selection techniques nearly always produced a subset containing at least one element of each transducer type, suggesting that the hybrid profiles were richer in information content. In most instances, the authors reported high within-group (transducers of the same type) correlations while between-group (transducers that were different in nature) responses showed little to no correlation. In most of the studies, the researchers demonstrated the value of the hybrid array by comparing results of data comprising the entire scope of sensors to subsets of transducers of the same type. Predictably, the results were consistently superior for the hybrid data. The high redundancy of the within-group sensors demonstrates the lack of selectivity that is often evidenced for the traditional EN array and the superior selectivity that can be achieved using a hybrid.

12.3   Multiple Measurements from a Single Sensing Element

The goal of using a hybrid approach is to increase the information content of the data and many researchers have found ways of increasing the dimensionality of the data without greatly increasing the complexity of the instrumentation. In terms of the hardware, these multidimensional sensors might not always be considered hybrid arrays because often at times a single type of transducer is used. Nevertheless, the desire to expand the richness of the data is in concordance with the goals of hybrid sensing, therefore we shall report examples of sensory data where multiple measurements were extracted from one sensor element.

12.3.1 Combination of Transient and Steady-State Measurements

Perhaps one of the simplest ways to increase the dimensionality of a sensing element without increasing the hardware complexity is to perform measurements at different temperatures [56–59]. In Ref. [59], an eight-element SnO_2 array was used and the temperature was cycled incrementally from 250 to 500°C, recording eight different parameters during each temperature change (maximum positive step change, maximum negative step change, time to maximum value, time to minimum value, maximum positive rate of change, time to maximum positive rate of change, maximum negative rate of change, and time to maximum negative rate of change), effectively increasing the dimensionality of the data from 8 to 208. For the analysis of three types of tea leaves (Ceylon, Earl Grey, and Kenya), a correct classification of 69% was achieved when using the eight-element steady-state array while the expanded array using the transient measurements improved the classification to 90%.

Use of both transient and steady-state measurements was also used for qualitative and quantitative analysis of ethanol, toluene, and o-xylene, whereby the conductance rise time and the overall change in conductance were measured from an array of four SnO_2 sensors [60]. Using ANN for classification, discrimination between the three VOCs improved from 66% using just steady-state measurements to 100% when the steady-state and transient measurements were combined. In another study, the kurtosis, variance, and skewness were extracted from the time-domain signal of a single SnO_2 sensor [61]. These variables, along with a human input variable, were used in a neural network to track the fermentation process of dough.

12.3.2 Higher Order Sensors

A "higher order" sensor is realized when more than one transducer is applied to the same sensing element, while a sensor (or sensor array) employing one transduction principle is considered zero order [62]. Hence, the information content which can be realized depends on the number of sensors used in the array and the order of each sensor. Janata et al. state that the higher-order signals should be orthogonal in their response [63].

Increasing the order of a sensor involves subjecting each sensing element to a variety of different measurements. For example, in a study concerning methanol vapor interaction with doped-polypyrrole thin films, changes in mass, work function, and optical absorbance were applied to each sensor [64]. In a different study involving polysiloxane-coated QMB sensors, changes in sensor thickness, mass, temperature, and capacitance were measured using an optical sensor, quartz microbalance oscillator, thermopile, and interdigital capacitor transducer, respectively [65]. These approaches may be useful for not only increasing the orthogonality of the data, but also to simplify the hardware requirements in terms of the number of sensing elements that are required in the array.

Extracting more than one type of measurement from each sensing element may improve the selectivity of the analysis. This was demonstrated in a study where both temperature and resistance were monitored for each element of a SnO_2-based sensor array for the identification of hydrogen (H_2) and various gaseous carbon compounds [66].

Another reason for increasing the order of a sensor is that different analytes may be more responsive to particular modes of sensing. In a study involving poly (ether urethane) coatings on quartz microbalances, changes in mass, capacitance, and temperature were measured for a variety of volatile organic compounds [63]. Capacitance produced a strong signal for compounds such as alcohols and alkanes, whose dielectric coefficients differed greatly from the polymer coatings, while calorimetric measurements were useful for compounds with high heats of vaporization, such as chlorinated hydrocarbons.

This approach was also taken in a recent study employing a multitransducer chip that incorporated three different sensing mechanisms: a mass-sensitive resonant cantilever, a capacitive sensor, and a calorimetric transducer [67]. Five such chips, each with different polymer coatings, were used to measure a variety of VOCs. The researchers noted the value of increasing the orthogonality of the array by providing measurements that give complementary information. A compound such as methanol produces a low response for a mass-sensitive transducer because of its low molecular mass and high saturation vapor pressure, while having a high dielectric constant responds highly to a capacitive measurement. Bar graphs for each of the eight compounds indicated that the three transducers gave distinct responses while the profiles between chips were similar for many cases. In other words, alterations of the transducing principle produced a richer data profile than simply changing the polymer coating.

12.3.3 Exploiting Higher Order Sensing Principles to Simplify Chip Designs

The concept of higher order sensors was demonstrated by Langereis et al. who constructed a multipurpose sensor-actuator structure consisting of platinum leads connected in a fingerlike pattern with four points of contact [68]. The structure operated in three different sensing modes (temperature, conductivity, or amperometry) in addition to two actuator modes (local heating or pH gradient control) by altering the connections of the leads and the type of stimulus applied to each contact point. The three sensing modes and two actuator modes gave a potential of six different measurement configurations that could be switched according to the desired experiment or application at hand. The principles behind this approach were used in a chip-based device that could be used to measure pH, penicillin concentration, diffusion coefficients of ions, temperature, flow velocity, and flow direction [69, 70]. Although only one type of transducer (field effect) was utilized as the sensing mechanism, the distinct ways of interrogating the sensing elements allowed for the measurement of physical parameters quite different in nature. In a micro-total analysis system (μ-TAS) device, parameters such as flow rate, conductivity, temperature, or pH often need to be monitored, but the integration of several types of transducers for these measurements is less practical and results in a higher cost of chip fabrication.

12.3.4 Conclusions

Sensor fabrication cost and complexity are important issues when considering a hybrid sensing scheme, however difficulties may be avoided by taking full advantage of the sensing elements incorporated in the device. Utilizing both transient and steady-state measurements, increasing the order of each sensor, or otherwise extracting multiple measurements from each sensor are ways to increase the information content of the data with only modest increases in the fabrication complexity. Interrogating a single sensor with several types of measurements can provide a richer data profile that can classify a wider range of compounds. This is accomplished by increasing the sensor order rather than extending the number of sensors used in the device.

12.4 Data Fusion Approaches

The concept of hybrid sensing is not new. Multisensory data fusion has long been used for military, civilian, and medical applications [71–74]. For chemical sensing, data fusion is slowly becoming an area of intrigue, but has yet to be widely accepted as an attractive approach to vapor sensing, most likely due to the presumption of complexity in the fabrication process and data analysis routines. Nevertheless, as more researchers share their successful implementations of hybrid sensing and data fusion in the area of chemical sensing, the use of such techniques is expected to grow. For now, the most commonly used data fusion approaches for gas sensing arrays involve use of more than one enose or an enose/etongue hybrid.

12.4.1 Examples of Basic Data Fusion Approaches

One example of a simple hybrid sensor involving data fusion was reported by Huyberechts et al. where a three-sensor array composed of two SnO_2 (one doped and one undoped) sensors and a humidity sensor were used to classify carbon monoxide and methane in humid air [75]. Although the authors did not use this terminology to describe their system, the concept of using multiple pieces of information (data for compound type combined with data for humidity level) can be seen as a data fusion approach. In theory, the confidence (and thus classification ability) of the two-sensor array used for analyte detection should be improved with the additional piece of information that the humidity sensor provides.

In another study, fusion of EN data with color and texture measurements was used to grade the freshness of fish [76]. The authors found that the results improved when more techniques were used in the analysis.

12.4.2 Enose-Enose Fusion

One simple way to implement data fusion is to analyze a vapor sample using more than one EN placed side-by-side. If a low level of data fusion was used, this type of hybrid could have been discussed in Sect. 12.2, however, we allow the EN-EN sensors to be placed in the data fusion category because the idea of combining two individual dedicated instruments in order to enrich the data profile is characteristic of a data fusion approach.

In one study, a hybrid array involving two enose sensors placed in parallel was used for the monitoring of hydrolyzates during wood fermentation [77]. The first consisted of 32 conductive polymers while the second was made up of ten MOSFET and six SnO_2 sensors. The latter sensor used five transient measurements for each sensor (response, on/off derivative, and on/off integral), producing in 112 of these, different response variables for the dual enose system. When sensor selection studies were performed, the reduced sensor sets nearly always contained elements from each type of transducer. For modeling ethanol production rate, one SnO_2 (off-derivative), four conductive polymer, and two MOSFET (response from one and off-integral for the second) sensors were chosen. Thus the array of 112 values was reduced to seven sensors, but the final selection included elements of each transducer type. Similarly, for ethanol concentration and sum of furfural and 5-(hydroxymethyl)furfural concentrations, the full range of transducers were represented in the reduced set of seven variables, but for furfural concentration only two types (six conductive polymers and one MOSFET response) were chosen.

Di Natale et al. found that data fusion from two types of enoses, one based on quartz resonators and another based on MOS sensors, was useful for the analysis of olive oils [78]. The researchers tested the fusion of data from the auto scaled sensor inputs (a low level abstraction) as well as fusion of the PCA scores (a higher level abstraction), the latter technique giving slightly better class resolution when the data was examined in bi-plots. The authors concluded that the integration of the two types of data significantly improved the classification of olive oils.

In another study, the fusion of data from four different ENs was found to improve the classification of the quality of pear juice compared to each system individually [79]. Data was combined using a low level of fusion for 10 features of *INRA*, 14 of *Roma*, 16 of *UPM*, and 32 of *Warwick* (enose systems are described in the reference). Using radial basis function neural networks, the combined inputs produced an 86.7% correct classification, while the individual systems produced 56.7, 40.0, 66.7, and 43.3% for *INRA*, *Roma*, *UPM*, and *Warwick*, respectively. Using probabilistic neural network (PNN), the individual results were 53.3, 33.3, 53.3, and 43.3% for *INRA*, *Roma*, *UPM*, and *Warwick*, respectively, while the fusion of all features produced 80.0% correct classification. The researchers demonstrated the need for greater selectivity in many enose systems and the improvements that can be realized when hybrid techniques are used.

12.4.3   Enose-Etongue fusion

Data fusion of EN and ET measurements has also been explored. Fusion of data from an enose and an etongue was implemented by Winquist et al., who used a combination of 14 enose and 18 etongue sensory outputs to classify age and type of fruit juice samples [80, 81]. The fusion of data improved the classification results compared to using the EN or ET systems separately.

Di Natale et al. combined data from an enose consisting of eight QMB sensors and an etongue made up of seven porphyrin electrodes for the analysis of urine samples as well as milk samples [82]. The researchers were interested in studying the appropriate abstraction levels for analysis (the level of feature extraction or data processing as well as the position in the classification scheme where each type of feature should be fused). They found that low levels of abstraction worked for enose and etongue data that was clearly distinct from each other (in other words, the variation between the two types of sensors was greater than the intrasensor variation). For the analysis of urine, the low abstraction level was found to be appropriate because PCA plots revealed the enose data was found to lie along the first principal component while the etongue data was well separated, lying along the second principal component axis. Scores of PCA analysis were found to be correlated to properties such as pH, specific weight, and blood cell concentration. For the analysis of milk, a higher level of abstraction was needed for the distinction between fresh and spoiled milk as well as two different pasteurization methods. The best clustering was found when PCA was performed separately for the enose and etongue data, and the first principal component of the enose was fused with the second principal component of the etongue.

Sensor fusion of data from an EN and ET has also been proposed for wine analysis [83]. In preliminary experiments, the fusion of an array of five MOS gas sensors and three liquid sensors was able to classify the alcohols into categories of beer, brandy, and vodka at a 94.4% correct classification, while the gas sensors alone or the taste sensors alone produced an 83.3% and 70.0% correct classification, respectively.

12.4.4   Conclusions

Relatively few examples concerning data fusion from multiple ENs or EN-ET configurations exist in the literature; however, for the studies mentioned above, the fusion approach proved to be useful in a variety of measurement settings. Reports of data fusion for gas sensors are expected to increase provided the conclusions for the studies indicate that the use of multiple sensors is worth the extra effort in the hardware cost and configuration. Naturally, this issue will depend on the intended application and whether or not current technologies are sufficient for the task at hand. There are several commercially available EN and ET sensors, many of which incorporate more than one type of transducer. Many manufacturers offer the hybrid and allow the user to decide which collection of sensory inputs is appropriate for the application. One commercially available gas sensor array that employs different types of hybrid instrumentation is the GDA 2 (Gas Detector Array II) manufactured by Airsense Analytics. This instrument incorporates ion mobility spectrometry, a photoionization detector, an electrochemical cell, and two metal-oxide sensors for the detection of chemical warfare agents and other gases. The signals from the various detectors are combined and pattern recognition methods are used to detect and classify gas species from a standard database of 45 compounds. The HAZMATCAD Plus manufactured by Microsensor Systems, Inc. uses surface acoustic wave sensors for blood and nerve agents and electrochemical sensors for toxic industrial chemicals, but a fusion of the two types of inputs is not used in the analysis. These two devices illustrate the high number of compounds that are able to be detected using a hybrid collection of sensors.

12.5 Conclusions and Prospects for Hybrid Array Sensing

Sections 12.2–12.4 introduced the concept of a hybrid array and the benefits it may offer in chemical sensing applications. Arguably, a hybrid sensing scheme may be seen as a data fusion approach. In Sect. 12.2 examples were given of EN sensors that employed data fusion at a low level of processing. The sensor readings from the hybrid array were able to be combined and analyzed in a fairly straightforward manner. Results repeatedly pointed to the use of different transducers as a means of enriching the data and improving classification of compounds. In Sect. 12.3, a different approach to enhancing the data content was discussed. Using transient measurements or interrogating each sensing element to measure more than one parameter is a way to increase the dimensionality of the data. In this way, the array can be exploited without increasing the number of sensing elements. Although some of the techniques may not be considered hybrid sensing in the conventional sense, the motivation for using a hybrid array is to increase the data content, therefore these techniques were presented as a means of doing so. Also, issues of sensor fabrication complexity (such as number of elements or hardware concerns) may be addressed using some of the methods highlighted in Sect. 12.3. In Sect. 12.4 we introduce examples of vapor sensors that employ data fusion in a more conventional sense. Sensing schemes utilizing ET-ET or ET-EN configurations were presented and examples of data analysis at a higher abstraction levels were given.

A common trend that was observed for all the research described in Sects. 12.2–12.4 was that the use of different transducers created a more powerful sensing array.

An important aspect of hybrid sensing is the attempt to obtain measurements that are orthogonal in nature. Arrays of semiselective sensors employing the same type of transducer were often seen to be highly correlated in their responses. This diminishes the potential of the array to produce unique fingerprints for various classes of compounds. The general consensus of the research covered in Sects. 12.2–12.4 was that using a collection of different transducers greatly improved the selectivity and sensing capabilities of the array. Although data fusion techniques have yet to see widespread use in chemical sensing applications, their implementations are expected to grow as more researchers confirm their success. For challenging analyses, where conventional EN technologies may fail to give the desired accuracy and reliability, hybrid sensing may provide an alternative that is cost-effective and feasible. A dedicated sensor for a specific application may be constructed from an array of spot sensors or that are commercially available at a reasonable cost. Sensors may be integrated onto a single platform or operate as separate entities followed by off-line fusion of the data. Integration of data off-line is often used in preliminary studies to select the appropriate array of sensors for a dedicated instrument. In the latter sections of this chapter we highlight some of the research conducted at the Naval Research Laboratory where hybrid sensing and data fusion approaches have been used for stand-off sensing applications such as situational assessment and target detection.

12.6   Implementations of Hybrid Array and Multisensor Data Fusion for US Naval Applications

In recent years, there has been much interest in developing sensors that provide automated monitoring and situational assessment of the surrounding environment. These "smart" sensors are designed to detect hazardous occurrences such as toxic chemical gas releases or liquid spills, flooding, fire, pipe ruptures, or other events that require damage control measures to be implemented. Sensory data is fed into mathematical algorithms that are modeled after a human-like thought processes such as classifying the data into predetermined patterns and/or making a logical decision about actions to be taken. To program a sensor to produce sophisticated conclusions or decisions, a multi-criteria approach using multiple pieces of information (i.e., an array of data) is generally necessary. Intuitively, it is expected that as the complexity of the analysis increases, such as identification of multiple events that are similar in nature, the selectivity of the array would need to be powerful. Hybrid sensing techniques that incorporate an array of orthogonal sensors are a powerful approach to passive situational assessment monitoring [84].

Studies conducted at the Naval Research Laboratory (NRL) have been geared toward sensors for the detection of toxic gases [85] (air quality monitors) as well as intelligent sensors that could detect and implement automated responses to hazardous events such as fires, pipe ruptures, gas leaks, or other occurrences that require damage control measures. These sensors allow reduced manning aboard naval vessels such as ships and submarines by providing automated response or early detection. One area of interest to the US Navy is the development of fire alarm systems that offer improved accuracy and reliability compared to traditional smoke detectors. Another topic of interest is the detection of UXO material that is buried beneath the ground at former bombing test sites [86]. Hybrid sensor arrays combined with multisensory data fusion have proven useful for addressing each of these concerns. Section 12.7 of this chapter describes a multicriteria sensor array called an Early Warning Fire Detector (EWFD) which contains a hybrid of smoke detectors and gas sensors. Section 12.8 describes a real-time, remote detection system for shipboard situational awareness known as the Volume Sensor. Section 12.9 describes wide-area assessment (WAA) techniques (sensory data from aircraft) for surveying potential UXO contamination. For all three projects, fusion of various types of data was necessary to provide an accurate detection with a low false alarm rate. Although the following sections describe research of interest to the navy, the methods used for sensor array design and data analysis are expected to be transferable to a wide range of applications.

12.7   Early Warning Multicriteria Fire Detector Based on Hybrid Sensing

12.7.1   Background

The goal of reducing manning aboard naval vessels means that fire detection capabilities must be accurate and reliable so that countermeasures such as sprinkler systems are not deployed due to false alarms. False alarms due to "nuisance" sources such as welding, grinding, torch-cutting, cooking, or other fire-like events are particularly problematic aboard naval vessels, where such activities are routinely performed. Another issue is the importance of detecting a fire as early as possible so that damage is minimized. Typical indoor fires may be characterized as either "flaming" or "smoldering," the latter of which may elude detection in its early stages.

Typical commercial fire alarm systems are spot-type (univariate) sensors designed to detect smoke. There are two common types, ionization and photoelectric smoke detectors. In order for the sensor to respond, smoke must diffuse to the chamber of the sensor. This may take many minutes, depending on the nature of the fire, air drift, and sensor location. Ionization detectors respond faster to flaming fires while photoelectric detectors respond faster to smoldering. In recent years, the development of a multicriteria fire detector, containing both smoke and thermal detectors has been advocated for decreasing nuisance alarms [87, 88].

Research suggests that the sensing capabilities of the multi-criteria fire detectors may be improved by adding gas sensors such as carbon monoxide (CO) and/or CO_2 detectors [89–91]. Section 12.7.2 describes a hybrid sensor comprised of smoke detectors and gas sensors for a fire alarm system that provides early warning capabilities as well as low false alarm rates. Classification using a PNN will be discussed as well as techniques utilizing multivariate statistical process control (MSPC).

12.7.2 Sensor Selection Studies

The selection of sensors for a dedicated fire alarm involved a two-tiered approach. First, a large collection of sensors were chosen based on the scientific knowledge about the system (i.e., what types of gasses or biproducts are generated by a fire or nuisance source) and which are the sensors that would be cost effective and easily implemented into a hybrid array. Second, a down-selection of sensors was performed based on experimental results of the full set.

Twenty sensors were selected based on a literature review, past experience, and available technologies. The choice of gas sensors were based on the expected effluents of a fire. Other detectors included temperature sensors, relative humidity detectors, photoelectric and ionization smoke detectors. A series of 120 experiments were conducted in a test compartment measuring $4.1 \times 6.5 \times 3.6$ m (96 m^3) wherein the set of 20 sensors were exposed to a variety of fire and nuisance scenarios. The fires included both flaming and smoldering fires using fuels such as a flammable liquids (propane, heptane, jet fuel, alcohol) and materials such as mattresses, paper (trash can fire), electrical cables, insulation, and wall panel materials. Nuisance sources included use of a toaster, welding, torch cutting (steel), grinding (steel or cinder block), cutting wood, burning popcorn in microwave, gasoline engine exhaust, electrical heaters and halogen lamps, and cigarette smoke. More details about the sensors and experiments can be found elsewhere [92, 93].

Response patterns were determined at discrete times corresponding to the high, medium, and low sensitivity alarm setting of the photoelectric smoke detector. Cross-validations using a PNN [94, 95] were used as the final measure of performance. The PNN is a nonparametric, nonlinear classification algorithm that has been successfully applied to a wide variety of multivariate data. In this study, patterns were trained and classified as three categories: clean air, nuisance, and fire.

Exploratory data analysis included PCA, hierarchical cluster analysis, and linear correlation coefficient analysis in order to visualize the natural clustering and determine redundancy among sensors. In addition, two other methods were used for sensor selection. First, a stepwise regression based on a chi-squared test of the classification model goodness of the fit was performed. Second, a forward selection that minimized the classification error of a PNN cross-validation was used to select the best set of sensors. Using cluster analysis as a guide, the subsets were further investigated by swapping sensors that were similar in nature, testing the consequences of leaving a sensor out or adding one in, and other ways of refining the array based on human deduction and sensor construct issues. The goal was to select a small subset of sensors that provided orthogonal content. A small subset was also desired to lessen the computational burden of the PNN, resulting in faster speed (important for real-time monitoring) and less memory requirements (important when using a dedicated processing chip).

The results indicated that five sensors provided the best overall classification. Increasing the number of sensors to eight did not improve the performance. The best results (98% correct) were achieved using oxygen, hydrogen sulfide, relative humidity, ionization, and photoelectric sensors at the response times that corresponded to the medium and low sensitivity levels of the photoelectric smoke detector. At the high sensitivity level, the same set of sensors provided 95% correct classification. Various other subsets also produced 95% correct classification. The classification results demonstrate an improvement over using one of the spot-type smoke detectors alone. The correct classifications for the ionization detector at the low, medium and high sensitivity alarm setting were 83%, 83%, and 76%, while the correct classification was 76%, 74%, and 85% for the photoelectric detectors. In addition, the cluster responses indicated that gas sensors could be varied with similar results. This is important when designing a fire detection system because long-term stability of the sensors is critical.

Similar experiments using the full set of sensors were performed aboard the ex-USS *Shadwell* (Mobile Bay, AL), where the sensors could be tested in a more realistic setting (various room configurations, background air conditions, etc.). After three test series involving 120 total fire or nuisance tests, the sensor array was reduced to four sensors: ionization, photoelectric, CO, and CO_2 detectors. A prototype system, EWFD, was built using the four-sensor array with the PNN classifier. The EWFD was evaluated in real time using full scale fire tests on the ex-USS *Shadwell*.

12.7.3 Conclusions and Reflections

The EWFD prototype provided improved detection for both flaming and smoldering fires. In general, the response times of the array were equal or better than the commercial smoke detectors. Subsequent studies have been performed with this hybrid array and the results are detailed elsewhere [96, 97]. Figure 12.1 shows a typical trend for various (scaled) sensor responses (a) as well as the PNN probability trend (b) for a typical fire. The figure was taken from Ref. [96]. The experiment involved a heptane pan fire and was performed aboard the ex-USS *Shadwell*. The source was ignited at 385 s and the trend shows a sharp increase at roughly 420 s which later falls due to extinguishing the fire.

Although most studies employed PNN, one study used a process monitoring approach to data evaluation [98]. MSPC is often used for fault detection, to identify when a process is out of control. For the EWFD, statistical diagnostics were

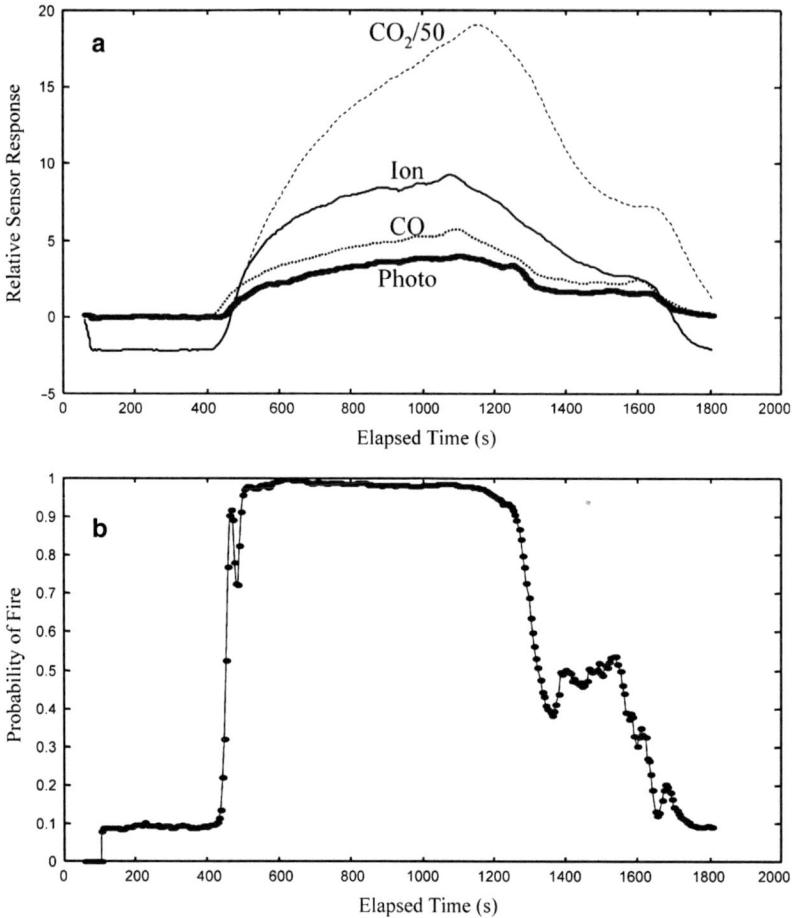

Fig. 12.1 Sensor responses (**a**) and PNN probabilities (**b**) generated during a heptane pan fire performed aboard the ex-USS *Shadwell*. The ignition time was 385 s. This figure was taken from Ref. [96]

computed for ambient, steady-state conditions using a PCA model of clean air data. Statistical parameters such as Hotelling's T^2-statistic (essentially the distance of a data point to the PCA model) and the Q-residual (sum of squared residuals after PCA projection) are commonly used to diagnose outliers or data that is statistically different from model (training) data. Critical limits such as a 95% or 99% confidence limit may be set as thresholds above which the data is considered to deviate from normal process conditions. Typically, the Q-residual is a more sensitive parameter. This was the case for the EWFD data collected aboard the ex-USS Shadwell. The Q-statistic was typically the first indicator of fire, while the T^2-statistic (referred to in this publication as the D-statistic) lagged behind before showing deviation above the critical threshold. The latter diagnostic was used for confirmatory analysis, but the use of both statistics was considered essential for accurate fire identification. Figure 12.2a shows a plot of the Q-statistic for a smoldering fire experiment that involved burning laundry clothes while Fig. 12.2b shows both the D- and Q-statistics for a welding experiment. These plots were taken from Ref. [98]. In plot a the fire was set at 313 s and the Q-statistic rose above the 99.9% upper confidence limit (UCL) at 782 s. The photoelectric detector alarmed at 1,133 s. In plot (b), both the photoelectric and ionization detectors alarmed for the nuisance source (welding) whereas the D- and Q- statistics did not reach the 95.0% UCL threshold. In this study, contribution plots were used

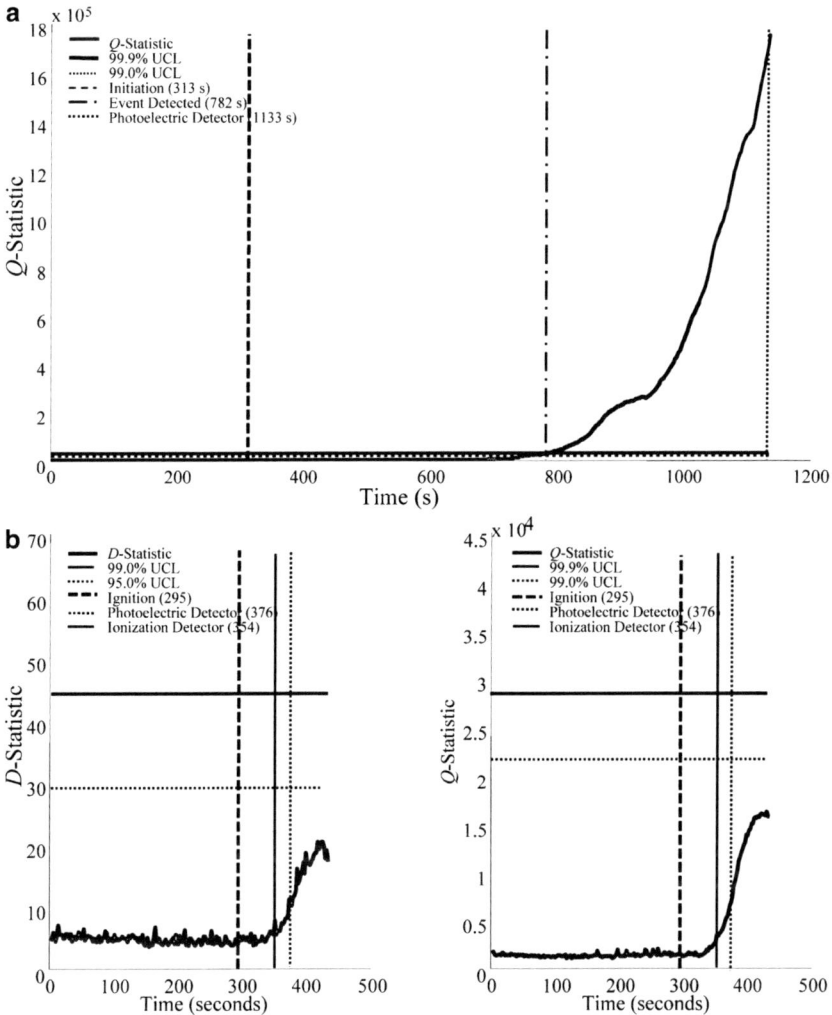

a

Legend:
— Q-Statistic
— 99.9% UCL
⋯ 99.0% UCL
--- Initiation (313 s)
—· Event Detected (782 s)
⋯ Photoelectric Detector (1133 s)

b

Legend:
— D-Statistic
— 99.0% UCL
⋯ 95.0% UCL
--- Ignition (295)
⋯ Photoelectric Detector (376)
— Ionization Detector (354)

Legend:
— Q-Statistic
— 99.9% UCL
⋯ 99.0% UCL
--- Ignition (295)
⋯ Photoelectric Detector (376)
— Ionization Detector (354)

Fig. 12.2 Trend of the Q-statistic during a smoldering fire experiment (**a**) and both D- and Q-statistics during a nuisance test involving welding steel (**b**). The figure was taken from Ref. [98]

to identify the source (location) of the fire and its rate of growth. This information is useful for deployment of countermeasures which (such as sprinkler systems or manpower) is targeted at the original source of the fire rather than adjacent compartments where effluent smoke may have drifted. The MSPC approach allows statistically meaningful diagnostics to be extracted from a time-domain data and represents a complimentary technique to a pattern recognition algorithm such as the PNN.

12.8 Data Fusion of Volume Sensor

In Sect. 12.8, a hybrid array called a *Volume Sensor* that uses a more sophisticated level of data fusion than the EWFD described in Sect. 12.7 is presented.

12.8.1 Background

While the EWFD provided improved fire detection, there were some limitations. Fire-like nuisances were still misclassified and the detection of smoldering fires was delayed because the smoke had to diffuse to spot-type or point detectors. Therefore, a new detection system was investigated based on real-time, remote detection system for shipboard situational awareness known as the Volume Sensor. The objective was to develop an affordable detection system that will identify shipboard damage control conditions and provide an alarm for events such as flaming and smoldering fires, explosions, pipe ruptures, and flooding. A multisensory approach that took advantage of the existing and emerging technology in the rapidly growing fields of optics, acoustics, image analysis, and computer processing was used. In addition, this technology utilizes conventional surveillance cameras, which are currently being incorporated into new ship designs, and therefore will provide multiple system functions with minimal new hardware. Intelligent data fusion algorithms were developed for event classification and situational assessment. The volume sensor prototype (VSP) was subject to full-scale testing aboard the ex-USS *Shadwell* and was evaluated by comparing its performance with two commercial video image detection (VID) and three spot-type fire detection systems.

12.8.2 Sensor Selection

The first phase of sensor selection consisted of a literature review and an industry review of current and emerging technologies for video, optical, and acoustic methods for the detection of smoke and fire [99]. Based on the study, several technologies were identified as having potential for meeting some of the objectives of the detection system development effort. A recent advance in fire detection is to exploit the data from video cameras that have already been installed for surveillance, employing special algorithms to detect smoke or fire. A full-scale laboratory evaluation of three VID systems using a variety of fire and nuisance sources indicated that the smoke alarm algorithms of these systems could provide fire detection capabilities equivalent to spot-type smoke detectors for most of the conditions evaluated [100]. However, nuisance rejection remained a problem, mainly due to bright nuisance events such as welding, torch cutting, and grinding that produce similar optical phenomenon in a VID image. Potential technologies were assessed to augment the VID. These included spectral sensors [101, 102], acoustic signatures [103], and long wavelength imaging [104, 105]. Each of the sensing technologies was evaluated using damage control events including fire, flooding and pipe ruptures, and against a variety of typical shipboard nuisance sources. The results indicated that each sensing technology provides unique information for use in the multisensory prototype. The VID systems were found to be effective for detecting smoke, while long wavelength video detection (LWVD) and spectral sensors successfully detected flame emission over the entire space without requiring a large number of cameras. In addition, the acoustic detection was useful for identifying pipe ruptures and flooding as well as for discriminating against nuisance sources, such as grinding.

Two VSPs have been assembled and demonstrated [106]. The sensor components were grouped into sensor suites, each of which contains a video camera, a long wavelength filtered video camera, spectral sensors, and a microphone. Data from multiple sensor suites were analyzed by the appropriate sensor systems. Figure 12.3 shows the data fusion architecture and labels the hardware and software components as "field monitoring" (the VSP), "sensor system computers," and "fusion system computer." The "supervisory control system" represents an interface to a higher level system. Each raw sensory output was subject to a specific algorithm before being sent to the fusion machine. The processed data was

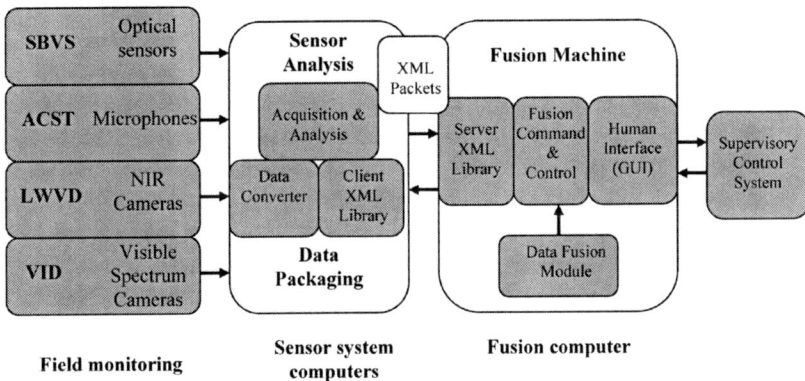

Fig. 12.3 Volume sensor system architecture and components

translated into the extensible markup language (XML) and transferred in message packets through a standard internet protocol (IP) network (i.e., Ethernet). The fusion computer (fusion machine) was a PC-based unit that contained the XML libraries used to encode and decode the message packets. The fusion machine also performed the data analysis routine and a graphical user interface (GUI) was developed to provide a visual aid and a human friendly interface.

12.8.3   Data Analysis

To build an event classifier, pattern recognition techniques such as PNN and linear discriminant analysis (LDA) were initially explored, but were found to produce higher than desired nuisance alarm rates and were less than successful when performed in real time. Results improved when an intelligent decision algorithm was written specifically for this application. Further improvements were found when Bayesian inference [107, 108] was used to guide the decision algorithm. For this approach, a set of training data was used to generate frequency tables based on the response profiles of each of the nine classes of events: fire, bright nuisance (i.e., welding or torch cutting), grinding, engine running, water (flooding), fire suppression system (mist), gas release, background, and people working/talking. The training data were gathered by picking event-specific signatures at known scenario times from the full-scale pool of tests. Each response profile consisted of 16 sensory outputs obtained from the preprocessing algorithms of the raw VSP data. For simplicity, the responses of each sensory input were binned into one of four states: low, mid/low, mid/high, and high. The states of the bins formulated the event-specific signatures that underwent Bayesian analysis to obtain a statistical probability that a certain event was occurring. The intelligent decision algorithm was necessary to provide a more robust analysis and prevent false alarms. For example, bright nuisance events (such as welding) may produce high sensory outputs for optical detectors, indicating a fairly high probability of fire. Using only spectral sensors, it is difficult to discern some of the bright nuisance event from fires. However, with the VSP acoustic sensors and the acoustic algorithm tailored to match welding, the Bayesian network may produce a very high likelihood that a welding event is occurring. With this additional piece of information, the likely conclusion is that the spectral phenomena are produced by welding rather than by fire. The intelligent decision algorithm can be programmed to place the system at a "prealarm" state or a warning state, but wait for further evidence of a fire. In this way, the decision algorithm was guided by the outputs of the Bayesian network, but was built upon a more complex set of instructions that modeled human inference. Both transient (past data that represented growth or trends) and current data from the Bayesian algorithm was used in the final analysis. The decision algorithm was written for the application at hand and was designed to provide accurate event detection with a minimum of false alarms.

12.8.4 Conclusion and Reflections

Performance of the VSP was compared to two commercial VID fire surveillance systems as well as three types of smoke detectors. The volume sensors outperformed all commercial systems. In addition to classification of fire sources, the volume sensors were able to categorize other events, correctly identifying most of the gas releases, flooding, and pipe rupture sources [109, 110].

The steps taken for the development and implementation of the volume sensor are applicable in many areas of chemical sensing. The approach to the problem of situational assessment was addressed by combining a diverse array of sensory elements that provided orthogonal information, producing event-specific data profiles that were able to be interpreted by intelligent fusion machine algorithms. Although microphones may not be the first detection mechanisms one would envision in a fire detector, they proved to be a surprisingly helpful addition for minimizing false alarms due to specific nuisance events. The VSPs were successful in fulfilling their goals for real-time monitoring in a shipboard environment. The systems provided timely and accurate detection of fires and other damage control events, outperforming a number of commercially available systems. The VSP systems were also designed with flexibility so that additional sensor components and software alterations could be accommodated if necessary or desired. The components were commercially available at a modest cost. The volume sensor demonstrated the performance gains that were realized when the multicriteria/multisensory approach was compared to the traditional univariate spot-type systems. Using a hybrid collection of sensors allowed for detecting the capabilities that expanded beyond the traditional systems, incorporating events such as gas releases, pipe ruptures, and flooding into the classification algorithms. The initial successes of this project should encourage the use of hybrid arrays as a feasible approach to addressing a host of challenging analytical problems.

12.9 UXO Analysis using a Data Fusion Approach

The following section describes some preliminary work performed at the Naval Research Laboratory concerning data fusion techniques to enhance UXO detection. The data fusion approach is similar to that described in Sect. 12.8 for the volume sensor, however the implementation is more complicated because the available data may be collected at different dates, contain different spatial resolutions, and be incomplete to one degree or another. This makes data registration, interpolation, and interpretation a much more involved task.

12.9.1 Background

An estimated 10 million acres of US land has been linked to former bombing test sites and have the potential to contain buried ordnance materials, although findings reveal that about 80% of a typical site is actually UXO-free. The most common method for detecting these materials is surface magnetometry which is slow and plagued with false alarms, adding to the high cost of remediation. The use of multiple sensing inputs combined with pattern recognition techniques has proven to be more effective at target detection and false alarm reduction than the use of a single sensing technology [111–113]. Recent efforts have been devoted to WAA using remote sensing techniques to pinpoint areas that may receive further scrutiny from close range methods. Preliminary results indicate that a data fusion approach to target detection would allow for a more rapid and reliable assessment of UXO-contaminated sites using WAA sensing technologies [114].

12.9.2   Data Fusion Architecture

For this preliminary investigation, low altitude (helicopter) airborne magneto-metry ("Helimag") and high-altitude airborne light detection and ranging (LiDAR) and high resolution aerial photography were analyzed from a survey taken at Pueblo Precision Bombing Range #2 in Colorado. In addition, auxiliary information such as historical records of site usage, data from previous surveys, information given by local residents, and other sources of expert knowledge were also gathered to supplement the WAA data. Figure 12.4 depicts an example of fusion architecture that is being implemented in this study. In a typical UXO remediation survey, there may be various types of data available in the three categories labeled "Wide Area," "Local Area," or "Auxiliary Information" and each type may be incomplete or differ in terms of data quality. The first step is the acquisition of sensory data and the second step is the data registration to a common grid system. Step three is the data fusion engine where the processing, fusion, and interpretation of the data occur. In this study, a Bayesian-based data fusion algorithm was used to establish confidence levels for each pixel location on the grid. In the fourth step, the probabilities are scaled and a color-coded map is generated based on the likelihoods of UXO existence. Step five allows for the refinement of the WAA map with additional survey scans that generate more detailed data in the regions reflecting high UXO likelihoods. Newly acquired data can be fed back into the fusion algorithm (steps 2–5) for a re-assessment of the probabilities, increasing the reliability of the feature map and establishing a more detailed and accurate dig sheet.

12.9.3   Data Analysis and Preliminary Findings

For the Pueblo site, the high resolution aerial photography data did not prove to be useful for UXO analysis; however the data could potentially assist in data registra-tion and interpretation of ground clutter or man-made artifacts. Geological survey

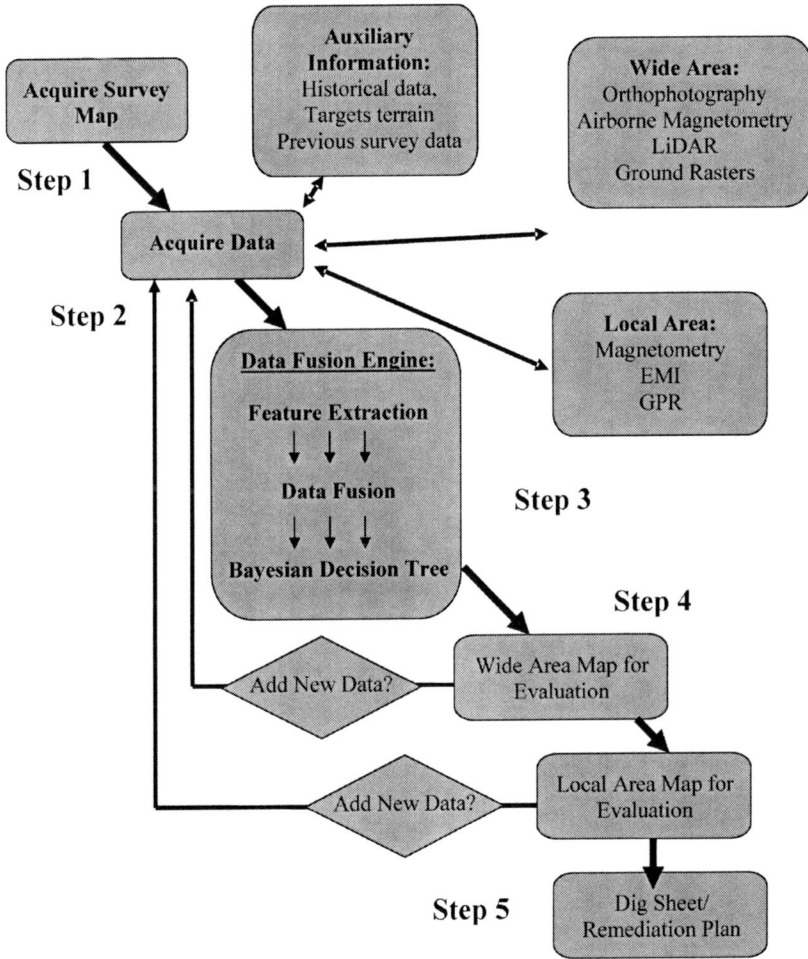

Fig. 12.4 Implementation of a data fusion framework for wide-area UXO assessment

data indicated minimal magnetic geological features as well as little foliage and other ground surface phenomena that may interfere with spectral or magnetometry measurements. Such features, however, are expected to pose a more difficult challenge for other sites. The principal feature of LiDAR data that may be correlated with UXO contamination is cratering of the ground surface. Figure 12.5 shows a grayscale feature map of LiDAR cratering side by side with a feature map based on total magnetometer signal. The coordinates span UTM Northing 4,169,390 to 4,178,634 and UTM Easting 614,590 to 618,627, and represented a rectangular survey area of approximately 37 km^2. A combined feature map was generated using both sources of data, as illustrated below. Taken individually, the results for each data source (expert information, orthophotography, LiDAR, and magnetometry)

Crater-like Feature Density Map (LiDAR)

Magnetometer Signal Density Map

+

Combined Feature Map

Auxiliary Data

Known and Suspected Targets (grey)

Man-made Structures

FUSION ALGORITHM

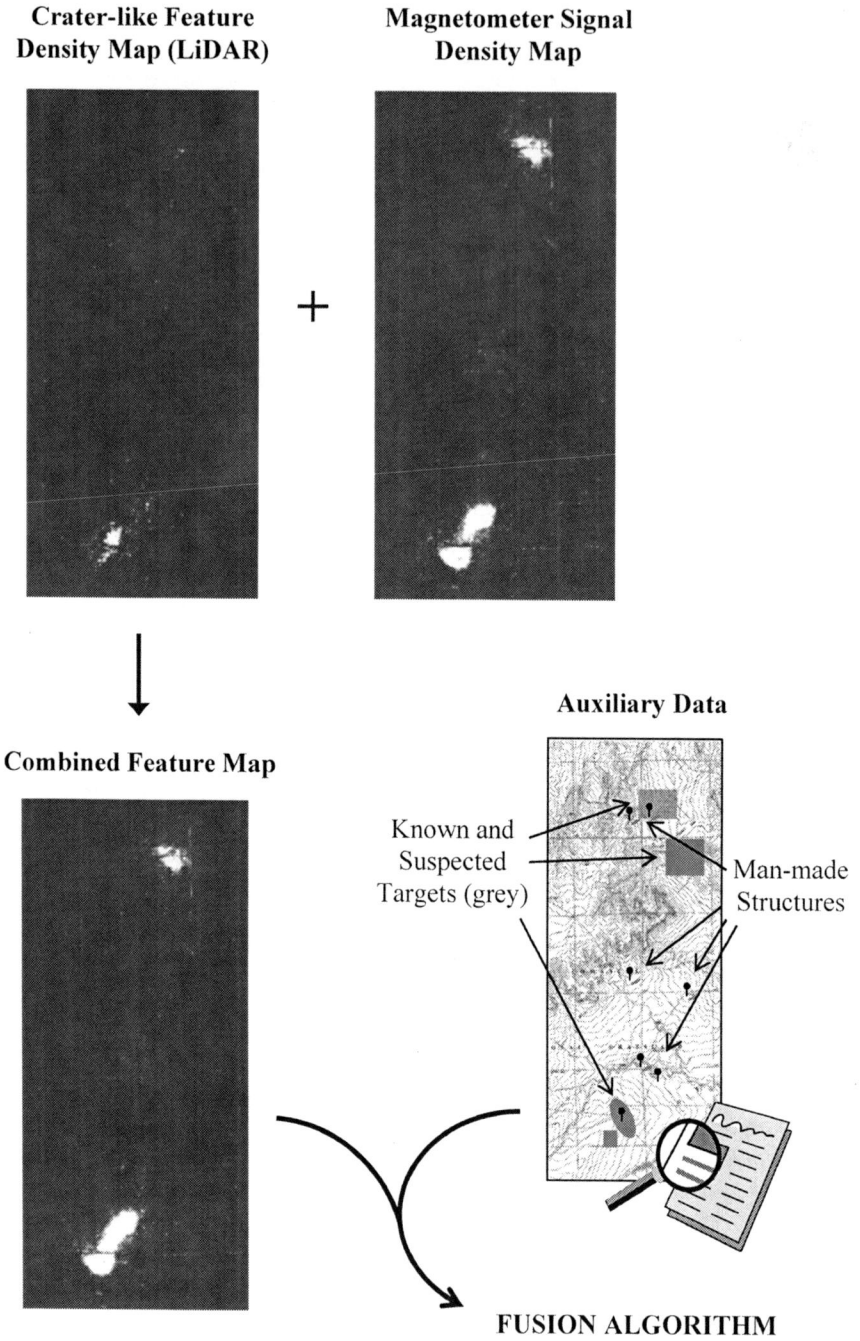

Fig. 12.5 Feature maps showing that a multisensory approach produces the most complete picture for UXO target detection

did not provide a complete and accurate profile of the Pueblo site. The combined feature map, however, provided a more reliable assessment. For instance, the crater-like feature density map generated with the LiDAR allowed for the gap to be filled between the two peaks in the southern portion of the magnetometer signal density map. This indicated that the target range known to be present in this area was used more extensively than suggested by the historical accounts. The density maps along with the combined map all confirmed the presence of a second, but less affected region in the northern portion of the site. However, no evidence in these maps supports the existence of the suspected 75-mm range. Ground truth data excavated in all three of these regions affirmed these conclusions.

12.9.4 Conclusions and Prospects

Determination of UXO contamination is a daunting task and the acquisition of sensory data is complicated by the diversities in terrain and ground cover of the various regions that are affected. For a sensing task that is plagued by false alarms, a multicriteria/multisensory approach using WAA techniques will ease the burden of data acquisition as well as lend a measure of reliability to the analysis. The data fusion architecture proposed above is flexible to accommodate a variety of data sources which may vary in nature, quantity, and quality. After appropriate scaling and transformation, the available pieces of information serve as inputs to a data fusion algorithm. More reliable measurements may be given greater weight in the analysis. For the Pueblo site, neither the magnetometry data nor the LiDAR alone was able to flag all suspected ordnance. However, the combined feature map revealed a more complete and accurate surveillance of UXO densities, confirmed by the ground truth data.

12.10 Conclusions and Reflections

In this chapter, the benefits of hybrid detector arrays are illustrated. Chemical vapor sensors, such as EN technologies, are typically comprised of semi-selective sensors based on a single transducer. There is much evidentiary support that a hybrid array can offer more a powerful detection capability than the more traditional designs. Array detectors are commonly developed to be application-specific and often at times, the desire to ease the manufacturing process, size, cost, or complexity precludes the investigation into hybrid sensing techniques. Hybrid designs may be deemed unworthy of the additional engineering efforts or overlooked due to the lack of published research that has proven them effective.

In Sects. 12.2–12.4, published reports of vapor sensing that utilized a hybrid array or higher order sensing principles were highlighted. Repeatedly, the authors stressed the benefits that were gained from the combination of two or more sensing mechanisms. Feature selection algorithms generally pointed to the redundancies among transducers of the same type and subsets based on a down-selection of sensors nearly always spanned the entire range of transducer types. The opening sections presented hybrid configurations where the sensory data was able to be combined in a straightforward manner, while the latter sections described more elaborate methods of data fusion. The current trend in hardware implementation is a modular approach to allow for design flexibility and alterations. Sensor elements of each class of transducer may be housed in a different module that may be addressed individually in terms of data acquisition and maintenance. The appropriate combination of data preprocessing, feature extraction, fusion architecture, and classification algorithm must be researched and optimized according to application.

In the second half of the chapter we discussed hybrid arrays and multisensory data fusion research conducted at the Naval Research Laboratory. Three projects were chosen to illustrate examples of data fusion using a relatively simple architecture (the EWFD) as well as those that were more complex (the VSP and UXO fusion schemes). The EWFD represents a hybrid array that may be the most closely associated with chemical vapor sensing. The VSP hybrid array was designed to detect multiple types of events, therefore the hardware components and software algorithms were correspondingly more complex. Nevertheless, the sensors were collocated and the data acquisition, fusion, and analysis were able to be performed in real time. The UXO analysis represents the most complicated system in terms of the data interpretation and fusion architecture. To generalize the process, the data fusion algorithms must be very flexible. It is expected that the ability to collect various types of data will depend on the time, cost, and instrumentation available and the fusion architecture may need to weight certain types of data according to reliability or quality. Similarly, abstract types of data (such as interviews with local residents or advisory reports from expert analysts) will need to be given a specific credence and somehow converted to a numerical representation if it is to be used in a fusion algorithm. Data registration to a common grid system (spatial alignment) is an important part of the process.

This chapter was aimed at presenting the research findings of those who have used hybrid arrays and data fusion for tasks such as chemical vapor sensing as well as for ongoing research conducted at the Naval Research Laboratory. The multi-criteria/multisensory approach is a powerful means of solving some challenging analytical problems. Although the use of hybrid arrays in chemical vapor sensing has not yet become commonplace, perhaps continuing research will encourage new developments and bring about more widespread usage. Oftentimes, the analytical benefits prove to be worthy of the modest increases in the hardware and software complexities.

Acknowledgments The Office of Naval Research provided funding for the Early Warning Fire Detector and the Volume Sensor. Funding for the UXO research was provided by the Strategic Environmental Research and Development Program (SERDP). Dr. Kirsten Kramer is a Post-Doctoral Fellow with the National Research Council.

References

1. Stetter, J. R.; Jurs, P. C.; Rose, S. L., Detection of hazardous gases and vapors: Pattern recognition analysis of data from electrochemical sensor array, *Anal. Chem.* **1986**, 58, 860–866

2. Müller, R.; Lange, E. Multidimensional sensor for gas analysis, *Sensors Actuat.* **1986**, 9, 39–48

3. Albert, K. J.; Lewis, N. S.; Schauer, C. L.; Sotzing, G. A.; Stitzel, S. E.; Vaid, T. P.; Walt, D. R., Cross-reactive chemical sensor arrays, *Chem. Rev.* **2000**, 100, 2595–2626

4. Jurs, P. C.; Bakken, G. A.; McClelland, H. E., Computational methods for the analysis of chemical sensor array data from volatile analytes, *Chem. Rev.* **2000**, 100, 2649–2678

5. Bourgeois, W.; Romain, A.; Nicolas, J.; Stuetz, R. M., The use of sensor arrays for environmental monitoring: interests and limitations, *J. Environ. Monit.* **2003**, 5, 852–860

6. Niessner, R., Chemical sensors for environmental analysis, *TrAC* **1991**, 10, 310–316

7. Di Natale, C.; Macagnano, A.; Paolesse, R.; D'Amico, A., Artificial olfaction systems: Principles and applications to food analysis, *Biotechnol. Agron. Soc. Environ.* **2001**, 5, 159–165

8. Deisingh, A. K.; Stone, D. C. Thompson, M., Applications of electronic noses and tongues in food analysis, *Int. J. Food Sci. Technol.* **2004**, 39, 587–604

9. Pavlou, A. K.; Turner, A. P., Sniffing out the truth: Clinical diagnosis using the electronic nose, *Clin. Chem. Lab. Med.* **2000**, 38, 99–112

10. Wilson, D. M.; Hoyt, S.; Janata, J.; Booksh, K.; Obando, L., Chemical sensors for portable, handheld field instruments, *IEEE Sensors J.* **2001**, 1, 256–274

11. Gardner, J. W.; Shurmer, H. V.; Corcoran, P., Integrated tin oxide odour sensors, *Sensors Actuat.* **1991**, B4, 117–121

12. Tomchenko, A. A.; Harmer, G. P.; Marquis, B. T.; Allen, J. W., Semiconducting metal oxide sensor array for the selective detection of combustion gases, *Sensors Actuat.* **2003**, B93, 126–134

13. Reinhoudt, D. N., Durable chemical sensors based on field-effect transistors, *Sensors Actuat.* **1995**, B24–25, 197–200

14. Ballantine, D. S., Jr.; Rose, S. L.; Grate, J. W.; Wohltjen, H., Correlation of surface acoustic wave device coating responses with solubility properties and chemical structure using pattern recognition, *Anal. Chem.* **1986**, 58, 3058–3066

15. Rose-Pehrsson, S. L.; Grate, J. W.; Ballantine, B. S., Jr.; Jurs, P. C., Detection of hazardous vapors including mixtures using pattern recognition analysis of responses from surface acoustic wave devices, *Anal. Chem.* **1988**, 60, 2801–2811

16. Grate, J. W., Acoustic wave microsensor arrays for vapor sensing, *Chem. Rev.* **2000**, 100, 2627–2648

17. Grate, J. W.; Patrash, S. J.; Kaganove, S. N.; Wise, B. M., Hydrogen bond acidic polymers for surface acoustic wave vapor sensors and arrays, *Anal. Chem.* **1999**, 71, 1033–1040

18. Grate, J. W.; Rose-Pehrsson, S.; Barger, W. R., Langmuir-blodgett films of a nickel dithiolene complex on chemical microsensors for the detection of hydrazine, *Langmuir* **1988**, 4, 1293–1301

19. Bartlett, P. N.; Archer, P. B. M.; Ling-Chung, S. K., Conducting polymer gas sensors - Part 1: Fabrication and characterization, *Sensors Actuat.* **1989**, 19, 125–140

20. Bartlett, P. N; Ling-Chung, S. K, Conducting polymer gas sensors - Part II: Response of polypyrrole to methanol vapor, *Sensors Actuat.* **1989**, 19, 141–150

21. Shurmer, H. V.; Corcoran, P.; Gardner, J. W., Integrated arrays of gas sensors using conducting polymers with molecular sieves, *Sensors Actuat.* **1991**, B4, 29–33

22. Walt, D. R.; Dickinson, T.; White, J.; Kauer, J.; Johnson, S.; Engelhardt, H.; Sutter, J.; Jurs, P., Optical sensor arrays for odor recognition, *Biosens. Bioelectron.* **1998**, 13, 697–699

23. Persaud, K.; Dodd, G. H., Analysis of discrimination mechanisms of the mammalian olfactory system using a model nose, *Nature* **1982**, 299, 352–355

24. Dickinson, T. A.; White, J.; Kauer, J. S.; Walt, D. R., Current trends in 'artificial-nose' technology, *Trends Biotechnol.* **1998**, 16, 250–258
25. Ziegler, C.; Gopel, W.; Hammerle, H.; Hatt, H.; Jung, G.; Laxhuber, L.; Schmidt, H. L.; Schutz, S.; Vogtle, F.; Zell, A., Bioelectronic noses: A status report. Part II, *Biosens. Bioelectron.* **1998**, 13, 539–571
26. Gardner, J. W.; Bartlett, P. N., Eds., Sensors and Sensory Systems for an Electronic Nose, Kluwer, Dordrecht, **1992**
27. Gardner, J. W.; Bartlett, P. N., A brief history of electronic noses, *Sensors Actuat.* **1994**, B18–19, 211–220
28. Strike, D. J.; Meijerink, M. G. H.; Koudelka-Hep, M., Electronic noses - A mini-review, *Fresenius J. Anal. Chem.* **1999**, 364, 499–505
29. Vlasov, Y.; Legin, A.; Rudnitskaya, A.; Di Natale, C.; D'Amico, A., Nonspecific sensor arrays ("electronic tongue") for chemical analysis of liquids, (IUPAC Technical Report), *Pure Appl. Chem.* **2005**, 77, 1965–1983
30. Shurmer, H. V.; Corcoran, P.; James, M. K., Sensitivity enhancement for gas sensing and electronic nose applications, *Sensors Actuat.* **1993**, B15–16, 256–259
31. Göpel, W., New materials and transducers for chemical sensors, *Sensors Actuat.* **1994**, B18–19, 1–21
32. He, L.; Toh, C., Review: Recent advances in analytical chemistry - A material approach, *Anal. Chim. Acta* **2006**, 556, 1–15
33. Walmsley, A. D.; Haswell, S. J.; Metcalfe, E., Methodology for the selection of suitable sensors for incorporation into a gas sensor array, *Anal. Chim. Acta* **1991**, 242, 31–36
34. Guadarrama, A.; Fernández, J. A.; Iniguez, M.; Souto, J.; de Saja, J. A., Discrimination of wine aroma using an array of conducting polymer sensors in conjunction with solid-phase micro-extraction (SPME) technique, *Sensors Actuat.* **2001**, B77, 401–408
35. James, D.; Scott, S. M.; Ali, Z; O'Hare, W. T., Review: Chemical sensors for electronic nose systems, *Microchim. Acta* **2005**, 149, 1–17
36. Craven, M. A.; Gardner, J. W; Bartlett, P. N., Electronic noses - development and future prospects, *TrAC* **1996**, 15, 486–493
37. Hall, D. L.; Llinas, J., An introduction to multisensor data fusion, *Proc. IEEE* **1997**, 85, 6–23
38. Winquist, F.; Hörnsten, E. G.; Sundgren, H.; Lundström, I., Performance of an electronic nose for quality estimation of ground meat, *Meas. Sci. Technol.* **1993**, 4, 1493–1500
39. Holmberg, M.; Winquist, F.; Lundström, I.; Gardner, J. W.; Hines, E. L., Identification of paper quality using a hybrid electronic nose, *Sensors Actuat.* **1995**, B26–27, 246–249
40. Börjesson, T.; Eklöv, T.; Jonsson, A.; Sundgren, H.; Schnürer, J., Electronic nose for odor classification of grains, *Cereal Chem.* **1996**, 73, 457–461
41. Mandenius, C.-F.; Hagman, A.; Dunås, F.; Sundgren, H.; Lundström, I., A multisensor array for visualizing continuous state transitions in biopharmaceutical processes using principal component analysis, *Biosens. Bioelectron.* **1998**, 13, 193
42. Mandenius, C.-F.; Eklöv, T.; Lundström, I., Sensor fusion with on-line gas emission multi-sensor arrays and standard process measuring devices in baker's yeast manufacturing process, *Biotechnol. Bioeng* **1997**, 55, 427–438
43. Lidén, H.; Mandenius, C.-F.; Gorton, L.; Meinander, N. Q.; Lundström, I.; Winquist, F., On-line monitoring of a cultivation using an electronic nose, *Anal. Chim. Acta* **1998**, 361, 223–231
44. Holmberg, M.; Gustafsson, F.; Hörnsten, E. G.; Winquist, F.; Nilsson, L. E.; Ljung, L.; Lundström, I., Bacteria classification based on feature extraction from sensor data, *Biotechnol. Tech.* **1998**, 12, 319–324
45. Mitrovics, J.; Weimar, U.; Göpel, W., Linearisation in multicomponent analysis based on a hybrid sensor array with 19 sensor elements, *Proc. Transducers '95* **1995**, 1, 25–29
46. Mitrovics, J.; Ulmer, H.; Weimar, U.; Göpel, W., Modular sensor systems for gas sensing and odor monitoring: The MOSES concept, *Acc. Chem. Res.* **1998**, 31, 307–315
47. Ulmer, H.; Mitrovics, J.; Noetzel, G.; Weimar, U.; Göpel, W., Odors and flavours identified with hybrid modular sensing systems, *Sensors Actuat.* **1997**, B43, 24–33

48. Sauter, D.; Weimar, U.; Noetzel, G.; Mitrovics, J.; Göpel, W., Development of modular ozone sensor system for application in practical use, *Sensors Actuat.* **2000**, B69, 1–9
49. Ulmer, H.; Mitrovics, J.; Weimar, U.; Göpel, W., Sensor arrays with only one or several transducer principles? The advantage of hybrid modular systems, *Sensors Actuat.* **2000**, B65, 79–81
50. Ulmer, H.; Mitrovics, J.; Weimar, U.; Göpel, W., Detection of off-odors using a hybrid modular sensor system, In Conference on Proceedings of Transducers '97, Chicago, USA, 555–558
51. Frank, M.; Ulmer, H; Ruiz, J.; Visani, P.; Weimar, U., Complementary analytical measurements based upon gas chromatography-mass spectrometry, sensor system and human sensory panel: a case study dealing with packaging materials, *Anal. Chim. Acta* **2001**, 431, 11–29
52. Pardo, M.; Kwong, L. G.; Sberveglieri, G.; Schneider, J.; Penrose, W. R.; Stetter, J. R., Detection of contraband food products with a hybrid chemical sensor system, *Proc. IEEE-Sensors* **2003**, 2, 1073–1076
53. Pardo, M.; Kwong, L. G.; Sberveglieri, G.; Brubaker, K.; Schneider, J. F.; Penrose, W. R.; Stetter, J. R., Data analysis for a hybrid sensor array, *Sensors Actuat.* **2005**, B106, 136–143
54. Benedetti, S.; Mannino, S.; Sabatini, A. G.; Marcazzan, G. L., Electronic nose and neural network use for the classification of honey, *Apidologie* **2004**, 35, 397–402
55. Morvan, M.; Talou, T.; Beziau, J.-F., MOS-MOSFET gas sensors array measurements versus sensory and chemical characterisation of VOC's emissions from car seat foams, *Sensors Actuat.* **2003**, B95, 212–223
56. Heilig, A.; Bârsan, N.; Weimar, U.; Schweizer-Berberich, M.; Gardner, J. W.; Göpel, W., Gas identification by modulating temperatures of SnO2-based thick film sensors, *Sensors Actuat.* **1997**, B43, 45–51
57. Sundgren, H.; Lundström, I.; Winquist, F.; Lukkari, I.; Carlsson, R.; Wold, S., Evaluation of a multiple gas mixture with a simple MOSFET gas sensor array and pattern recognition, *Sensors Actuat.* **1990**, B2, 115–123
58. Wilson, D. M.; Roppel, T.; Kalim, R., Aggregation of sensory input for robust performance in chemical sensing Microsystems, *Sensors Actuat.* **2000**, B64, 107–117
59. Corcoran, P.; Lowery, P.; Anglesea, J., Optimal configuration of a thermally cycled gas sensor array with neural network pattern recognition, *Sensors Actuat.* **1998**, B48, 448–455
60. Llobet, E.; Brezmes, J.; Vilanova, X.; Sueiras, J. E.; Correig, X., Qualitative and quantitative analysis of volatile organic compounds using transient and steady-state responses of a thick-film tin oxide gas sensor array, *Sensors Actuat.* **1997**, B41, 13–21
61. Wide, P., A human-knowledge-based sensor implemented in an intelligent fermentation-sensor system, *Sensors Actuat.* **1996**, B32, 227–231
62. Janata, J.; Josowicz, M.; Vanysek, P.; Devaney, D. M., Chemical sensors, *Anal. Chem.* **1998**, 70, 179R–208R
63. Zhou, R.; Hierlemann, A.; Weimar, U.; Göpel, W., Gravimetric, dielectric and calorimetric methods for the detection of organic solvent vapours using poly(ether urethane) coatings, *Sensors Actuat.* **1996**, B34, 356–360
64. Topart, P.; Josowicz, M., Transient effects in the interaction between polypyrrole and methanol vapor, *J. Phys. Chem.* **1992**, 96, 8662–8666
65. Haug, M.; Schierbaum, K. D.; Gauglitz, G.; Göpel, W., Chemical sensors based upon polysiloxanes: Comparison between optical, quartz microbalance, calorimetric, and capacitance sensors, *Sensors Actuat.* **1993**, B11, 383–391
66. Heilig, A.; Bârsan, N.; Weimar, U.; Göpel, W., Selectivity enhancement of SnO2 gas sensors: Simultaneous monitoring of resistances and temperatures, *Sensors Actuat.* **1999**, B58, 302–309
67. Kurzawski, P.; Hagleitner, C.; Hierlemann, A., Detection and discrimination capabilities of a multitransducer single-chip gas sensor system, *Anal. Chem.* **2006**, 78, 6910–6920
68. Langereis, G. R.; Olthuis, W.; Bergveld, P., Using a single structure for three sensor operations and two actuator operations, *Sensors Actuat.* **1998**, B53, 197–203

69. Poghossian, A.; Schultze, J. W.; Schöning, M. J., Multi-parameter detection of (bio-)chemical and physical quantities using an identical transducer principle, *Sensors Actuat.* 2003, B91, 83–91

70. Poghossian, A.; Lüth, H.; Schultze, J. W.; Schöning, M. J., (Bio-)chemical and physical microsensor arrays using an identical transducer principle, *Electrochim. Acta* 2001, 47, 243–249

71. Hall, D. L; Llinas, J., Eds., Handbook of Multisensor Data Fusion, CRC, Boca Raton, FL, 2001

72. Hall, D. L; McMullen, S. A., Mathematical Techniques in Multisensor Data Fusion, 2nd edn.; Artech House, Inc., Norwood, MA, 2004

73. Klein, L. A., Sensor and Data Fusion: A Tool for Information Assessment and Decision Making, SPIE, Bellingham, WA, 2006

74. Naidu, P. S., Sensor Array Signal Processing, CRC, Boca Raton, FL, 2000

75. Huyberechts, G.; Szecówka, P.; Roggen, J.; Licznerski, B. W., Simultaneous quantification of carbon monoxide and methane in humid air using a sensor array and an artificial neural network, *Sensors Actuat.* 1997, B45, 123–130

76. Macagnano, A.; Careche, M.; Herrero, A.; Paolesse, R.; Martinelli, E.; Pennazza, G.; Carmona, P.; D'Amico, A.; Di Natale, C., A model to predict fish quality from instrumental features, *Sensors Actuat* 2005, B111–112, 293–298

77. Mandenius, C.-F.; Lidén, H.; Eklöv, T.; Taherzadeh, M. J.; Lidén G., Predicting fermentability of wood hydrolyzates with responses from electronic noses, *Biotechnol. Prog.* 1999, 15, 617–621

78. Di Natale, C.; Macagnano, A.; Nardis, S.; Paolesse, R.; Falconi, C.; Proietti, E.; Siciliano, P.; Rella, R.; Taurino, A.; D'Amico, A., Comparison and integration of arrays of quartz resonators and metal-oxide semiconductor chemoresistors in the quality evaluation of olive oils, *Sensors Actuat.* 2001, B78, 303–309

79. Boilot, P.; Hines, E. L.; Gongora, M. A.; Folland, R. S., Electronic noses inter-comparison, data fusion and sensor selection in discrimination of standard fruit solutions, *Sensors Actuat.* 2003, B88, 80–88

80. Winquist, F.; Wide, P.; Lundström, I., The combination of an electronic tongue and an electronic nose for improved classification of fruit juices, In Technical Digest of Eurosensors XII Conference, Southampton, UK, IOP, Bristol, 1998

81. Wide, P.; Winquist, F.; Bergsten, P.; Petriu, E. M., The human-based multisensor fusion method for artificial nose and tongue sensor data, *IEEE Trans. Instrum. Measure.* 1998, 47, 531–536

82. Di Natale, C.; Paolesse, R.; Macagnano, A.; Mantini, A.; D'Amico, A.; Legin, A.; Lvova, L.; Rudnitskaya, A.; Vlasov, Y., Electronic nose and electronic tongue integration for improved classification of clinical and food samples, *Sensors Actuat.* 2000, B64, 15–21

83. Rong, L.; Ping, W.; Wenlei, H., A novel method for wine analysis based on sensor fusion technique, *Sensors Actuat.* 2000, B66, 246–250

84. Luo, R. C.; Yih, C. C.; Su, K. L., Multisensor fusion and integration: Approaches, applications, and future research directions, *IEEE Sensors J.* 2002, 2, 107–119

85. Hammond, M. H.; Johnson, K. J.; Rose-Pehrsson, S. L.; Ziegler, J.; Walker, H.; Caudy, K.; Gary, D.; Tillett, D., A novel chemical detector using cermet sensors and pattern recognition methods for toxic industrial chemicals, *Sensors Actuat.* 2006, B116, 135–144

86. Hart, S. J.; Shaffer, R. E.; Rose-Pehrsson, S. L.; McDonald, J. R., Using physics-based modeler outputs to train probabilistic neural networks for unexploded ordnance (uxo) classification in magnetometry surveys, *IEEE Trans. Geosci. Remote Sensing* 2001, 39, 797–804

87. Bernstein, D. S., The end of false alarms?" National Fire Protection Association Magazine, Jan/Feb 1998

88. Pfister, G., Multisensor/multicriteria fire detection: A new trend rapidly becomes state of the art, *Fire Technol.* 1997, 33, 115–139

89. Jackson, M. A.; Robins, I., Gas sensing for fire detection: Measurements of CO, CO2, H2, O2 and smoke density in European standard fire tests, *Fire Safety J.* **1994**, 23, 181–205

90. Milke, J. A., Monitoring multiple aspects of fire signatures for discriminating fire detection, *Fire Technol.* **1999**, 35, 195–209

91. Milke, J. A.; Hulcher, M. E.; Worrel, C. L.; Gottuk, D. T.; Williams, F. W., Investigation of multi-sensor algorithms for fire detection, *Fire Technol.* **2003**, 39, 363–382

92. Gottuk, D. T.; Hill, S. A.; Schemel, C. F.; Strehlen, B. D.; Rose-Pehrsson, S. L.; Shaffer, R. E.; Tatem, P. A.; Williams, F. A., Identification of fire signatures for shipboard mulit-criteria fire detection systems, In NRL Memorandum Report NRL/MR/6180–99–8386, June 18, **1999**

93. Rose-Pehrsson, S. L.; Shaffer, R. E.; Hart, S. J.; Williams, F. W.; Gottuk, D. T.; Strehlen, B. D.; Hill, S. A., Multi-criteria fire detection systems using a probabilistic neural network, *Sensors Actuat.* **2000**, B69, 325–335

94. Shaffer, R. E.; Rose-Pehrsson, S. L., Improved probabilistic neural network algorithm for chemical sensor array pattern recognition, *Anal. Chem.* **1999**, 71, 4263–4271

95. Hammond, M. H.; Riedel, J. C.; Rose-Pehrsson, S. L.; Williams, F. W., Training set optimization methods for a probabilistic neural network, *Chemom. Intell. Lab. Syst.* **2004**, 71, 73–78

96. Hart, S. J.; Hammond, M. H.; Wong, J. T.; Wright, M. T.; Gottuk, D. T.; Rose-Pehrsson, S. L.; Williams, F. W., Real-time classification performance and failure mode analysis of a physical/chemical sensor array and probabilistic neural network, *Field Anal. Chem. Technol.* **2001**, 5, 244–258

97. Rose-Pehrsson, S. L.; Hart, S. J.; Street, T. T.; Williams, F. W.; Hammond, M. H.; Gottuk, D. T.; Wright, M. T.; Wong, J. T., Early warning fire detection system using a probabilistic neural network, *Fire Technol.* **2003**, 39, 147–171

98. JiJi, R. D.; Hammond, M. A.; Williams, F. W.; Rose-Pehrsson, S. L., Multivariate statistical process control for continuous monitoring of networked early warning fire detection (EWFD) systems, *Sensors Actuat.* **2003**, B93, 107–116

99. Rose-Pehrsson, S. L.; Owrutsky, J. C.; Gottuk, D. T.; Geiman, J. A.; Williams, F. W.; Farley, J. P., Phase I: FY01 investigative study for the advanced volume sensor, In NRL Memorandum Report NRL/MR/6110–03–8688, June 30, **2003**

100. Gottuk, D. T.; Lynch, J. A.; Rose-Pehrsson, S. L.; Owrutsky, J. C.; Williams, F. W., Video image fire detection for shipboard use, *Fire Safety J.* **2006**, 41, 321–326

101. Owrutsky, J. C.; Steinhurst, D. A.; Nelson, H. H.; Williams, F. W., Spectral based volume sensor component, In NRL Memorandum Report NRL/MR/6110–03–8694, July 30, **2003**

102. Steinhurst, D. A.; Lynch, J. A.; Gottuk, D. T.; Owrutsky, J. C.; Nelson, H. H.; Rose-Pehrsson, S. L.; Williams, F. W., Spectral-based volume sensor testbed algorithm development, test series VS2, In NRL Memorandum Report NRL/MR/6110–05–8856, January 12, **2005**

103. Wales, S. C.; McCord, M. T.; Lynch, J. A.; Rose-Pehrsson, S. L.; Williams, F. W., Acoustic event signatures for damage control: Water events and shipboard ambient noise, In NRL Memorandum Report NRL/MR/7120–04–8445, October 12, **2004**

104. Steinhurst, D. A.; Minor, C. P.; Owrutsky, J. C.; Rose-Pehrsson, S. L. Gottuk, D. T.; Williams, F. W., Long wavelength video-based event detection, preliminary results from the CVNX and VS1 test series, ex-USS SHADWELL, April 7–25, 2003, In NRL Memorandum Report NRL/MR/6110–03–8733, December 31, **2003**

105. Owrutsky, J. C.; Steinhurst, D. A.; Minor, C. P.; Rose-Pehrsson, S. L.; Gottuk, D. T.; Williams, F. W., Long wavelength video detection of fire in ship compartments, *Fire Safety J.* **2006**, 41, 315–320

106. Rose-Pehrsson, S. L.; Minor, C. P.; Steinhurst, D. A.; Owrutsky, J. C.; Lynch, J. A.; Gottuk, D. T.; Wales, S. C.; Farley, J. P.; Williams, F. W., Volume sensor for damage assessment and situational awareness *Fire Safety J.* **2006**, 41, 301–310

107. James, P. S., Bayesian Statistics: Principles, Models, and Applications, Wiley, New York, **1989**

108. Roussel, S.; Bellon-Maurel, V.; Roger, J.-M.; Grenier, P., Fusion of aroma, FT-IR and UV sensor data based on Bayesian inference. Application to the discrimination of white grape varieties, *Chemom. Intell. Lab. Syst.* **2003**, 65, 209–219

109. Minor, C. P.; Johnson, K. J.; Rose-Pehrsson, S. L; Owrutsky, J. C.; Wales, S. C.; Steinhurst, D. A.; Gottuk, D. T., A full-scale prototype multisensor system for damage control and situational awareness, *Fire Technol.*, in press

110. Lynch, J. A.; Gottuk, D. T.; Owrutsky, J. C.; Steinhurst, D. A.; Minor, C. P.; Wales, S. C.; Farley, J. P.; Rose-Pehrsson, S. L; Williams, F. W., Volume sensor development test series 5 – Multi-compartment system, In NRL Memorandum Report NRL/MR/6180–05–8931, December 30, **2005**

111. Collins, L. M.; Zhang, Y.; Li, J.; Wang, H.; Carin, L.; Hart, S. J.; Rose-Pehrsson, S. L.; Nelson, H. H.; McDonald, H. H., A comparison of the performance of statistical and fuzzy algorithms for unexploded ordnance detection, IEEE Trans. Fuzzy Systems **2001**, 9, 17–30

112. Barrow, B.; Nelson, H. H., Model-based characterization of electromagnetic induction signatures obtained with the MTADS electromagnetic array, *IEEE Trans. Geosci. Remote Sensing* **2001**, 39, 1279–1285

113. Nelson, H. H.; McDonald, J. R., Multisensor towed array detection system for UXOdetection, *IEEE Trans. Geosci. Remote Sensing* **2001**, 39, 1139–1145

114. Rose-Pehrsson, S. L.; Johnson, K. J.; Minor, C. P., Intelligent data fusion for wide-area assessment of UXO contamination. SERDP Project MM-1510. FY06 Annual Report, In NRL Memorandum Report NRL/MR/6181–07–9039, April 20, **2007**

Part IV
Future Directions

Chapter 13
Future Directions

Margaret A. Ryan and Abhijit V. Shevade

Abstract In this volume, several computational methods which may be used for evaluation and selection of sensing materials have been discussed. These computational methods have ranged from first principles or de novo methods, such as those discussed in Part 1, to semi-empirical and statistical methods as discussed in Parts 2 and 3. Some chapters have focused on designing sensing materials to respond to specific analytes and some on combining sensors to create arrays to detect a suite of chemical species. Nevertheless, challenges in computational evaluation of chemical sensing materials remain. We see two principal challenges. One challenge is in refining the methods discussed to yield accurate prediction of sensor response [1, 2]; methods as presented here may certainly be used to evaluate and rank candidate materials and to determine which materials to test. The question is whether these approaches can be used to discover new materials for sensing applications and whether these approaches will be able to predict the response of known sensing materials to new analytes accurately. The second challenge is in constructing arrays, that is, selecting which combination of materials and sensor types to use in an array. In this emerging field, the question is how to select a suite of sensors for a suite of analytes and how to analyze and understand the information gathered from the array.

Computationally, there are several approaches discussed here, which may be used singly and in combination to develop new sensing materials, to design arrays, to predict how an existing array will respond to particular stimuli, and to create a database of sensor responses to analytes. New computational tools that provide a more detailed understanding of the mechanisms and action of chemical sensing from molecular to system level are required. Developing these tools involves scalable and

M.A. Ryan (✉) and A.V. Shevade
Jet Propulsion Laboratory, California Institute of Technology, 4800 Oak Grove Drive, Pasadena, CA 91109, USA
e-mail: mryan@jpl.nasa.gov

M.A. Ryan et al. (eds.), *Computational Methods for Sensor Material Selection*, 301
Integrated Analytical Systems,
DOI 10.1007/978-0-387-73715-7_13, © Springer Science+Business Media, LLC 2009

robust computational algorithms to enable integration of materials modeling and simulation approaches from the atomistic through the meso, to the continuum scale. A validated approach based on these formulations would then provide tools with which to evaluate sensing materials and their properties based on fundamental understanding of sensing at the electronic, atomic, and molecular levels.

Advancement of validated computational approaches and tools will assist experiments, as computational studies could be done prior to experiments to evaluate sensing materials. A priori sensor response predictions using computational approaches can also be used to generate parameters for quantification and identification algorithms. This will facilitate the generation of virtual response data sets for any given sensor/sensor array for analytes that may not easily be tested, such as highly explosive or toxic compounds. Subsequently, fewer experimental tests will be needed. Sensing experiments are time consuming, owing to the need to measure and catalog sensor responses for various target analytes at different concentrations and under a variable set of environmental conditions (humidity, temperature, and pressure). Only when these experiments are completed can data analysis algorithms be tested.

Information content in chemical sensor arrays can be created by a multitude of combinations of materials, structures and methods, and that variety is only now coming to be appreciated. Chapters 10 and 11 discuss selecting different sensors of a single type to construct an array; similar methods may be used to construct an array for multiple sensor types. However, it is as important to develop computational methods to arrange, store, and analyze complex data sets; such methods are making a significant impact on progress in this field of research and in applications, and will continue to do so. Future work will involve combining many of the aspects of computation and arrays discussed throughout this volume. Hybrid Arrays as discussed in Chaps. 1 and 12 are, perhaps, the future of sensing arrays. The computational methods discussed here can be applied to development of such hybrid arrays.

Future computational research to design, select, and optimize sensing materials will also focus on developing a prediction database for sensor materials to aid in developing a discovery frame work of new sensing materials for desired applications. Such databases could be developed for organic, inorganic, and a combination of material types. This calls for the development of novel, mathematical, statistical, and optimization approaches and their integration with advanced computational materials modeling approaches.

Combination of quantum mechanics with first-principles molecular dynamics affords a great deal of information that it is useful in designing and selecting materials for specific analytes, as shown in Chaps. 2 and 3. In future work, a newly developed Molecular Dynamics method for the direct estimation of free energies [3] may be applied to such computation. Computational methods such as the Grand Canonical Monte Carlo method discussed in Chap. 4 may also lead to estimation of energies involved in the sensing process, such as electrostatic forces.

Combinatorial and high-throughput experimentation methodologies, as discussed in Chap. 7, provide an opportunity to generate new experimental data to discover new sensing materials and/or to optimize existing material compositions.

Design of new sensing materials is the important cornerstone in the effort to develop new sensors. Computer-aided design of materials, as discussed in Chap. 5, is one approach to sensor design. However, it is often found that sensing materials are too complex to predict their performance quantitatively in the design stage. Applications of data mining approaches to sensing materials [4–7] will assist in understanding prospective sensing material performance in the design stage. In addition, new chemometric methods are under development to extract chemical information from array responses in terms of solvation and other parameters, as described in Chaps. 8 and 9, as descriptors of the detected vapor and of the sensing materials.

Hybrid biomimetic nanosensors, discussed in Chap. 6, which use selective polymeric and biological materials that integrate flexible recognition moieties with nanometer size transducers, are new sensing approaches currently under investigation. Their potential biocompatibility combined with advanced mechanistic modeling studies could lead to applications such as unobtrusive implantable medical sensors for disease diagnostics, light weight multi-purpose sensing devices for aerospace applications, ubiquitous environmental monitoring devices in urban and rural areas, and inexpensive smart packaging materials for active in situ food safety labeling.

Inclusion of biomimetic influences into development of sensing materials and signal processing for a sensor array is a rising area of research, and progress in this area will result in improvement of the functionality and fidelity of sensor devices. To strengthen pattern recognition capability of gas sensors, aggregating sensors with overlapping specificities is an approach under study [8]. This process mimics the biological olfaction process that is known to aggregate the raw sensory information collected by large numbers of olfactory receptors into a smaller number of aggregate inputs before these signals reach the olfactory bulb. Biologically inspired models have been investigated to remove concentration effects from the multivariate response of a chemical sensor array [9]. Chemosensory adaptation of sensing arrays could be enhanced by using algorithms [10] that reduce array sensitivity to odors previously detected in the environment. This approach is inspired by adaptation in biological olfactory processes that remove constant noninformative stimuli, so that new ones are detected.

Finally, it must be noted that the benefits of a multisensory or hybrid sensing approach are not automatically achieved. Multivariate data analysis techniques such as those employed in the field of chemometrics have become more important in analyzing sensor array data. Depending on the nature of the acquired data, a number of chemometric algorithms may prove useful in the analysis and interpretation of data from hybrid sensor arrays. The multicriteria/multisensory approach is a powerful means of solving some challenging analytical problems. Although the use of hybrid arrays in chemical vapor sensing has not yet become commonplace, perhaps continuing research will encourage new developments and bring about more widespread usage. Often, the analytical benefits prove to be worthy of the modest increases in the hardware and software complexities.

References

1. Ryan, M. A.; Shevade A. V., Computational approaches to design and evaluation of chemical sensing materials, In Combinatorial Methods for Chemical and Biological Sensors; Potyrailo, R. A.; Mirsky, V. M., Eds.; Springer Science, New York, **2009**, 455–463
2. Goddard, W. A.; Cagin, T.; Blanco, M.; Vaidehi, N.; Dasgupta, S.; Floriano, W.; Belmares, M.; Kua, J.; Zamanakos, G.; Kashihara, S.; Iotov, M.; Gao, G. H., Strategies for multiscale modeling and simulation of organic materials: Polymers and biopolymers, *Comput. Theor. Polym. Sci.* **2001**, 11, 329–343
3. Becke, A. D., Density-functional thermochemistry. 3. The role of exact exchange, *J. Chem. Phys.* **1993**, 98, 5648–5652
4. Potyrailo, R. A.; McCloskey, P. J.; Wroczynski, R. J.; Morris, W. G., High-throughput determination of quantitative structure-property relationships using resonant multisensor system: Solvent-resistance of bisphenol a polycarbonate copolymers, *Anal. Chem.* **2006**, 78, 3090–3096
5. Frenzer, G.; Frantzen, A.; Sanders, D.; Simon, U.; Maier, W. F., Wet chemical synthesis and screening of thick porous oxide films for resistive gas sensing applications, *Sensors* **2006**, 6, 1568–1586
6. Villoslada, F. N.; Takeuchi, T., Multivariate analysis and experimental design in the screening of combinatorial libraries of molecular imprinted polymers, *Bull. Chem. Soc. Jpn* **2005**, 78, 1354–1361
7. Mijangos, I.; Navarro-Villoslada, F.; Guerreiro, A.; Piletska, E.; Chianella, I.; Karim, K.; Turner, A.; Piletsky, S., Influence of initiator and different polymerisation conditions on performance of molecularly imprinted polymers, *Biosens. Bioelectron.* **2006**, 22, 381–387
8. Wilson, D. M.; Roppel, T.; Kalim, R., Aggregation of sensory input for robust performance in chemical sensing microsystems, *Sensors Actuat. B* **2000**, 64,107–117
9. Raman, B.; Gutierrez-Osuna, R., Concentration normalization with a model of gain control in the olfactory bulb, *Sensors Actuat. B* **2006**, 116, 36–42
10. Gutierrez-Osuna, R.; Powar, N. U., Odor mixtures and chemosensory adaptation in gas sensor arrays, *Int.J. Artif. Intell. Tools* **2003**, 12, 1–16

Index

Breinigsville, PA USA
20 October 2009
226132BV00004B/32/P